London Mathematical Society Lecture Note Series: 366

Highly Oscillatory Problems

Edited by

BJORN ENGQUIST
University of Texas, Austin

ATHANASIOS FOKAS
University of Cambridge

ERNST HAIRER
Université de Genève

ARIEH ISERLES
University of Cambridge

CAMBRIDGE
UNIVERSITY PRESS

CAMBRIDGE UNIVERSITY PRESS
Cambridge, New York, Melbourne, Madrid, Cape Town, Singapore, São Paulo, Delhi

Cambridge University Press
The Edinburgh Building, Cambridge CB2 8RU, UK

Published in the United States of America by Cambridge University Press, New York

www.cambridge.org
Information on this title: www.cambridge.org/9780521134439

First published 2009

Printed in the United Kingdom at the University Press, Cambridge

A catalogue record for this publication is available from the British Library

ISBN 978-0-521-13443-9 paperback

LONDON MATH

All the titles listed below can be obtained from good booksellers or from Cambridge University Press. For a complete series listing visit www.cambridge.org/mathematics

218 Surveys in combinatorics, 1995, P. ROWLINSON (ed.)
220 Algebraic set theory, A. JOYAL & I. MOERDIJK
221 Harmonic approximation, S. J. GARDINER
222 Advances in linear logic, J.-Y. GIRARD, Y. LAFONT & L. REGNIER (eds)
223 Analytic semigroups and semilinear initial boundary value problems, KAZUAKI TAIRA
224 Computability, enumerability, unsolvability, S. B. COOPER, T. A. SLAMAN & S. S. WAINER (eds)
225 A mathematical introduction to string theory, S. ALBEVERIO *et al.*
226 Novikov conjectures, index theorems and rigidity I, S. FERRY, A. RANICKI & J. ROSENBERG (eds)
227 Novikov conjectures, index theorems and rigidity II, S. FERRY, A. RANICKI & J. ROSENBERG (eds)
228 Ergodic theory of $\mathbf{Z}^d$ actions, M. POLLICOTT & K. SCHMIDT (eds)
229 Ergodicity for infinite dimensional systems, G. DA PRATO & J. ZABCZYK
230 Prolegomena to a middlebrow arithmetic of curves of genus 2, J. W. S. CASSELS & E. V. FLYNN
231 Semigroup theory and its applications, K. H. HOFMANN & M. W. MISLOVE (eds)
232 The descriptive set theory of Polish group actions, H. BECKER & A. S. KECHRIS
233 Finite fields and applications, S. COHEN & H. NIEDERREITER (eds)
234 Introduction to subfactors, V. JONES & V. S. SUNDER
235 Number theory 1993–94, S. DAVID (ed.)
236 The James forest, H. FETTER & B. G. DE BUEN
237 Sieve methods, exponential sums, and their applications in number theory, G. R. H. GREAVES *et al.*
238 Representation theory and algebraic geometry, A. MARTSINKOVSKY & G. TODOROV (eds)
240 Stable groups, F. O. WAGNER
241 Surveys in combinatorics, 1997, R. A. BAILEY (ed.)
242 Geometric Galois actions I, L. SCHNEPS & P. LOCHAK (eds)
243 Geometric Galois actions II, L. SCHNEPS & P. LOCHAK (eds)
244 Model theory of groups and automorphism groups, D. EVANS (ed.)
245 Geometry, combinatorial designs and related structures, J. W. P. HIRSCHFELD *et al.*
246 p-Automorphisms of finite p-groups, E. I. KHUKHRO
247 Analytic number theory, Y. MOTOHASHI (ed.)
248 Tame topology and o-minimal structures, L. VAN DEN DRIES
249 The atlas of finite groups: ten years on, R. CURTIS & R. WILSON (eds)
250 Characters and blocks of finite groups, G. NAVARRO
251 Gröbner bases and applications, B. BUCHBERGER & F. WINKLER (eds)
252 Geometry and cohomology in group theory, P. KROPHOLLER, G. NIBLO, R. STÖHR (eds)
253 The q-Schur algebra, S. DONKIN
254 Galois representations in arithmetic algebraic geometry, A. J. SCHOLL & R. L. TAYLOR (eds)
255 Symmetries and integrability of difference equations, P. A. CLARKSON & F. W. NIJHOFF (eds)
256 Aspects of Galois theory, H. VÖLKLEIN *et al.*
257 An introduction to noncommutative differential geometry and its physical applications 2ed, J. MADORE
258 Sets and proofs, S. B. COOPER & J. TRUSS (eds)
259 Models and computability, S. B. COOPER & J. TRUSS (eds)
260 Groups St Andrews 1997 in Bath, I, C. M. CAMPBELL *et al.*
261 Groups St Andrews 1997 in Bath, II, C. M. CAMPBELL *et al.*
262 Analysis and logic, C. W. HENSON, J. IOVINO, A. S. KECHRIS & E. ODELL
263 Singularity theory, B. BRUCE & D. MOND (eds)
264 New trends in algebraic geometry, K. HULEK, F. CATANESE, C. PETERS & M. REID (eds)
265 Elliptic curves in cryptography, I. BLAKE, G. SEROUSSI & N. SMART
267 Surveys in combinatorics, 1999, J. D. LAMB & D. A. PREECE (eds)
268 Spectral asymptotics in the semi-classical limit, M. DIMASSI & J. SJÖSTRAND
269 Ergodic theory and topological dynamics, M. B. BEKKA & M. MAYER
270 Analysis on Lie Groups, N. T. VAROPOULOS & S. MUSTAPHA
271 Singular perturbations of differential operators, S. ALBEVERIO & P. KURASOV
272 Character theory for the odd order theorem, T. PETERFALVI
273 Spectral theory and geometry, E. B. DAVIES & Y. SAFAROV (eds)
274 The Mandelbrot set, theme and variations, TAN LEI (ed.)
275 Descriptive set theory and dynamical systems, M. FOREMAN *et al.*
276 Singularities of plane curves, E. CASAS-ALVERO
277 Computational and geometric aspects of modern algebra, M. D. ATKINSON *et al.*
278 Global attractors in abstract parabolic problems, J. W. CHOLEWA & T. DLOTKO
279 Topics in symbolic dynamics and applications, F. BLANCHARD, A. MAASS & A. NOGUEIRA (eds)
280 Characters and automorphism groups of compact Riemann surfaces, T. BREUER
281 Explicit birational geometry of 3-folds, A. CORTI & M. REID (eds)
282 Auslander–Buchweitz approximations of equivariant modules, M. HASHIMOTO
283 Nonlinear elasticity, Y. FU & R. OGDEN (eds)
284 Foundations of computational mathematics, R. DEVORE, A. ISERLES & E. SÜLI (eds)
285 Rational points on curves over finite fields, H. NIEDERREITER & C. XING
286 Clifford algebras and spinors 2ed, P. LOUNESTO
287 Topics on Riemann surfaces and Fuchsian groups, E. BUJALANCE *et al.*
288 Surveys in combinatorics, 2001, J. HIRSCHFELD (ed.)
289 Aspects of Sobolev-type inequalities, L. SALOFF-COSTE

290 Quantum groups and Lie theory, A. PRESSLEY (ed.)
291 Tits buildings and the model theory of groups, K. TENT (ed.)
292 A quantum groups primer, S. MAJID
293 Second order partial differential equations in Hilbert spaces, G. DA PRATO & J. ZABCZYK
294 Introduction to operator space theory, G. PISIER
295 Geometry and integrability, L. MASON & YAVUZ NUTKU (eds)
296 Lectures on invariant theory, I. DOLGACHEV
297 The homotopy category of simply connected 4-manifolds, H.-J. BAUES
298 Higher operands, higher categories, T. LEINSTER
299 Kleinian groups and hyperbolic 3-manifolds, Y. KOMORI, V. MARKOVIC & C. SERIES (eds)
300 Introduction to Möbius differential geometry, U. HERTRICH-JEROMIN
301 Stable modules and the D(2)-problem, F. E. A. JOHNSON
302 Discrete and continuous nonlinear Schrödinger systems, M. J. ABLOWITZ, B. PRINARI & A. D. TRUBATCH
303 Number theory and algebraic geometry, M. REID & A. SKOROBOGATOV (eds)
304 Groups St Andrews 2001 in Oxford Vol. 1, C. M. CAMPBELL, E. F. ROBERTSON & G. C. SMITH (eds)
305 Groups St Andrews 2001 in Oxford Vol. 2, C. M. CAMPBELL, E. F. ROBERTSON & G. C. SMITH (eds)
306 Peyresq lectures on geometric mechanics and symmetry, J. MONTALDI & T. RATIU (eds)
307 Surveys in combinatorics 2003, C. D. WENSLEY (ed.)
308 Topology, geometry and quantum field theory, U. L. TILLMANN (ed.)
309 Corings and comodules, T. BRZEZINSKI & R. WISBAUER
310 Topics in dynamics and ergodic theory, S. BEZUGLYI & S. KOLYADA (eds)
311 Groups: Topological, combinatorial and arithmetic aspects, T. W. MÜLLER (ed.)
312 Foundations of computational mathematics, Minneapolis 2002, FELIPE CUCKER *et al.* (eds)
313 Transcendantal aspects of algebraic cycles, S. MÜLLER-STACH & C. PETERS (eds)
314 Spectral generalizations of line graphs, D. CVETKOVIĆ, P. ROWLINSON & S. SIMIĆ
315 Structured ring spectra, A. BAKER & B. RICHTER (eds)
316 Linear logic in computer science, T. EHRHARD *et al.* (eds)
317 Advances in elliptic curve cryptography, I. F. BLAKE, G. SEROUSSI & N. SMART
318 Perturbation of the boundary in boundary-value problems of partial differential equations, DAN HENRY
319 Double affine Hecke algebras, I. CHEREDNIK
320 L-functions and Galois representations, D. BURNS, K. BUZZARD & J. NEKOVÁŘ (eds)
321 Surveys in modern mathematics, V. PRASOLOV & Y. ILYASHENKO (eds)
322 Recent perspectives in random matrix theory and number theory, F. MEZZADRI, N. C. SNAITH (eds)
323 Poisson geometry, deformation quantisation and group representations, S. GUTT *et al.* (eds)
324 Singularities and computer algebra, C. LOSSEN & G. PFISTER (eds)
325 Lectures on the Ricci flow, P. TOPPING
326 Modular representations of finite groups of Lie type, J. E. HUMPHREYS
328 Fundamentals of hyperbolic manifolds, R. D. CANARY, A. MARDEN & D. B. A. EPSTEIN (eds)
329 Spaces of Kleinian groups, Y. MINSKY, M. SAKUMA & C. SERIES (eds)
330 Noncommutative localization in algebra and topology, A. RANICKI (ed.)
331 Foundations of computational mathematics, Santander 2005, L. PARDO, A. PINKUS, E. SULI & M. TODD (eds)
332 Handbook of tilting theory, L. ANGELERI HÜGEL, D. HAPPEL & H. KRAUSE (eds)
333 Synthetic differential geometry 2ed, A. KOCK
334 The Navier–Stokes equations, P. G. DRAZIN & N. RILEY
335 Lectures on the combinatorics of free probability, A. NICA & R. SPEICHER
336 Integral closure of ideals, rings, and modules, I. SWANSON & C. HUNEKE
337 Methods in Banach space theory, J. M. F. CASTILLO & W. B. JOHNSON (eds)
338 Surveys in geometry and number theory, N. YOUNG (ed.)
339 Groups St Andrews 2005 Vol. 1, C. M. CAMPBELL, M. R. QUICK, E. F. ROBERTSON & G. C. SMITH (eds)
340 Groups St Andrews 2005 Vol. 2, C. M. CAMPBELL, M. R. QUICK, E. F. ROBERTSON & G. C. SMITH (eds)
341 Ranks of elliptic curves and random matrix theory, J. B. CONREY, D. W. FARMER, F. MEZZADRI & N. C. SNAITH (eds)
342 Elliptic cohomology, H. R. MILLER & D. C. RAVENEL (eds)
343 Algebraic cycles and motives Vol. 1, J. NAGEL & C. PETERS (eds)
344 Algebraic cycles and motives Vol. 2, J. NAGEL & C. PETERS (eds)
345 Algebraic and analytic geometry, A. NEEMAN
346 Surveys in combinatorics, 2007, A. HILTON & J. TALBOT (eds)
347 Surveys in contemporary mathematics, N. YOUNG & Y. CHOI (eds)
348 Transcendental dynamics and complex analysis, P. RIPPON & G. STALLARD (eds)
349 Model theory with applications to algebra and analysis Vol 1, Z. CHATZIDAKIS, D. MACPHERSON, A. PILLAY & A. WILKIE (eds)
350 Model theory with applications to algebra and analysis Vol 2, Z. CHATZIDAKIS, D. MACPHERSON, A. PILLAY & A. WILKIE (eds)
351 Finite von Neumann algebras and masas, A. SINCLAIR & R. SMITH
352 Number theory and polynomials, J. MCKEE & C. SMYTH (eds)
353 Trends in stochastic analysis, J. BLATH, P. MÖRTERS & M. SCHEUTZOW (eds)
354 Groups and analysis, K. TENT (ed)
355 Non-equilibrium statistical mechanics and turbulence, J. CARDY, G. FALKOVICH & K. GAWEDZKI, S. NAZARENKO & O. V. ZABORONSKI (eds)
356 Elliptic curves and big Galois representations, D. DELBOURGO
357 Algebraic theory of differential equations, M. A. H. MACCALLUM & A. V. MIKHAILOV (eds)
359 Moduli spaces and vector bundles, L. BRAMBILA-PAZ, S. B. BRADLOW, O. GARCÍA-PRADA & S. RAMANAN (eds)
361 Words: Notes on verbal width in groups, D. SEGAL
363 Foundations of computational mathematics, Hong Kong 2008, F. CUCKER, A. PINKUS & M. J. TODD (eds)

Contents

Preface

High oscillation is everywhere and it is difficult to compute. The conjunction of these two statements forms the rationale of this volume and it is therefore appropriate to deliberate further upon them.

Rapidly oscillating phenomena occur in electromagnetics, quantum theory, fluid dynamics, acoustics, electrodynamics, molecular modelling, computerised tomography and imaging, plasma transport, celestial mechanics – and this is a partial list! The main reason to the ubiquity of these phenomena is the presence of signals or data at widely different scales. Typically, the slowest signal is the carrier of important information, yet it is overlayed with signals, usually with smaller amplitude but with considerably smaller wavelength (cf. the top of Fig. 1). This presence of different frequencies renders both analysis and computation considerably more challenging. Another example of problems associated with high oscillation is provided by the wave packet at the bottom of Fig. 1 and by other phenomena which might appear dormant (or progress sedately, at measured pace) for a long time, only to demonstrate suddenly (and often unexpectedly) much more hectic behaviour.

The difficulty implicit in high oscillation becomes a significant stumbling block once we attempt to produce reliable numerical results. In principle, the problem can be alleviated by increasing the resolution of the computation (the step size, spatial discretization parameter, number of modes in an expansion, the bandwidth of a filter), since high oscillation is, after all, an artefact of resolution: zoom in sufficiently and all signals oscillate slowly. Except that such 'zooming in' requires huge computer resources and the sheer volume of computations, even were it possible, would have led to an unacceptable increase of error because of the roundoff error accumulation.

The situation is reminiscent, yet very different, of the phenomenon of *transient behaviour,* commonly associated with stiff ordinary differential equations or with boundary layers for singularly perturbed partial differential equations. In those cases a differential system undergoes a brief but very intensive change over a small domain of the independent variables. Away from this domain the solution settles down to its asymptotic behaviour, typically at an exponential rate. In that case it is enough to

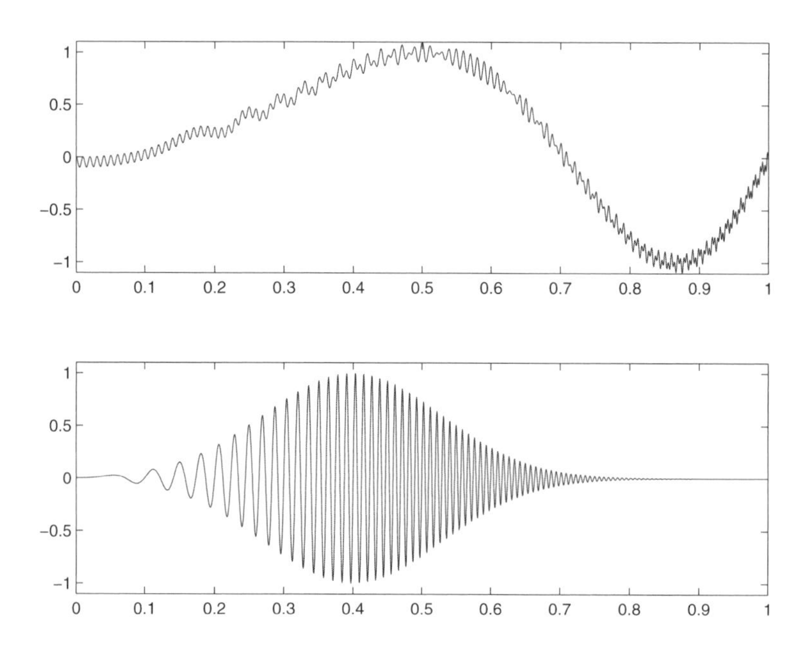

Fig. 1 Two highly oscillatory signals.

filter the contribution of transiency (or use a numerical method that dampens transient components). Not so with high oscillation, because this is a persistent phenomenon which stretches out over a large part of the computational domain. We cannot banish it by mesh refinement in the domain of rapid change and then just apply a stable algorithm. In a sense, the entire computational domain is in a transient phase.

Numerical analysis of differential equations is, at its very core, based upon Taylor expansions. Although often disguised by the formalism of order or of Sobolev-space inequalities, Taylor expansions are the main organising principle in the design of numerical methods and a criterion for their efficacy. Thus, typically numerical error scales as a derivative (or an elementary differential, or a norm of the derivative...). And this is precisely why standard numerical methods experience severe problems in the presence of high oscillation.

Our point is illustrated in Fig. 2 by the function

$$f(x) = \frac{\sin 2\pi x}{1+2x} + \tfrac{1}{10}\mathrm{Ai}(-100x),$$

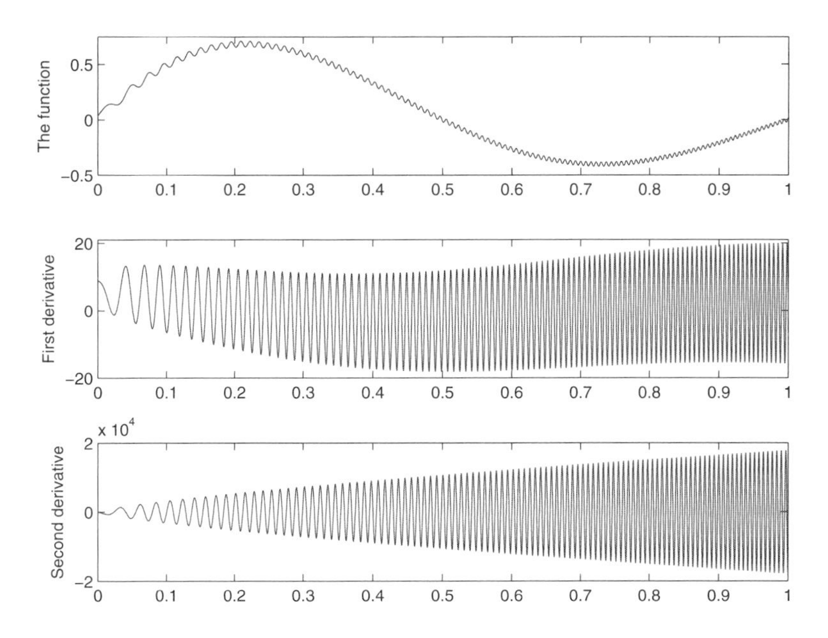

Fig. 2 A highly oscillatory function and its derivatives.

where Ai(·) is the Airy function. Note that the function f itself is basically a gently decaying sinusoid, with small-amplitude, yet increasingly rapid oscillations superimposed. For all intents and purposes, it is the sinusoid that is likely to describe whatever natural phenomenon we attempt to model. Yet, once we start differentiating f, the amplitude associated with the highly oscillating Airy function is magnified rapidly. Suppose that f is a solution of a differential equation which we attempt to discretize by any classical numerical method: multistep or Runge–Kutta, say. The error of a pth-order method scales like a $(p+1)$st power of the step size times the $(p+1)$st derivative (or a linear combination of elementary differentials including the derivative in question). This imposes severe restrictions on the step size, which rapidly lead to unacceptable computational cost.

No wonder, thus, that high oscillation in computation has been at worst disregarded and eliminated from the mathematical model by fiat, at best approached by a fairly unstructured bag of tricks and *ad hoc* ideas. The motivation for the six-months' long programme on "Highly oscillatory problems: Computation, theory and applications" at the Isaac Newton Institute of Mathematical Sciences (January–July 2007) was to provide an overarching setting for research focusing on high oscillation. Thus, rather than dealing with highly oscillatory phenomena separately in each application, we have attempted to provide a broader, synergistic approach, with high oscillation – its analysis and computation – at the centre of attention.

At the outset of the programme we have identified a number of significant threads and challenges in high oscillation research to provide a focus for our deliberations:

Homogenization Many large scale physical problems involve highly oscillatory solutions that involve many non-separable scales. Examples include applications from environmental and geosciences, combustion, fluid dynamics, plasma physics, materials science, and biological applications. A key mathematical difficulty is that the problem has no scale separation, so traditional asymptotic methods fail to yield useful models. New systematic multiscale analysis needs to be developed that can account for interaction of infinitely many non-separable scales. Such analysis will shed useful light in designing accurate and reliable multiscale models. One approach is to develop a new homogenization theory that applies to nonlinear dynamic problems without scale separation. This can be achieved by using the two-scale analysis iteratively in space and incrementally in time.

Future advances in multiscale modeling and computation require rigorous mathematical analysis to quantify modelling error across scales, approximation error within each scale, uncertainty error, and to address the well-posedness of the multiscale model. An important issue is to derive a rigorous microscopic interface condition that connects large scales to small scales, or connects continuum model to subgrid microscopic model. It is in general easier to derive a local interface condition for elliptic or diffusion equations, because there is a strong localization of small scale interaction for diffusion dominated processes. Deriving accurate interface condition for convection-dominated transport or hyperbolic problems is more difficult due to the nonlocal memory effect: in this case, small scales are propagated along characteristics. Upscaling the saturation equation for two-phase flow in heterogeneous porous media is considerably more difficult mathematically than solving the pressure equation. While various *ad hoc* upscaling models have been proposed by engineers, there is still lack of a systematic derivation of upscaling models for two-phase flow with rigorous error control. Upscaling the nonlinear convection-dominated transport problem provides an excellent prototype problem for developing a dynamic multiscale computational method with error control.

Asymptotic theory In recent years substantial progress has been made in the analysis of *Riemann–Hilbert (RH)* problems containing highly oscillatory integrals. Such RH problems arise in a variety of mathematical formulations including inverse scattering, initial-boundary value problems for linear and integrable nonlinear partial differential equations, the isomonodromy method, orthogonal polynomials and random matrices. The question of extracting useful asymptotic information from these formulations, reduces to the question of studying the asymptotic behaviour of RH problems involving highly oscillatory integrals. The latter question can be rigorously investigated using the nonlinear steepest descent asymptotic technique introduced in the beginning of the 90's by P. Deift and X. Zhou.

Another important development in asymptotic theory, which was the subject of a highly successful INI programme in 1995, is exponential asymptotics. It leads to considerably tighter and more powerful asymptotic estimates and tremendous speedup in the convergence of asymptotic series. In the specific context of scientific computation, this creates the prospect of techniques originating in exponential asymptotics for better computation of special functions, differential equations and integrals.

Symplectic algorithms The numerical solution of Hamiltonian differential equations is one of the central themes in *geometric numerical integration.* By interpreting the numerical approximation as the exact solution of a modified differential equation (backward error analysis), much insight into the long-time behaviour of symplectic and symmetric integrators is obtained. Unfortunately, in the presence of highly oscillatory solutions, this theory breaks down and new techniques have to be found. For some important situations (Fermi–Pasta–Ulam-like models for molecular dynamics simulations), where the high frequencies stem from a linear part in the differential equation, it is known that the exact flow has adiabatic invariants corresponding to actions (oscillatory energies divided by frequencies). The technique of *modulated Fourier expansions* has been developed to study the preservation of such adiabatic invariants by simulations. Numerical methods that nearly preserve the total energy and such adiabatic invariants are very important, because in realistic applications the solution is very sensitive to perturbations in initial values, so that it is not possible to control the global error of the approximation. Difficulties arise with resonant frequencies, numerical resonances, high dimensions as obtained with discretizations of nonlinear wave equations, and one is interested to get more insight in such situations. Problems where the high oscillations come from a time-dependent or nonlinear part of the differential equation are still more difficult to analyse and to treat numerically.

Integral expansion methods Changing locally the variables with respect to a rapidly-rotating frame of reference results in differential systems of the form $\partial u/\partial t = A(t)u + g(u)$, where the elements of the matrix A themselves oscillate rapidly, while g is, in some sense, small. Such systems can be computed very effectively by time-stepping methods that expand the solution by means of integral series: the main idea is that, integration being a smoothing operator, integral series converges significantly *faster* in the presence of oscillations. Two types of such methods have been recently the object of much attention. Firstly, Magnus, Cayley and Neumann expansions: the first two have been developed in the context of Lie-group methods and they possess remarkable structure-preserving features. The latter, known in physics as the Dyson expansion, is commonly dismissed as ineffective but it comes into its own in the presence of oscillations. The second (and related!) type of integral methods are *exponential integrators,* which represent the solution using variation of constants and employ, sometimes in succession, different discretization methods on the "linear but large" and "nonlinear but small" parts. Such methods have proved themselves in practice in the last few years but much work remains in harnessing them for the highly oscillatory setting.

Highly oscillatory quadrature Practical computation of integral expansions and exponential integrators with highly oscillatory kernels calls for efficient quadrature methods in one or several dimensions. Historically, this was considered very difficult and expensive – wrongly. Using appropriate asymptotic expansions and new numerical methods, it is possible to compute highly oscillating integrals very inexpensively indeed: rapid oscillation, almost paradoxically, renders quadrature much cheaper and more precise! The implications of such methods, which are presently subject to a very active research effort, are clear in the context of the above integral expansions. Wider implications are a matter of conjecture, as are the connections of these methods with exponential asymptotics, harmonic analysis and, in a multivariate setting, degree theory. Insofar as applications are concerned, highly oscillatory quadrature is central to calculations in electrodynamic and acoustic scattering and with wide range of other applications in fluid dynamics, molecular modelling and beyond.

The purpose of the INI programme was to bring together specialists in all these specialities (and beyond), to weave the distinct threads into a seamless theory of high oscillation. We have never laboured under the illusion that six months of an intensive exchange of ideas will somehow produce such a theory: rather more modestly, we have hoped at the first instance to open channels of communication, establish a dialogue and lay the foundations for future work. This, we believe, has been accomplished in a most outstanding manner.

The setting of the Isaac Newton Institute – its location, architecture, facilities and, perhaps most importantly, its very efficient and helpful staff – make it an ideal venue for collaborative projects. We have taken advantage of this to the fullest. For the first time ever high oscillation has been approached in a concerted manner not as an appendix to another activity or to an application but as a subject matter of its own. The eight review papers in this volume revisit in their totality the main themes of the programme. Carefully reading between the lines, it is possible to discern the impact of our six-months'-long conversation on different aspects of high oscillation research. We have every reason to believe that this impact is bound to progress and grow in leaps and bounds. The INI programme was a first step in what is a long journey toward the goal of *understanding* high oscillation in its analytic, computational and applied aspects. This volume is a first way-station on this journey.

This is the moment to thank the staff of the Isaac Newton Institute for their unstinting help and the 120 participants of the programme for their many contributions to its success.

We hope that this volume acts as a window to the fascinating world of computational high oscillation, sketching challenges, describing emerging methodologies and pointing the way to future research.

Björn Engquist

Athanasios Fokas

Ernst Hairer

Arieh Iserles

1

Oscillations over long times in numerical Hamiltonian systems

Ernst Hairer
Dept. de Mathématiques, Université de Genève
CH-1211 Genève 4
Switzerland
Email: Ernst.Hairer@math.unige.ch

Christian Lubich
Mathematisches Institut
Universität Tübingen
D-72076 Tübingen
Germany
Email: lubich@na.uni-tuebingen.de

1 Introduction

The numerical treatment of ordinary differential equations has continued to be a lively area of numerical analysis for more than a century, with interesting applications in various fields and rich theory. There are three main developments in the design of numerical techniques and in the analysis of the algorithms:

- *Non-stiff differential equations.* In the 19th century (Adams, Bashforth, and later Runge, Heun and Kutta), numerical integrators have been designed that are efficient (high order) and easy to apply (explicit) in practical situations.
- *Stiff differential equations.* In the middle of the 20th century one became aware that earlier developed methods are impractical for a certain class of differential equations (stiff problems) due to stability restrictions. New integrators (typically implicit) were needed as well as new theories for a better understanding of the algorithms.
- *Geometric numerical integration.* In long-time simulations of Hamiltonian systems (molecular dynamics, astronomy) neither classical explicit methods nor implicit integrators for stiff problems give satisfactory results. In the last few decades, special numerical methods have been designed that preserve the geometric structure of the exact flow and thus have an improved long-time behaviour.

The basic developments (algorithmic and theoretical) of these epochs are documented in the monographs [HNW93], [HW96], and [HLW06]. Within geometric numerical integration we can also distinguish between non-stiff and stiff situations. Since here the main emphasis is on conservative Hamiltonian systems, the term "stiff" has to be interpreted as "highly oscillatory".

The present survey is concerned with geometric numerical integration with emphasis on theoretical insight for the long-time behaviour of numerical solutions. There are several degrees of difficulty:

- *Non-stiff Hamiltonian systems — backward error analysis.* The main theoretical tool for a better understanding of the long-time behaviour of numerical methods for structured problems is backward error analysis (Sect. 2). Rigorous statements over exponentially long times have been obtained in [BG94, HL97, Rei99] for symplectic integrators. Unfortunately, the analysis is restricted to the non-stiff situation, and does not provide any information for problems with high oscillations.
- *Highly oscillatory problems — modulated Fourier expansion.* The main part of this survey treats Hamiltonian systems of the form

$$\ddot{q} + \Omega^2 q = -\nabla U(q), \tag{1.1}$$

  where Ω is a diagonal matrix with real entries between 0 and a large ω, and $U(q)$ is a smooth potential function. The additional difficulty is the presence of two time scales, and the crucial role of harmonic actions in the long-time analysis. Basic work for the analytic solution is in [BGG87]. Section 3 presents the technique of modulated Fourier expansions which permits to prove simultaneously the conservation of energy and actions for the analytic and the numerical solution (where the product of the time step size and ω is of size one or larger). This is developed in [HL01, CHL03] for one high frequency and in [CHL05] for several high frequencies.
- *Non-linear wave equations.* An extension to infinite dimension with arbitrarily large frequencies permits to treat the long-time behaviour of one-dimensional semi-linear wave equations. Long-time conservation of harmonic actions along the analytic solution is studied in [Bou96, Bam03]. The technique of modulated Fourier expansion yields new insight into the long-time behaviour of the analytic solution [CHL08b], of pseudo-spectral semi-discretizations [HL08], and of full discretizations [CHL08a]. This is discussed in Sect. 4.

In Sect. 5, an interesting analogy between highly oscillatory differential equations and linear multistep methods for non-stiff problems $\ddot{q} = -\nabla U(q)$ is established (see [HL04]). The inverse of the step size plays the role of ω, and the parasitic solutions of the multistep method correspond to high oscillations in the solution of (1.1). The near conservation of the harmonic actions thus yields the bounded-ness of the parasitic solutions over long times, and permits to prove that special linear multistep methods are suitable for the long-time integration of Hamiltonian systems (like those arising in the computation of planetary motion).

2 Backward error analysis

An important tool for a better understanding of the long-time behaviour of numerical methods for ordinary differential equations is backward error analysis. We present the main ideas, some important consequences, and also its limitations in the case of highly oscillatory problems.

2.1 General idea

The principle applies to general ordinary differential equations $\dot{y} = f(y)$ and to general (numerical) one-step methods $y_{n+1} = \Phi_h(y_n)$, such as Runge–Kutta, Taylor series, composition and splitting methods. It consists in searching for a modified differential equation

$$\dot{z} = f_h(z) = f(z) + hf_2(z) + h^2 f_3(z) + \cdots , \qquad z(0) = y_0, \tag{2.1}$$

where the vector field is written as a formal series in powers of the step size h, such that the numerical solution for the original problem is equal (in the sense of formal power series) to the exact solution of the modified differential equation (see Fig. 1).

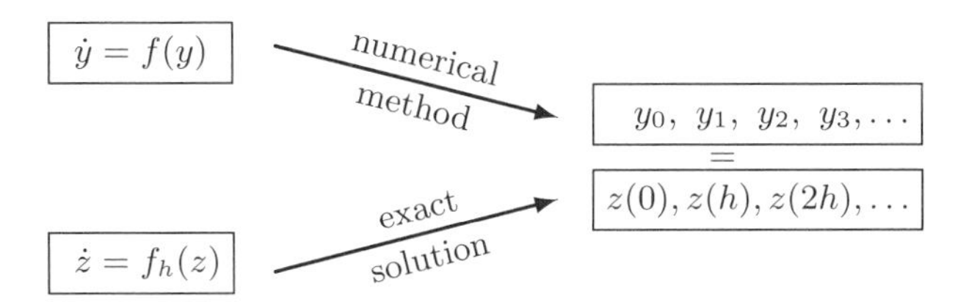

Fig. 1. Idea of backward error analysis

To obtain the coefficient functions $f_j(y)$, we note that we have the relation $z(t+h) = \Phi_h\big(z(t)\big)$ for the (formal) solution of (2.1). Expanding both sides of this relation into a power series of h and comparing equal powers of h, permits us to compute the functions $f_j(y)$ in a recursive manner.

The importance of backward error analysis resides in the fact that for differential equations with certain structures (Hamiltonian, reversible, divergence-free, etc.) solved with suitable geometric integrators (symplectic, symmetric, volume-preserving, etc.), the modified differential equation has the same structure as the original problem. The study of the modified differential equation then gives insight into the numerical solution. The rest of this section is devoted to make these statements more precise for the important special case of the Störmer–Verlet (leapfrog) discretisation.

2.2 Störmer–Verlet discretisation

For ease of presentation we restrict our considerations to the special Hamiltonian system

$$\ddot{q} = f(q) \quad \text{with} \quad f(y) = -\nabla U(q), \tag{2.2}$$

where $U(q)$ is a smooth potential function. Its most obvious discretisation (augmented with an approximation to the velocity $p = \dot{q}$) is

$$\begin{aligned} q^{n+1} - 2q^n + q^{n-1} &= h^2 f(q^n) \\ q^{n+1} - q^{n-1} &= 2h\, p^n . \end{aligned} \tag{2.3}$$

Due to pioneering work on higher order variants by Störmer, and due to its importance in molecular dynamics simulations recognised by Verlet, it is often called Störmer–Verlet method. In the literature on partial differential equations it is known as the leapfrog discretisation.

Introducing $p^{n+1/2} := (q^{n+1} - q^n)/h$ as an intermediate slope, this method can be written as

$$\begin{aligned} p^{n+1/2} &= p^n + \frac{h}{2} f(q^n) \\ q^{n+1} &= q^n + h\, p^{n+1/2} \\ p^{n+1} &= p^{n+1/2} + \frac{h}{2} f(q^{n+1}) \end{aligned} \tag{2.4}$$

which is clearly recognised as a symmetric one-step method for (2.2). It is a geometric integrator par excellence: the numerical flow $(q^n, p^n) \mapsto (q^{n+1}, p^{n+1})$ is symplectic when $f(q) = -\nabla U(q)$, it is volume preserving in the phase space, and it is time reversible (see [HLW03]). It is also the basic scheme for various extensions to higher order methods: composition and splitting methods, partitioned Runge–Kutta methods, and symmetric multistep methods.

2.3 Formal backward error analysis

We search for a modified differential equation such that its solution $\big(q(t), p(t)\big)$, which should not be confused with the solution of (2.2), formally interpolates the numerical solution of (2.3), i.e.,

$$\begin{aligned} q(t+h) - 2q(t) + q(t-h) &= h^2 f\big(q(t)\big) \\ q(t+h) - q(t-h) &= 2h\, p(t). \end{aligned} \tag{2.5}$$

Expanding the left hand sides into Taylor series around $h = 0$, eliminating higher derivatives by successive differentiation, and expressing the resulting differential equations in terms of q and p, yields

$$\begin{aligned} \dot p &= f(q) + \frac{h^2}{12}\Big(f''(q)(p,p) + f'(q)f(q)\Big) - \frac{h^4}{720}\Big(f''''(q)(p,p,p,p) \\ &\quad + 6\,f'''(q)\big(f(q),p,p\big) + 24\,f''(q)\big(f'(q)p,p\big) + 3\,f''(q)\big(f(q),f(q)\big) \\ &\quad + 6\,f'(q)f''(q)(p,p) + 6\,f'(q)f'(q)f(q)\Big) + \mathcal{O}(h^6) \\ \dot q &= p - \frac{h^2}{6}\,f'(q)p + \frac{h^4}{180}\Big(f'''(q)(p,p,p) + 3\,f''(q)\big(f(q),p\big) \\ &\quad + 6\,f'(q)f'(q)p\Big) + \mathcal{O}(h^6). \end{aligned} \tag{2.6}$$

Due to the symmetry of the method, the modified differential equation becomes a series in even powers of h.

For the case of a Hamiltonian system (2.2), i.e., $f(q) = -\nabla U(q)$, the modified differential equation (2.6) is also Hamiltonian

$$\dot p = -\nabla_q H_h(p,q), \qquad \dot q = \nabla_p H_h(p,q)$$

with modified Hamiltonian

$$\begin{aligned} H_h(p,q) &= \frac{1}{2}\|p\|^2 + U(q) + \frac{h^2}{24}\Big(2\,U''(q)(p,p) - \|U'(q)\|^2\Big) \\ &\quad - \frac{h^4}{720}\Big(U^{(4)}(q)(p,p,p,p) - 6\,U'''(q)\big(U'(q),p,p\big) \\ &\qquad + 3\,U''(q)\big(U'(q),U'(q)\big) - 12\,\|U''(q)p\|^2\Big) + \mathcal{O}(h^6). \end{aligned} \tag{2.7}$$

An important consequence of this observation is the following: since the numerical solution of the Störmer–Verlet discretisation is (at least formally) equal to the exact solution of the modified differential equation, we have that $H_h(p^n, q^n) = \text{const}$. As long as the numerical solution stays in a compact set, this implies that the energy $H(p,q) = \frac{1}{2}\|p\|^2 + U(q)$ remains close to a constant, i.e., $H(p^n, q^n) = \text{const} + \mathcal{O}(h^2)$ without any drift.

The next section shows how this statement can be made rigorous.

2.4 Rigorous backward error analysis

For a rigorous analysis, the modified differential equation constructed in the previous sections has to be truncated suitably:

$$\dot{z} = f_{h,N}(z) = f(z) + hf_2(z) + \cdots + h^{N-1} f_N(z), \qquad z(0) = y_0. \quad (2.8)$$

Obviously, equality does not hold any more in Fig. 1 and an error of size $\mathcal{O}(h^{N+1})$ is introduced. More precisely, if $y_{n+1} = \Phi_h(y_n)$ denotes the one-step method, and $\varphi_{N,t}(y)$ the flow of the truncated differential equation (2.8), we have $\|\Phi_h(y_0) - \varphi_{N,h}(y_0\| \le C_N h^{N+1}$ for arbitrary N. The freedom of choosing the truncation index N can be used to minimise this estimate. For analytic $f(y)$ and for standard numerical integrators (such as partitioned Runge–Kutta methods including the Störmer–Verlet discretisation), the choice $N \sim h^{-1}$ yields an estimate

$$\|\Phi_h(y_0) - \varphi_{N,h}(y_0\| \le h\gamma M \,\mathrm{e}^{-\alpha/\omega h}, \qquad (2.9)$$

where α and γ are constants that only depend on the numerical method, M is an upper bound of $f(y)$ on a disc of radius $2R$ around the initial value y_0, and $\omega = M/R$ is related to a Lipschitz constant of $f(y)$. A detailed proof can be found in [HLW06, Chap. IX].

Notice that (2.9) yields an estimate for one step only (local error). To get estimates for the global error and information on the long-time behaviour, knowledge on the propagation of perturbations is needed.

- *Conservation of energy.* In the case of a symplectic method applied to a Hamiltonian system, the modified equation is Hamiltonian (see Section 2.3). The truncated modified Hamiltonian $H_{N,h}(p,q)$ is exactly conserved along the solution of (2.8). Therefore, local deviations in $H_{N,h}(p^n, q^n)$ are just summed up, and one obtains from (2.9) that this modified Hamiltonian is conserved along the numerical solution up to exponentially small errors $\mathcal{O}(\mathrm{e}^{-\gamma/2\omega h})$ on exponentially long time intervals $0 \le t \le \mathcal{O}(\mathrm{e}^{\gamma/2\omega h})$. This implies the absence of any drift in the numerical Hamiltonian $H(p^n, q^n)$.

- *Integrable Hamiltonian systems.* Symplectic integrators applied to a nearly integrable Hamiltonian system give raise to a modified equation that is a perturbed Hamiltonian system. The celebrated KAM theory can be used get insight into the long-time behaviour of numerical integrators, e.g., linear growth of the global error.
- *Chaotic systems.* In the presence of positive Lyapunov exponents, the numerical solution remains close to the exact solution of the truncated modified equation only on time intervals of length $\mathcal{O}(h^{-1})$. Energy is well conserved by symplectic integrators also in this situation.

2.5 Limitation in the presence of high oscillations

The estimate (2.9) does not give any useful information if the product ωh is of size one or larger. Recall that ω is a kind of Lipschitz constant of the vector field $f(y)$ which, in the case of a stable Hamiltonian system, can be interpreted as the highest frequency in the solution. This means that for highly oscillatory differential equations the step size is restricted to unrealistic small values.

From the example of the harmonic oscillator $H(p,q) = \frac{1}{2}(p^2 + \omega^2 q^2)$ it can be seen that the estimate (2.9) cannot qualitatively be improved. In fact, for all reasonable integrators, the scaled numerical solution $(\omega q^n, p^n)$ depends on the step size h only via the product ωh.

The aim of the next section is to present a theory that permits to analyse the long-time behaviour of numerical time integrators in the presence of high oscillations.

3 Modulated Fourier expansion

In this section we consider Hamiltonian systems

$$\ddot{q} + \Omega^2 q = g(q), \qquad g(q) = -\nabla U(q), \tag{3.1}$$

where, for ease of presentation, Ω is a diagonal matrix and $U(q)$ is a smooth potential function. Typically, Ω will contain diagonal entries ω with large modulus. We are interested in the long-time behaviour of numerical solutions when ω times the step size h is not small, so that classical backward error analysis cannot be applied.

3.1 Modulated Fourier expansion of the analytic solution

We start with the situation, where Ω contains only diagonal entries which are either 0 or ω, and we split the components of q accordingly, i.e., $q = (q_0, q_1)$ and $\Omega = \mathrm{diag}\,(0, \omega I)$. Both, q_0 and q_1 are allowed to be vectors. There are two time scales in the solution of equation (3.1):

- fast time ωt in oscillations of the form $\mathrm{e}^{\mathrm{i}\omega t}$;
- slow time t due to the zero eigenvalue and the non-linearity.

In the absence of the non-linearity $g(q)$, the solution of (3.1) is a linear combination of 1, t, and $\mathrm{e}^{\pm\mathrm{i}\omega t}$. For the general case we make the ansatz

$$q(t) = \sum_{k\in\mathbb{Z}} z^k(t)\,\mathrm{e}^{\mathrm{i}k\omega t}, \tag{3.2}$$

where $z^k(t)$ are smooth functions with derivatives bounded uniformly in ω. The function $z^0(t)$ is real-valued, and $z^{-k}(t)$ is the complex conjugate of $z^k(t)$. Inserting (3.2) into the differential equation (3.1), expanding the non-linearity into a Taylor series around z^0, and comparing coefficients of $\mathrm{e}^{\mathrm{i}k\omega t}$ yields

$$\begin{pmatrix} \ddot z_0^k + 2\mathrm{i}k\omega \dot z_0^k - k^2\omega^2 z_0^k \\ \ddot z_1^k + 2\mathrm{i}k\omega \dot z_1^k + (1-k^2)\,\omega^2 z_1^k \end{pmatrix} = \sum_{m\ge 0} \frac{1}{m!} \sum_{s(\alpha)=k} g^{(m)}(z^0)\, z^\alpha, \tag{3.3}$$

where $\alpha = (\alpha_1,\ldots,\alpha_m)$ is a multi-index, $s(\alpha) = \sum_{j=1}^m \alpha_j$, and $g^{(m)}(z^0)\,z^\alpha = g^{(m)}(z^0)\big(z^{\alpha_1},\ldots,z^{\alpha_m}\big)$. The second sum is over multi-indices $\alpha = (\alpha_1,\ldots,\alpha_m)$ with $\alpha_j \neq 0$.

To obtain smooth functions $z_j^k(t)$ with derivatives bounded uniformly for large ω, we separate the dominating term in the left-hand side of (3.3), and eliminate higher derivatives by iteration. This gives a second order differential equation for z_0^0, first order differential equations for z_1^1 and z_1^{-1}, and algebraic relations for all other variables. Under the "bounded energy" assumption on the initial values

$$\|\dot q(0)\|^2 + \|\Omega\, q(0)\|^2 \le E, \tag{3.4}$$

it is possible to prove that the coefficient functions are bounded (on intervals of size one) as follows: $z_0^0(t) = \mathcal{O}(1)$, $z_1^{\pm1}(t) = \mathcal{O}(\omega^{-1})$, $\dot z_1^{\pm1}(t) = \mathcal{O}(\omega^{-2})$, and $z_j^k(t) = \mathcal{O}(\omega^{-|k|-2})$ for the remaining indices (j,k), see [HLW06, Sect. XIII.5].

The time average of the potential $U(q)$ along the analytic solution (3.2) only depends on the smooth coefficient functions $z^k(t)$ and is (formally) given by (with $\mathbf{z} = (\ldots, z^{-1}, z^0, z^1, z^2, \ldots)$)

$$\mathcal{U}(\mathbf{z}) = U(z^0) + \sum_{m\geq 1} \frac{1}{m!} \sum_{s(\alpha)=0} U^{(m)}(z^0) z^\alpha. \tag{3.5}$$

It is an interesting fact and crucial for the success of the expansion (3.2) that the functions $y^k(t) = z^k(t)\,\mathrm{e}^{\mathrm{i}k\omega t}$ are solution of the infinite dimensional Hamiltonian system

$$\ddot{y}^k + \Omega^2 y^k = -\nabla_{-k}\,\mathcal{U}(\mathbf{y}), \tag{3.6}$$

where ∇_{-k} indicates the derivative with respect to the component "$-k$" of the argument $\mathbf{y}$. Its Hamiltonian

$$\mathcal{H}(\mathbf{y}, \dot{\mathbf{y}}) = \frac{1}{2} \sum_{k\in\mathbb{Z}} \Big((\dot{y}^{-k})^T \dot{y}^k + (y^{-k})^T \Omega^2 y^k \Big) + \mathcal{U}(\mathbf{y}) \tag{3.7}$$

is therefore a conserved quantity of the system (3.6), and hence also of (3.3). Since $q_0(t) = z_0^0(t) + \mathcal{O}(\omega^{-3})$, $\dot{q}_0(t) = \dot{z}_0^0(t) + \mathcal{O}(\omega^{-2})$, $q_1(t) = z_1^1(t)\,\mathrm{e}^{\mathrm{i}\omega t} + z_1^{-1}(t)\,\mathrm{e}^{-\mathrm{i}\omega t} + \mathcal{O}(\omega^{-2}) = y_1^1(t) + y_1^{-1}(t) + \mathcal{O}(\omega^{-2})$ and $\dot{q}_1(t) = \mathrm{i}\omega\big(y_1^1(t) - y_1^{-1}(t)\big) + \mathcal{O}(\omega^{-2})$ by the estimates for z_j^k, the quantity (3.7) is $\mathcal{O}(\omega^{-1})$ close to the total energy of the system

$$H\big(q(t), \dot{q}(t)\big) = \frac{1}{2}\big(\|\dot{q}(t)\|^2 + \|\Omega q(t)\|^2\big) + U\big(q(t)\big). \tag{3.8}$$

The averaged potential $\mathcal{U}(\mathbf{y})$ is invariant under the one-parameter group of transformations $y^k \to \mathrm{e}^{\mathrm{i}k\tau} y^k$. Therefore, Noether's theorem yields the additional conserved quantity

$$\mathcal{I}(\mathbf{y}, \dot{\mathbf{y}}) = -\mathrm{i}\omega \sum_{k\in\mathbb{Z}} k\,(y^{-k})^T \dot{y}^k \tag{3.9}$$

for the system (3.6). It is $\mathcal{O}(\omega^{-1})$ close to the harmonic energy

$$I\big(q(t), \dot{q}(t)\big) = \frac{1}{2}\big(\|\dot{q}_1(t)\|^2 + \omega^2 \|q_1(t)\|^2\big) \tag{3.10}$$

of the highly oscillatory part of the system.

The analysis of this section can be made rigorous by truncating the arising series and by patching together estimates on short intervals to get information on intervals of length ω^{-N} (with arbitrary N). In this way one can prove that the harmonic energy (3.10) remains constant up to oscillations of size $\mathcal{O}(\omega^{-1})$ on intervals of length ω^{-N}, a result first obtained by [BGG87].

3.2 Exponential integrators

Since $q^{n+1} - 2\cos(h\Omega)\, q^n + q^{n-1} = 0$ is an exact discretisation of the equation $\ddot q + \Omega^2 q = 0$, it is natural to consider the numerical scheme

$$q^{n+1} - 2\cos(h\Omega)\, q^n + q^{n-1} = h^2 \Psi g(\Phi q^n) \tag{3.11}$$

as discretisation of (3.1). Here, $\Psi = \psi(h\Omega)$ and $\Phi = \phi(h\Omega)$, where the filter functions $\psi(\xi)$ and $\phi(\xi)$ are even, real-valued functions satisfying $\psi(0) = \phi(0) = 1$. Special cases are the following:

(A)	$\psi(h\Omega) = \mathrm{sinc}^2(\frac{1}{2}h\Omega)$	$\phi(h\Omega) = 1$	[Gau61]
(B)	$\psi(h\Omega) = \mathrm{sinc}\,(h\Omega)$	$\phi(h\Omega) = 1$	[Deu79]
(C)	$\psi(h\Omega) = \mathrm{sinc}^2(h\Omega)$	$\phi(h\Omega) = \mathrm{sinc}\,(h\Omega)$	[GASS99]

where $\mathrm{sinc}\,(\xi) = \sin\xi/\xi$. It is also natural to complete formula (3.11) with a derivative approximation p^n given by

$$q^{n+1} - q^{n-1} = 2h\, \mathrm{sinc}\,(h\Omega)\, p^n, \tag{3.12}$$

because, for $q(t) = \exp(\mathrm{i}\Omega t)\, q^0$, the derivative $p(t) = \dot q(t)$ satisfies this relation without error.

Written as a one-step method, we obtain

$$\begin{aligned} \widetilde{p}^n &= p^n + \frac{h}{2}\, \Psi_1 g(\Phi q^n) \\ q^{n+1} &= \cos(h\Omega)\, q^n + \Omega^{-1}\sin(h\Omega)\, \widetilde{p}^n \\ p^{n+1} &= \Omega\sin(h\Omega)\, q^n + \cos(h\Omega)\, \widetilde{p}^n + \frac{h}{2}\, \Psi_1 g(\Phi q^{n+1}), \end{aligned} \tag{3.13}$$

where $\Psi_1 = \psi_1(h\Omega)$ with $\psi_1(\xi) = \psi(\xi)/\,\mathrm{sinc}\,(\xi)$. Notice that, for $\Omega \to 0$, this integrator reduces to the Störmer–Verlet discretisation (2.4).

3.3 Modulated Fourier expansion of numerical solution

We are interested in the long-time behaviour of numerical approximations to the highly oscillatory Hamiltonian system (3.1). Our focus will be on the near conservation of the total energy (3.8) and of the harmonic energy (3.10) over long times.

In complete analogy to what we did in Sect. 3.1 for the analytic solution, we separate the fast and slow modes by the ansatz

$$q^n = \widetilde{q}(t_n) \qquad \text{with} \qquad \widetilde{q}(t) = \sum_{k\in\mathbb{Z}} z^k(t)\, \mathrm{e}^{\mathrm{i}k\omega t}, \tag{3.14}$$

where $t_n = nh$, and the coefficient functions are again assumed to be smooth with derivatives bounded uniformly in ω. Inserting this ansatz into the numerical scheme (3.11), expanding the functions $z^k(t \pm h)$ into Taylor series around $h = 0$ and the non-linearity into a Taylor series around z^0, and finally comparing coefficients of $e^{ik\omega t}$ yields

$$\mathcal{L}^k(hD)\,z^k = h^2 \sum_{m\geq 0} \frac{1}{m!} \sum_{s(\alpha)=k} \Psi\, g^{(m)}(z^0)\,(\Phi z)^\alpha, \tag{3.15}$$

where, with the abbreviations $s_k = \sin(\frac{1}{2}kh\omega)$ and $c_k = \cos(\frac{1}{2}kh\omega)$, the differential operator is given by

$$\mathcal{L}^k(hD)\,z^k = \begin{pmatrix} -s_k^2 z_0^k + 2\mathrm{i}hs_{2k}\dot{z}_0^k + h^2 c_{2k}\ddot{z}_0^k + \frac{1}{3}\mathrm{i}h^3 s_{2k}\dddot{z}_0^k + \cdots \\ -s_{k-1}s_{k+1}z_1^k + 2\mathrm{i}hs_{2k}\dot{z}_1^k + h^2 c_{2k}\ddot{z}_1^k + \frac{1}{3}\mathrm{i}h^3 s_{2k}\dddot{z}_1^k + \cdots \end{pmatrix}$$

As in (3.3) we separate the dominating term in the left-hand expression of (3.15). This gives a second order differential equation for z_0^0, first order differential equations for z_1^1 and z_1^{-1}, and algebraic relations for the other functions, provided that s_k^2 for $k \neq 0$ and $s_{k-1}s_{k+1}$ for $k \neq \pm 1$ are bounded away from zero. To achieve this, we assume the numerical non-resonance condition

$$|\sin(\tfrac{1}{2}kh\omega)| \geq c\sqrt{h} \qquad \text{for} \qquad k = 1, 2, \ldots, N \tag{3.16}$$

with some fixed $N \geq 2$.

The functions $\psi_1(\xi)$ and $\phi(\xi)$ in the one-step formulation (3.13) of the exponential integrator have the role to suppress or weaken numerical resonance, when $h\omega$ is close to an integral multiple of π. For the product $h\omega$ we therefore suppose

$$|\psi_1(h\omega)| \leq C\,|\operatorname{sinc}(\tfrac{1}{2}h\omega)|, \qquad |\phi(h\omega)| \leq C\,|\operatorname{sinc}(\tfrac{1}{2}h\omega)| \tag{3.17}$$

with some moderate constant C. Moreover, in addition to the standard assumption of a small step size h, we restrict our considerations to large frequencies ω and, in particular, to

$$h\omega \geq c_0 \qquad \text{for some given} \quad c_0 > 0, \tag{3.18}$$

which is precisely the situation where standard backward error analysis (Sect. 2) is not applicable.

These assumptions permit us to carry over the analysis from the analytic solution to the numerical solution of the exponential integrator. The coefficient functions z_j^k can be bounded in a similar way and they decay rapidly with increasing k. With the averaged potential $\mathcal{U}(\mathbf{z})$ of (3.5), the system (3.15) can be written as

$$\mathcal{L}^k(hD)z^k = -h^2 \Psi\, \nabla_{-k}\, \mathcal{U}(\Phi\mathbf{z}), \tag{3.19}$$

where $\Phi\mathbf{z}$ is the vector composed of Φz^k. We would like to extract from this relation a conserved quantity. Notice that the Hamiltonian (3.7) can be obtained from (3.6) by taking the scalar product of the Hamiltonian equation (3.6) with $\dot{z}^{-k}$, by summing up over all $k \in \mathbb{Z}$, and by writing the appearing expressions as total differentials. Here, we can try the same. Taking the scalar product of the equation (3.19) with $\Psi^{-1}\Phi\,\dot{z}^{-k}$ and summing up over all $k \in \mathbb{Z}$, the right-hand side is recognised as the total derivative of $\mathcal{U}(\Phi\mathbf{z})$. On the left-hand side, this procedure yields linear combinations of expressions of the form

$$\Re\langle \dot{z}^{-k}, (z^k)^{(2\ell)}\rangle, \qquad \Im\langle \dot{z}^{-k}, (z^k)^{(2\ell+1)}\rangle$$

which all, by miracle, can be written as a total derivative, e.g.,

$$\begin{aligned} \Re\langle \dot{\bar{z}}, \ddot{z}\rangle &= \frac{d}{dt}\Re\Big(\frac{1}{2}\langle \dot{\bar{z}}, \dot{z}\rangle\Big), & \Re\langle \dot{\bar{z}}, \ddddot{z}\rangle &= \frac{d}{dt}\Re\Big(\langle \dot{\bar{z}}, \dddot{z}\rangle - \frac{1}{2}\langle \ddot{\bar{z}}, \ddot{z}\rangle\Big), \\ \Im\langle \dot{\bar{z}}, \dddot{z}\rangle &= \frac{d}{dt}\Im\Big(\langle \dot{\bar{z}}, \ddot{z}\rangle\Big), & \Im\langle \dot{\bar{z}}, z^{(5)}\rangle &= \frac{d}{dt}\Im\Big(\langle \dot{\bar{z}}, \ddddot{z}\rangle - \langle \ddot{\bar{z}}, \dddot{z}\rangle\Big). \end{aligned}$$

This yields a conserved quantity for the system (3.15) that is $\mathcal{O}(h)$ close to the Hamiltonian (3.8) of (3.1).

If we take the scalar product of (3.19) with $ik\omega\, z^{-k}$, we get in a similar manner a second conserved quantity. This one turns out to be $\mathcal{O}(h)$ close the harmonic energy (3.10).

To make the analysis rigorous we truncate all appearing series after N terms, and we then patch together the estimates on small time intervals. Under the assumptions of this section this proves that the numerical solution of the exponential integrator satisfies

$$\begin{aligned} H(q^n, p^n) &= H(q^0, p^0) + \mathcal{O}(h) \\ I(q^n, p^n) &= I(q^0, p^0) + \mathcal{O}(h) \end{aligned}$$

on intervals of length $\mathcal{O}(h^{-N+1})$, where the constant symbolising the $\mathcal{O}(h)$ reminder may depend on N.

3.4 Several high frequencies and resonance

The results of the previous sections can be extended to more than one high frequencies. We consider the Hamiltonian system (3.1), where the entries of the diagonal matrix Ω are $\omega_j = \lambda_j\omega$ (for $j = 0, \ldots, \ell$) with $\lambda_0 = 0$ and $\lambda_j \geq 1$ for $j \geq 1$. As before, ω is a large parameter. The

Hamiltonian of this system is

$$H(q,\dot q) = \frac{1}{2}\sum_{j=0}^{\ell}\Big(\|\dot q_j\|^2 + \omega_j^2\|q_j\|^2\Big) + U(q), \tag{3.20}$$

where the components q_j of q can themselves be vectors in different dimensions. With the aim of extending the analysis of Sect. 3.1 we are led to consider oscillators $\mathrm{e}^{\pm \mathrm{i}\omega_j t}$ and products thereof, i.e., $\mathrm{e}^{\mathrm{i}(\mathbf{k}\cdot\omega)t}$, where $\mathbf{k}\cdot\omega$ is the scalar product of $\mathbf{k}=(k_1,\ldots,k_\ell)\in\mathbb{Z}^\ell$ and $\omega=(\omega_1,\ldots,\omega_\ell)$. To avoid redundancy in a linear combination of such expressions, we introduce the resonance module

$$\mathcal{M} = \{\mathbf{k}\in\mathbb{Z}^\ell\ ;\ k_1\lambda_1+\cdots+k_\ell\lambda_\ell = 0\}, \tag{3.21}$$

and we consider the equivalence relation $\mathbf{k}\sim\mathbf{j}$ defined by $\mathbf{k}-\mathbf{j}\in\mathcal{M}$. We choose a set $\mathcal{K}$ of representatives which is such that $|\mathbf{k}| = |k_1|+\cdots+|k_\ell|$ is minimal within the equivalence class $\mathbf{k}+\mathcal{M}$, and such that with $\mathbf{k}\in\mathcal{K}$ also $-\mathbf{k}\in\mathcal{K}$. Extending (3.2) we make the ansatz

$$q(t) = \sum_{\mathbf{k}\in\mathcal{K}} z^{\mathbf{k}}(t)\,\mathrm{e}^{\mathrm{i}(\mathbf{k}\cdot\omega)t} \tag{3.22}$$

for the solution of (3.1). Here, the smooth function $z^{\mathbf{k}}$ is partitioned into $z_j^{\mathbf{k}}$ in the same as q into q_j.

The whole programme of Sects. 3.1 and 3.3 can now be repeated to get information on the long-time behaviour of the analytic and the numerical solution in the case of several high frequencies. Let us just mention a few crucial steps.

Inserting (3.22) into the differential equation (3.1), a Taylor series expansion and a comparison of the coefficients of $\mathrm{e}^{\mathrm{i}(\mathbf{k}\cdot\omega)t}$ yields

$$\ddot z_j^{\mathbf{k}} + 2\mathrm{i}(\mathbf{k}\cdot\omega)\dot z_j^{\mathbf{k}} + \big(\omega_j^2-(\mathbf{k}\cdot\omega)^2\big)z_j^{\mathbf{k}} = \sum_{m\ge 0}\frac{1}{m!}\sum_{s(\boldsymbol{\alpha})\sim\mathbf{k}} g^{(m)}(z^{\mathbf{0}})\,z^{\boldsymbol{\alpha}}, \tag{3.23}$$

where $\boldsymbol{\alpha}=(\boldsymbol{\alpha}_1,\ldots,\boldsymbol{\alpha}_m)$ is a multi-index of elements in $\mathcal{K}$, $\boldsymbol{\alpha}_j\neq\mathbf{0}$, and $s(\boldsymbol{\alpha})=\sum_{j=1}^m\boldsymbol{\alpha}_j$. Since there is only a fixed finite number of frequencies and since we restrict all considerations to indices $\mathbf{k}\in\mathcal{K}$ with $|\mathbf{k}|\le N$, the situation is very similar to that for one high frequency. We obtain a second order differential equation for $z_0^{\mathbf{0}}$, first order differential equations for $z_j^{\pm\langle j\rangle}$ where $\langle j\rangle\in\mathcal{K}$ is the vector with value 1 at position j and 0 else, and algebraic relations for the remaining coefficient functions. As in Sect. 3.1, the equations (3.23) can be interpreted as a complex

Hamiltonian system with potential

$$\mathcal{U}(\mathbf{z}) = U(z^{\mathbf{0}}) + \sum_{m\geq 1} \frac{1}{m!} \sum_{s(\boldsymbol{\alpha})\sim \mathbf{0}} U^{(m)}(z^{\mathbf{0}}) z^{\boldsymbol{\alpha}}, \tag{3.24}$$

where $\mathbf{z} = (z^{\mathbf{k}})_{\mathbf{k}\in\mathcal{K}}$. This is again a time average of the potential $U(q)$.

This time, the potential (3.24) is invariant under the one-parameter group of transformations $z^{\mathbf{k}} \to \mathrm{e}^{\mathrm{i}(\mathbf{k}\cdot\boldsymbol{\mu})\tau} z^{\mathbf{k}}$ for all $\boldsymbol{\mu} \perp \mathcal{M}$. We always have $(\lambda_1, \ldots, \lambda_\ell) \perp \mathcal{M}$, but there may be many more vectors perpendicular to $\mathcal{M}$. For example, if the λ_j are rationally independent, then $\mathcal{M} = \{\mathbf{0}\}$ and all vectors $\boldsymbol{\mu}$ are perpendicular to $\mathcal{M}$. Therefore, Noether's theorem yields for every $\boldsymbol{\mu} \perp \mathcal{M}$ a conserved quantity which, with $y^{\mathbf{k}}(t) = z^{\mathbf{k}}(t)\,\mathrm{e}^{\mathrm{i}(\mathbf{k}\cdot\omega)t}$ is given by

$$\mathcal{I}_{\boldsymbol{\mu}}(\mathbf{y}, \dot{\mathbf{y}}) = -\mathrm{i} \sum_{\mathbf{k}\in\mathcal{K}} (\mathbf{k}\cdot\boldsymbol{\mu})(y^{-\mathbf{k}})^T \dot{y}^{\mathbf{k}}. \tag{3.25}$$

This invariant turns out to be $\mathcal{O}(\omega^{-1})$ close to

$$I_{\boldsymbol{\mu}}\big(q(t), \dot{q}(t)\big) = \sum_{j=1}^{\ell} \frac{\mu_j}{\lambda_j}\, I_j\big(q(t), \dot{q}(t)\big)$$

where

$$I_j\big(q(t), \dot{q}(t)\big) = \frac{1}{2}\big(\|\dot{q}_j(t)\|^2 + \omega_j^2 \|q_j(t)\|^2\big)$$

is the harmonic energy corresponding to the frequency ω_j.

It is possible to continue the analysis as for the case with one high frequency. This yields the long-time near conservation of the quantities $I_{\boldsymbol{\mu}}\big(q(t), \dot{q}(t)\big)$. These ideas can be extended to the numerical solution of exponential integrators, and they lead to statements on the near conservation of total and harmonic energies over long time intervals (see [CHL05] for an elaboration of the details).

3.5 Störmer–Verlet as exponential integrator

Applying the Störmer–Verlet discretisation (2.3) to the highly oscillatory differential equation (3.1) yields

$$\begin{aligned} q^{n+1} - 2q^n + q^{n-1} &= -(h\Omega)^2 q^n + h^2 g(q^n) \\ q^{n+1} - q^{n-1} &= 2h\, p^n. \end{aligned} \tag{3.26}$$

This can be written as an exponential integrator

$$\begin{aligned} \tilde{q}^{n+1} - 2\cos(h\tilde{\Omega})\,\tilde{q}^{n} + \tilde{q}^{n-1} &= h^2 \Psi g(\Phi \tilde{q}^{n}) \\ \tilde{q}^{n+1} - \tilde{q}^{n-1} &= 2h\,\mathrm{sinc}\,(h\tilde{\Omega})\,\tilde{p}^{n}, \end{aligned}$$

where $\Psi = \psi(h\tilde{\Omega})$, $\Phi = \phi(h\tilde{\Omega})$ with $\psi(\xi) = \phi(\xi) = 1$, the diagonal matrix $\tilde{\Omega}$ is related to Ω via

$$I - \frac{1}{2}(h\Omega)^2 = \cos(h\tilde{\Omega}) \qquad \text{or} \qquad \sin\Big(\frac{h\tilde{\Omega}}{2}\Big) = \frac{h\Omega}{2},$$

and the numerical approximations are related by

$$\tilde{q}^{n} = q^{n} \qquad \text{and} \qquad \mathrm{sinc}\,(h\tilde{\Omega})\,\tilde{p}^{n} = p^{n}.$$

This interpretation permits us to apply all the results that we know for the long-time behaviour of exponential integrators (3.11) to get information for the numerical solution of the Störmer–Verlet discretisation.

A linear stability analysis (i.e., vanishing non-linearity in (3.26)) shows that a necessary condition for stability is $h\omega_j \le 2$ which corresponds to $h\tilde{\omega}_j \le \pi$. Assumption (3.17) is therefore automatically satisfied for $h\tilde{\omega}_j$.

The results of Sect. 3.3 imply that $\tilde{H}(\tilde{q}^{n},\tilde{p}^{n})$ and $\tilde{I}(\tilde{q}^{n},\tilde{p}^{n})$ are conserved up to an error of size $\mathcal{O}(h)$ on intervals of length $\mathcal{O}(h^{-N+1})$. Here, $\tilde{H}$ and $\tilde{I}$ are defined in the same way as H and I, but with $\tilde{\omega}_j$ in place of ω_j. Rewritten in the original variables, this implies that on intervals of size $\mathcal{O}(h^{-N+1})$

$$\begin{aligned} H(q^{n},p^{n}) + \frac{\gamma(h\omega)}{2}\,\|p^{n}\|^2 &= \text{const} \; + \mathcal{O}(h) \\ I(q^{n},p^{n}) + \frac{\gamma(h\omega)}{2}\,\|p^{n}\|^2 &= \text{const} \; + \mathcal{O}(h), \end{aligned}$$

where

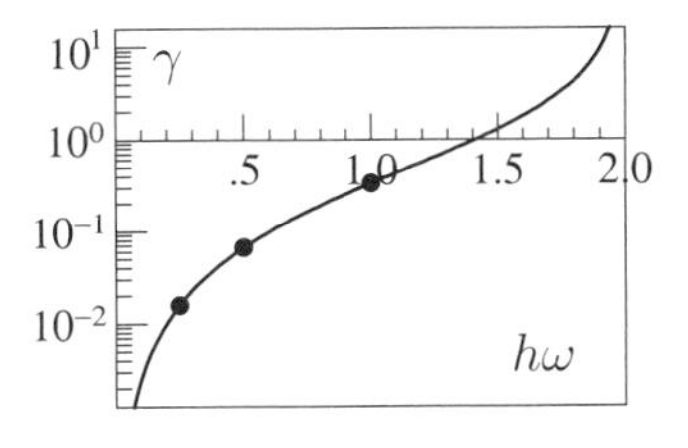

$$\gamma(h\omega) = \frac{(h\omega/2)^2}{1-(h\omega/2)^2}.$$

If $h\omega$ is not too close to 2, say $h\omega \le 1/2$, the perturbation in the above formula is small and, what is even more important, bounded without any drift.

The extension to several high frequencies is more delicate, and one has to pay attention. We still have that $\tilde{I}_{\boldsymbol{\mu}}(\tilde{q}^{n},\tilde{p}^{n})$ is well conserved over long times when $\boldsymbol{\mu}$ is perpendicular to the resonance module $\tilde{\mathcal{M}}$ corresponding to the frequencies $\tilde{\omega}_j$. However, $\boldsymbol{\mu}$ need not be perpendicular to $\mathcal{M}$. This means that a quantity is nearly conserved over long times by the numerical solution, but not by the analytic solution of the problem. Also the converse situation is possible.

4 Nonlinear wave equation

We consider the one-dimensional wave equation (the non-linear Klein–Gordon equation)

$$\partial_t^2 u - \partial_x^2 u + \rho\, u + g(u) = 0 \tag{4.1}$$

for $t \geq 0$ and $-\pi \leq x \leq \pi$ subject to periodic boundary conditions. We assume $\rho > 0$, a smooth non-linearity $g(u)$ satisfying $g(0) = g'(0) = 0$, and initial data that are small in a Sobolev norm of sufficiently high differentiation order s:

$$\Big(\big\| u(\cdot,0)\big\|_{s+1}^2 + \big\|\partial_t u(\cdot,0)\big\|_s^2 \Big)^{1/2} \leq \varepsilon\,, \qquad 0 < \varepsilon \ll 1. \tag{4.2}$$

We expand the solution into a Fourier series

$$u(x,t) = \sum_{j=-\infty}^{\infty} u_j(t) \mathrm{e}^{\mathrm{i}jx}, \qquad \partial_t u(x,t) = \sum_{j=-\infty}^{\infty} v_j(t) \mathrm{e}^{\mathrm{i}jx},$$

so that, in terms of the Fourier coefficients, the wave equation (4.1) becomes

$$\partial_t^2 u_j + \omega_j^2 u_j + \mathcal{F}_j g(u) = 0 \qquad \text{for} \qquad j \in \mathbb{Z}, \tag{4.3}$$

where $\omega_j = \sqrt{j^2 + \rho}$ and $\mathcal{F}_j v$ denotes the jth Fourier coefficient of a function $v(x)$. We consider only real solutions, so that $u_{-j} = \overline{u_j}$. The system (4.3) can be viewed as an infinite-dimensional version of (3.1) with an infinite number of frequencies.

4.1 Modulated Fourier expansion of the analytic solution

For the analytic solution we an make an ansatz analogous to (3.2) and (3.22),

$$u_j(t) \approx \sum_{\|\mathbf{k}\| \leq 2N} z_j^{\mathbf{k}}(\varepsilon t)\, \mathrm{e}^{\mathrm{i}(\mathbf{k}\cdot\boldsymbol{\omega})t},$$

where we use the multi-index notation $\mathbf{k} = (k_1, k_2, k_3, \ldots)$ with $k_j \in \mathbb{Z}$, $\|\mathbf{k}\| = |k_1| + |k_2| + |k_3| + \cdots$, and $\mathbf{k}\cdot\boldsymbol{\omega} = k_1\omega_1 + k_2\omega_2 + k_3\omega_3 + \cdots$. We insert this ansatz by a modulated Fourier expansion into the wave equation and compare the coefficients of $\mathrm{e}^{\mathrm{i}(\mathbf{k}\cdot\boldsymbol{\omega})t}$. This yields relations of the form

$$(\omega_j^2 - |\mathbf{k}\cdot\boldsymbol{\omega}|^2)z_j^{\mathbf{k}} + 2\mathrm{i}(\mathbf{k}\cdot\boldsymbol{\omega})\varepsilon\dot{z}_j^{\mathbf{k}} + \varepsilon^2\ddot{z}_j^{\mathbf{k}} + \cdots = 0, \qquad (4.4)$$

where $\dot{z}_j^{\mathbf{k}}$ and $\ddot{z}_j^{\mathbf{k}}$ are derivatives with respect to the slow time $\tau = \varepsilon t$ and the three dots indicate the contribution due to the non-linearity. We separate the dominant term, i.e.,

for $(\mathbf{k}, j)$ with $\mathbf{k}\cdot\omega = \pm\omega_j$ the second term in (4.4);
for $(\mathbf{k}, j)$ with $\big|\omega_j - |\mathbf{k}\cdot\omega|\big| \geq \varepsilon^{1/2}$ the first term in (4.4);
for $(\mathbf{k}, j)$ with $\big|\omega_j - |\mathbf{k}\cdot\omega|\big| < \varepsilon^{1/2}$ it is undecidable.

We put $z_j^{\mathbf{k}} = 0$ in the third case, and estimate the defect with the help of a non-resonance condition, which imposes a restriction on the choice of ρ, but holds for almost all ρ. We refer the reader to [CHL08b] for more details.

The theory then follows the following steps:

- Proving existence of smooth functions $z_j^{\mathbf{k}}$ with derivatives bounded independently of ε (on intervals of length ε^{-1}).
- Establishing a Hamiltonian structure and the existence of formal invariants in the differential and algebraic equations for the $z_j^{\mathbf{k}}$.
- Proving closeness (on intervals of length ε^{-1}) of the formal invariants to actions

$$I_j(t) = \frac{\omega_j}{2}|u_j(t)|^2 + \frac{1}{2\omega_j}|v_j(t)|^2$$

  and to the total energy

$$H(t) = \frac{1}{2\pi}\int_{-\pi}^{\pi}\left(\frac{1}{2}\Big((\partial_t u)^2 + (\partial_x u)^2 + \rho\, u^2\Big) + U\big(u\big)\right)dx,$$

  and the momentum

$$K(t) = \frac{1}{2\pi}\int_{-\pi}^{\pi} \partial_x u\ \partial_t u\, dx.$$

  (Unlike the actions I_j, energy and momentum are exactly conserved along solutions of the wave equation (4.1).)
- Stretching from short to long intervals of length ε^{-N+1} by patching together previous results along an invariant.

Carrying out this programme yields, in particular, an estimate on the long-time near-conservation of the harmonic actions:

$$\sum_{\ell=0}^{\infty}\omega_\ell^{2s+1}\,\frac{|I_\ell(t) - I_\ell(0)|}{\varepsilon^2} \leq C\,\varepsilon \qquad \text{for} \qquad 0 \leq t \leq \varepsilon^{-N+1}. \qquad (4.5)$$

The proof of this result is given in full detail in [CHL08b]. Related results on the long-time near-conservation of actions were previously obtained in [Bou96, Bam03] by different techniques.

We note that N can be chosen arbitrarily (but with s and C depending on N via the non-resonance condition). The near-conservation of harmonic actions is thus valid on time intervals that are much longer than the natural time scale of the problem. Moreover, the result implies spatial regularity over long times

$$\|u(\cdot,t)\|_{s+1}^2 + \|\partial_t u(\cdot,t)\|_s^2 \le \varepsilon^2(1+C\varepsilon) \qquad \text{for} \qquad 0 \le t \le \varepsilon^{-N+1}.$$

4.2 Pseudo-spectral semi-discretisation

We approximate the solution of (4.1) by a trigonometric polynomial

$$u^M(x,t) = \sum_{|j|\le M}{}' q_j(t) \mathrm{e}^{\mathrm{i}jx},$$

where the prime on the sum indicates a factor $\frac{1}{2}$ in the first and last terms. The $2M$-periodic sequence $q = (q_j)$ is solution of the system

$$\frac{d^2 q_j}{dt^2} + \omega_j^2 q_j = f_j(q) \qquad \text{with} \qquad f(q) = -\mathcal{F}_{2M}\, g(\mathcal{F}_{2M}^{-1} q), \tag{4.6}$$

where $\mathcal{F}_{2M}$ stands for the discrete Fourier transform. This is in fact a finite-dimensional Hamiltonian system with Hamiltonian

$$H_M(q,p) = \frac{1}{2} \sum_{|j|\le M}{}' \Big(|p_j|^2 + \omega_j^2 |q_j|^2 \Big) + V(q),$$

$$V(q) = \frac{1}{2M} \sum_{k=-M}^{M-1} U\big((\mathcal{F}_{2M}^{-1} q)_k \big).$$

The above programme can again be carried out in the semi-discrete case. It yields the near-conservation over long times, uniformly in the discretization parameter M, of the harmonic actions as in (4.5) and of the continuous total energy H (which stays close to H_M) and the momentum K along semi-discrete solutions [HL08].

4.3 Full discretisation

The system (4.6), which we also write as $\frac{d^2 q}{dt^2} + \Omega^2 q = f(q)$, is an ordinary differential equation, and we can apply the exponential integrators of Section 3.2 and the Störmer–Verlet method.

Combining the techniques addressed in Sections 3.3 and 4, we obtain under a numerical non-resonance condition that along the numerical solution of a symplectic exponential integrator (3.13), for $0 \le t_n \le \varepsilon^{-N+1}$,

$$\begin{aligned} \frac{|H(t_n) - H(0)|}{\varepsilon^2} &\le C\,\varepsilon \\ \frac{|K(t_n) - K(0)|}{\varepsilon^2} &\le C(\varepsilon + M^{-s} + t_n\,\varepsilon\,M^{-s+1}) \\ \sum_{\ell=0}^{\infty} \omega_\ell^{2s+1}\,\frac{|I_\ell(t_n) - I_\ell(0)|}{\varepsilon^2} &\le C\,\varepsilon. \end{aligned}$$

Here, the functions $u(x, t_n)$ and $\partial_t u(x, t_n)$ in $H(t)$, $K(t)$, and $I_\ell(t)$ have to be replaced by the trigonometric interpolation polynomial with Fourier coefficients q^n and p^n, respectively.

For the Störmer–Verlet discretisation (2.4) this holds with an additional $\mathcal{O}(h^2)$ term on the right-hand side of these estimates, if in addition the step size is restricted by the CFL condition $h\omega_M \le c < 2$.

These results, which are proved in [CHL08a], are apparently the first rigorous results on the long-time near-conservation of energy (and momentum and actions) for symplectic discretisations of a non-linear partial differential equation.

5 Linear multistep methods

There is an interesting connection between the numerical solution of special linear multistep methods for non-stiff Hamiltonian equations $\ddot{q} = -\nabla U(q)$ and the analytic solution of the highly oscillatory problem (3.1). An analogue of the technique of modulated Fourier expansion (Sect. 3) will provide new insight into the long-time behaviour of such methods. In particular, the long-time conservation of harmonic actions corresponds to the bounded-ness of parasitic solutions in the multistep discretisation.

5.1 Multistep methods for second order problems

For the second order differential equation

$$\ddot{q} = f(q) \qquad \text{with} \qquad f(q) = -\nabla U(q) \tag{5.1}$$

we consider linear multistep methods of the form

$$\sum_{j=0}^{k} \alpha_j \, q^{n+j} = h^2 \sum_{j=0}^{k} \beta_j \, f\big(q^{n+j}\big) \tag{5.2}$$

together with an approximation to the derivative $p(t) = \dot{q}(t)$ which is obtained by a finite difference formula and does not affect the propagation of errors:

$$p^n = \frac{1}{h} \sum_{j=-k}^{k} \gamma_j q^{n+j} . \tag{5.3}$$

A special case is the Störmer–Verlet discretisation (2.3). The general method (5.2) is characterised by its generating polynomials

$$\rho(\zeta) = \sum_{j=0}^{k} \alpha_j \, \zeta^j , \qquad \sigma(\zeta) = \sum_{j=0}^{k} \beta_j \, \zeta^j . \tag{5.4}$$

The classical theory of Dahlquist tells us that zero-stability (all roots of $\rho(\zeta) = 0$ satisfy $|\zeta| \leq 1$ and those on the unit circle have at most multiplicity two) and consistency ($\rho(1) = \rho'(1) = 0$, $\rho''(1) = 2\sigma(1) \neq 0$) imply convergence to the analytic solution on intervals of length $\mathcal{O}(1)$.

For the case of a Hamiltonian system (5.1) we are often interested in the long-time behaviour of numerical solutions and, in particular, in the near conservation of the total energy

$$H(t) = \frac{1}{2} \dot{q}^T \dot{q} + U(q) \tag{5.5}$$

along numerical solutions. For methods that are

- *symmetric,* i.e., $\alpha_{k-j} = \alpha_j$ and $\beta_{k-j} = \beta_j$ for all j,
- *s-stable,* i.e., all roots of $\rho(\zeta) = 0$ are on the unit circle and they are simple roots with the exception of $\zeta = 1$ which is a double root,
- *order r,* i.e., $\rho(\mathrm{e}^h) - h^2 \sigma(\mathrm{e}^h) = \mathcal{O}(h^{p+2})$ asymptotically for $h \to 0$,

we shall prove that with sufficiently accurate starting approximations the total energy is nearly conserved over long times

$$H(q_n, p_n) = H(q_0, p_0) + \mathcal{O}(h^r) \qquad \text{for} \qquad nh \leq \mathcal{O}(h^{-r-2}). \tag{5.6}$$

For methods, where no zero of $\rho(\zeta)$ other than 1 can be written as the product of two other zeros, the estimate holds even on intervals of length $\mathcal{O}(h^{-2r-3})$.

A similar statement holds for quadratic first integrals of the form $p^T Cq$ such as the angular momentum in N-body problems, and for all action variables in nearly integrable Hamiltonian systems.

5.2 Parasitic solutions

We consider a linear multistep method (5.2) where $\zeta_0 = 1$ is a double root of $\rho(\zeta) = 0$, and $\zeta_j, \zeta_{-j} = \overline{\zeta}_j$ are pairs of complex conjugate roots for $j = 1, \ldots, \ell$. In the limit $h \to 0$, (5.2) becomes a linear difference relation with characteristic polynomial $\rho(\zeta)$. Its solution is a linear combination of 1, n, and ζ_j^n, ζ_{-j}^n for $j = 1, \ldots, \ell$. Similar to the analytic solution of (3.1), we are also here confronted with two time scales:

- fast time t/h in oscillations of the form $\zeta_j^n = \zeta_j^{t/h}$;
- slow time t due to the zero root and the dynamics of (5.1).

In the general situation, ζ_j will be slightly perturbed leading to a modulation of the coefficients, and the non-linearity in the differential equation provokes the presence of products of ζ_j^n in the numerical solution. This motivates the following ansatz with smooth functions $z^j(t)$

$$q^n = \widetilde{q}(nh), \qquad \widetilde{q}(t) = \sum_{j \in \mathcal{I}} z^j(t)\, \zeta_j^{t/h} \tag{5.7}$$

which is in complete analogy to (3.2). Here, the sum is not only over $|j| \leq \ell$, but the index-set $\mathcal{I}$ includes also finite products of roots of $\rho(\zeta) = 0$ which are denoted by $\zeta_{\ell+1}, \zeta_{\ell+2}, \ldots$ and $\zeta_{-j} = \overline{\zeta}_j$. The set $\mathcal{I}$ can be finite (e.g., if the roots of $\rho(\zeta) = 0$ are all roots of unity) or infinite. In the latter case the sum will be truncated suitably for a rigorous analysis. In the representation (5.7) only the function $z^0(t)$ contributes to an approximation of the solution of (5.1), the other coefficient functions are called parasitic solutions.

For determining the smooth functions $z^j(t)$, we insert q^n from (5.7) into the multistep formula (5.2), we expand the non-linearity around z^0, and we compare the coefficients of ζ_j^n. This yields

$$\rho(\zeta_j\, \mathrm{e}^{hD})\, z^j = h^2 \sigma(\zeta_j\, \mathrm{e}^{hD}) \sum_{m \geq 0} \frac{1}{m!} \sum_{p(\alpha)=j} f^{(m)}(z^0)\, z^\alpha,$$

where D represents differentiation with respect to time, so that the Taylor series expansion of a function $z(t+h)$ is given by $\mathrm{e}^{hD} z(t)$, and the second sum is over multi-indices $\alpha = (\alpha_1, \ldots, \alpha_m)$ such that the product $\zeta_{\alpha_1} \cdot \ldots \cdot \zeta_{\alpha_m}$ equals ζ_j. This is symbolised by $p(\alpha) = j$. We divide this relation formally by $\sigma(\zeta_j\, \mathrm{e}^{hD})$ and we introduce the notation

$$\frac{\rho(\zeta_j \mathrm{e}^{\mathrm{i}x})}{\sigma(\zeta_j \mathrm{e}^{\mathrm{i}x})} = \nu_{j,0} + \nu_{j,1}\, x + \nu_{j,2}\, x^2 + \nu_{j,3}\, x^3 + \cdots . \tag{5.8}$$

Due to the symmetry of the method (i.e., $\rho(\zeta) = \zeta^k \rho(\zeta^{-1})$ and $\sigma(\zeta) =$

$\zeta^k\sigma(\zeta^{-1}))$ the coefficients $\nu_{j,0}, \nu_{j,1}, \ldots$ are all real. The functions $z^j(t)$ of (5.7) are thus determined by

$$\nu_{j,0}\, z^j + \nu_{j,1}(-\mathrm{i}h)\, \dot z^j + \nu_{j,2}(-\mathrm{i}h)^2\, \ddot z^j + \cdots \\ = h^2 \sum_{m\ge 0} \frac{1}{m!} \sum_{p(\alpha)=j} f^{(m)}(z^0)\, z^\alpha . \tag{5.9}$$

For $j=0$, we have from $\rho(1)=\rho'(1)=0$ and $\rho''(1)=2\sigma(1)\neq 0$ that $\nu_{0,0}=\nu_{0,1}=0$ and $\nu_{0,2}=-1$ so that we get a second order differential equation for $z^0(t)$ which is a perturbation of (5.1). For $0<|j|\le \ell$, where ζ_j is a simple root of $\rho(\zeta)=0$, we get a first order differential equation for $z^j(t)$. Finally, for $|j|>\ell$, where $\rho(\zeta_j)\neq 0$ we get an algebraic relation for $z^j(t)$ (to avoid technical difficulties, we assume $\sigma(\zeta_j)\neq 0$ also in this case).

5.3 Long-term stability

From now on we assume that the differential equation (5.1) is Hamiltonian, i.e., $f(q)=-\nabla U(q)$. We introduce the extended potential

$$\mathcal{U}(\mathbf{z}) = U(z^0) + \sum_{m\ge 1} \frac{1}{m!} \sum_{p(\alpha)=0} U^{(m)}(z^0) z^\alpha \tag{5.10}$$

in complete analogy to (3.5), so that the relation (5.9) becomes

$$\nu_{j,0}\, z^j + \nu_{j,1}(-\mathrm{i}h)\, \dot z^j + \nu_{j,2}(-\mathrm{i}h)^2\, \ddot z^j + \cdots = -\, h^2 \nabla_{-k}\, \mathcal{U}(\mathbf{z}). \tag{5.11}$$

Here, the situation is very similar to formula (3.19). In the left-hand side the derivatives of z^j are multiplied with the corresponding power of h, and the coefficients are real for even derivatives and purely imaginary for odd derivatives. To get a conserved quantity close to the Hamiltonian of the system, we take the scalar product of (5.11) with $\dot z^{-j}$ and sum over all $j\in\mathcal{I}$. The same miracle which helped us in Sect.3.3 to obtain a conserved quantity, applies also here. We thus get a formally conserved quantity $\mathcal{H}(\mathbf{z},\dot{\mathbf{z}})$ that is close to the Hamiltonian of the system.

Concerning further conserved quantities of the system (5.11) we encounter a serious difficulty. The extended potential (5.10) is invariant with respect to the transformation $z^j \to \zeta_j^n z^j$ only for integral values of n, so that Noether's theorem cannot be applied. Nevertheless, we take the scalar product of (5.11) with $\mathrm{i}jz^{-j}$, but this time we sum up only the relation for j and that for $-j$ (for $1\le j\le \ell$). Since $\nu_{j,0}=\nu_{-j,0}=0$ and $\nu_{j,1}=-\nu_{-j,1}\neq 0$, we obtain as in Sect. 3 that the left-hand side is the total derivative of an expression which is close to $ch\|z^j\|^2$ with a constant $c \neq 0$. The dominant term of the right-hand side which, up to the factor $-h^2/2$, equals $U''(z^0)(\mathrm{i}jz^{-j}, z^j) + U''(z^0)(-\mathrm{i}jz^j, z^{-j})$, vanishes due to the symmetry of the Hessian $U''(z^0)$ of the potential. Consequently, all terms of the right-hand side contain at least three times a factor z^j with $j \neq 0$. As long as $\|z^j\| \leq \delta$ for $j \neq 0$, we thus get an expression $\mathcal{I}_j(\mathbf{z}, \dot{\mathbf{z}})$ close to $\|z_j\|^2$ whose time derivative is bounded by $\mathcal{O}(h\delta^3)$.

If the starting approximations $q^0, q^1, \ldots, q^{k-1}$ for the multistep formula (5.2) are such that $\|z^j(0)\| \leq \delta$, then the parasitic solutions $z^j(t)$ for $j \neq 0$ remain bounded by 2δ on a time interval of length $\mathcal{O}(h^{-1}\delta^{-1})$. In a typical situation, when the multistep method is of order r and the starting approximations are obtained by a one-step method of order r, we have $\delta = \mathcal{O}(h^{r+1})$. The parasitic solutions are then bounded by $\mathcal{O}(h^{r+1})$ on intervals of length $\mathcal{O}(h^{-r-2})$. This implies that also the Hamiltonian is nearly conserved on a time interval of this length.

A rigorous elaboration of these ideas can be found in the publication [HL04] and in [HLW06, Chap. XV].

Bibliography

[Bam03] D. Bambusi (2003). Birkhoff normal form for some nonlinear PDEs, *Comm. Math. Phys.* **234**, 253–285.

[BG94] G. Benettin & A. Giorgilli (1994). On the Hamiltonian interpolation of near to the identity symplectic mappings with application to symplectic integration algorithms, *J. Statist. Phys.* **74**, 1117–1143.

[BGG87] G. Benettin, L. Galgani & A. Giorgilli (1987). Realization of holonomic constraints and freezing of high frequency degrees of freedom in the light of classical perturbation theory. Part I, *Comm. Math. Phys.* **113**, 87–103.

[Bou96] J. Bourgain (1996). Construction of approximative and almost periodic solutions of perturbed linear Schrödinger and wave equations, *Geom. Funct. Anal.* **6**, 201–230.

[CHL03] D. Cohen, E. Hairer & C. Lubich (2003). Modulated Fourier expansions of highly oscillatory differential equations, *Foundations of Comput. Maths* **3**, 327–345.

[CHL05] D. Cohen, E. Hairer & C. Lubich (2005). Numerical energy conservation for multi-frequency oscillatory differential equations, *BIT* **45**, 287–305.

[CHL08a] D. Cohen, E. Hairer & C. Lubich (2008). Conservation of energy, momentum and actions in numerical discretizations of nonlinear wave equations, *Numer. Math.*, **110** 113–143.

[CHL08b] D. Cohen, E. Hairer, & C. Lubich (2008). Long-time analysis of nonlinearly perturbed wave equations via modulated Fourier expansions, *Arch. Ration. Mech. Anal.* **187**, 341–368.

[Deu79] P. Deuflhard (1979). A study of extrapolation methods based on multistep schemes without parasitic solutions, *Z. Angew. Math. Phys.* **30**, 177–189.

[GASS99] B. García-Archilla, J. M. Sanz-Serna, & R. D. Skeel (1999). Long-time-step methods for oscillatory differential equations, *SIAM J. Sci. Comput.* **20**, 930–963.

[Gau61] W. Gautschi (1961). Numerical integration of ordinary differential equations based on trigonometric polynomials, *Numer. Math.* **3** 381–397.

[HL97] E. Hairer & C. Lubich (1997). The life-span of backward error analysis for numerical integrators, *Numer. Math.* **76**, 441–462 (Erratum: `http://www.unige.ch/math/folks/hairer/`).

[HL01] E. Hairer & C. Lubich (2001). Long-time energy conservation of numerical methods for oscillatory differential equations, *SIAM J. Numer. Anal.* **38**, 414–441.

[HL04] E. Hairer & C. Lubich (2004). Symmetric multistep methods over long times, *Numer. Math.* **97**, 699–723.

[HL08] E. Hairer & C. Lubich (2008). Spectral semi-discretisations of nonlinear wave equations over long times, *Foundations of Comput. Math.* **8** 319–334.

[HLW03] E. Hairer, C. Lubich, & G. Wanner (2003). Geometric numerical integration illustrated by the Störmer–Verlet method, *Acta Numerica* **12**, 399–450.

[HLW06] E. Hairer, C. Lubich, & G. Wanner (2006). *Geometric Numerical Integration. Structure-Preserving Algorithms for Ordinary Differential Equations* (2nd edition), Springer Series in Computational Mathematics 31 (Springer-Verlag, Berlin).

[HNW93] E. Hairer, S. P. Nørsett, & G. Wanner (1993). *Solving Ordinary Differential Equations I. Nonstiff Problems* (2nd edition), Springer Series in Computational Mathematics 8 (Springer, Berlin).

[HW96] E. Hairer & G. Wanner (1996). *Solving Ordinary Differential Equations II. Stiff and Differential-Algebraic Problems* (2nd edition), Springer Series in Computational Mathematics 14 (Springer-Verlag, Berlin).

[Rei99] S. Reich (1999). Backward error analysis for numerical integrators, *SIAM J. Numer. Anal.* **36**, 1549–1570.

2
Highly oscillatory quadrature

Daan Huybrechs
Department of Computer Science
K.U. Leuven
Celestijnenlaan 200A
BE-3001 Leuven
Belgium
Email: daan.huybrechs@cs.kuleuven.be

Sheehan Olver
St John's College
Oxford University
St Giles'
Oxford OX1 3JP
United Kingdom
Email: Sheehan.Olver@sjc.ox.ac.uk

Abstract

Oscillatory integrals are present in many applications, and their numerical approximation is the subject of this paper. Contrary to popular belief, their computation can be achieved efficiently, and in fact, the more oscillatory the integral, the more accurate the approximation. We review several existing methods, including the asymptotic expansion, Filon method, Levin collocation method and numerical steepest descent. We also present recent developments for each method.

1 Introduction

The aim of this paper is to review recent methods for the evaluation of the oscillatory integral

$$I[f] = \int_\Omega f(\boldsymbol{x}) e^{i\omega g(\boldsymbol{x})} \, dx,$$

where f and g are nonoscillatory functions, the frequency of oscillations ω is large and Ω is some piecewise smooth domain. By taking the real and imaginary parts of this integral, we obtain integrals with trigonometric kernels:

$$\Re I[f] = \int_\Omega f(\boldsymbol{x}) \cos \omega g(\boldsymbol{x}) \,\mathrm{d}V \quad \text{and} \quad \Im I[f] = \int_\Omega f(\boldsymbol{x}) \sin \omega g(\boldsymbol{x}) \,\mathrm{d}V.$$

Highly oscillatory integrals of this form play a valuable role in applications. Using the *modified Magnus expansion* [Ise04], highly oscillatory differential equations of the form $y'' + g(t)y = 0$, where $g(t) \to \infty$ while the derivatives of g are moderate, can be expressed in terms of an infinite sum of highly oscillatory integrals. Differential equations of this form appear in many areas, including special functions, e.g., the Airy function. Oscillatory integrals also typically appear in scientific disciplines that involve the modelling of wave phenomena. For example in acoustics, the boundary element method requires the evaluation of highly oscillatory integrals, in order to solve integral equations with oscillatory kernels [HV07b]. *Modified Fourier series* use highly oscillatory integrals to obtain a function approximation scheme that converges faster than the standard Fourier series ([IN06a], reviewed in [HO08]). There are many other applications, including fluid dynamics, image analysis and more.

To understand why we need special methods for oscillatory integrals, it is important to study where traditional quadrature methods fail. Most nonoscillatory quadrature methods approximate an integral by a weighted sum sampling the integrand at n discrete points $\{x_1, \ldots, x_n\}$, and averaging the samples with suitable weights $\{w_1, \ldots, w_n\}$:

$$\int_a^b w(x) f(x) \,\mathrm{d}x \approx \sum_{k=1}^n w_k f(x_k), \tag{1.1}$$

where w is some nonnegative weight function. Regardless of the particular method used, (1.1) fails as a quadrature scheme for high frequency oscillation when $w(x) \equiv 1$, unless n grows with ω. To see this, consider the integral

$$\int_a^b f(x) \mathrm{e}^{\mathrm{i}\omega x} \,\mathrm{d}x \approx \sum_{k=1}^n w_k f(x_k) \mathrm{e}^{\mathrm{i}\omega x_k},$$

where n, w_k and x_k are all fixed for increasing ω. Assuming that this sum is not identically zero, it cannot decay as ω increases. This can be seen in Fig. 1 for the integral

$$\int_0^1 x^2 \mathrm{e}^{\mathrm{i}\omega x} \,\mathrm{d}x.$$

A simple application of integration by parts—which will be investigated

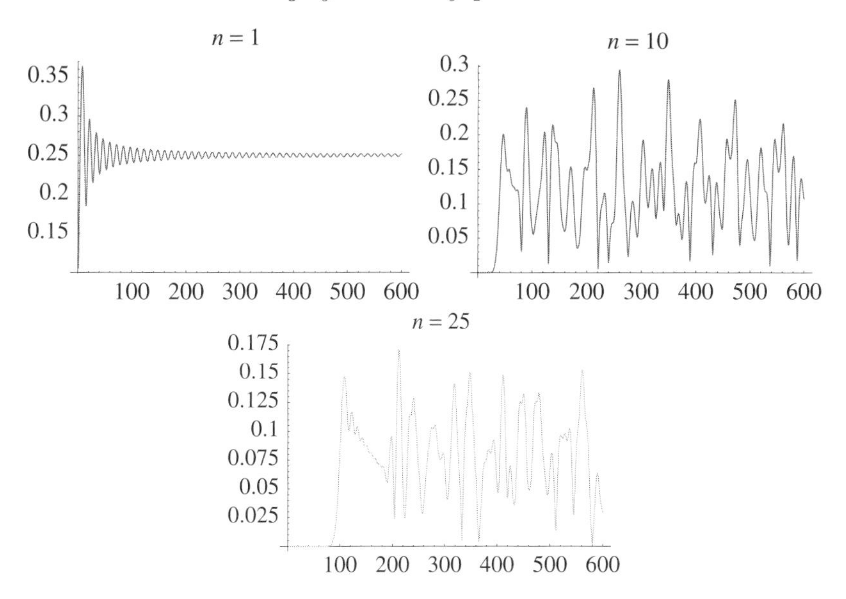

Fig. 1. The absolute error in approximating $\int_0^1 x^2 \mathrm{e}^{\mathrm{i}\omega x}\,\mathrm{d}x$ as a function of ω by an n-point Gauss–Legendre quadrature scheme, for $n = 1$, 10 and 25.

further in the next section—reveals that the integral itself decays like $\mathcal{O}(\omega^{-1})$. Thus the error of any weighted sum is $\mathcal{O}(1)$, which compares to an error of order $\mathcal{O}(\omega^{-1})$ when approximating the integral by zero! It is safe to assume that a numerical method which is less accurate than equating the integral to zero is of little practical use. On the other hand, letting n be proportional to the frequency can result in considerable computational costs. This is magnified significantly when we attempt to integrate over multivariate domains. Even nonoscillatory quadrature is computationally difficult for multivariate integrals, and high oscillations would only serve to further exasperate the situation. Thus we must look for alternative methods to approximate such integrals.

In this paper, we focus on four methods of approximation: the asymptotic expansion, Filon method, Levin collocation method and numerical steepest descent. An enormous amount of research has been conducted on the asymptotics of oscillatory integrals, hence we investigate the derivation of an *asymptotic expansion* in Section 2. Here there is a distinction between integrals with *stationary points*—points where $g'(x)$ vanishes—and those without. We can also derive asymptotic results for multivariate integrals.

With the asymptotic groundwork in place, we consider in depth methods based on the *Filon method* in Section 3. These include Filon-type methods which achieve higher asymptotic orders and Moment-free Filon-type methods for irregular oscillators with stationary points. We can also derive a multivariate Filon-type method, though only for linear oscillators over a simplex.

Like Moment-free Filon-type methods, *Levin collocation methods* of Section 4 can be constructed for irregular oscillators, though only for oscillators without stationary points. We also develop Levin-type methods which obtain higher asymptotic orders. Alternatively, we can use a Chung, Evans and Webster construction which frames the method in a more traditional quadrature context. Levin-type methods' generality lends itself to multivariate integration, which can be accomplished over more general domains than Filon-type methods, as long as a *nonresonance condition* is satisfied.

Steepest descent is a classical method for determining the asymptotic expansion of oscillatory integrals with stationary points. Recently, it has been used as a numerical quadrature scheme, which is reviewed in Section 5. These methods achieve twice the asymptotic decay for the same amount of function evaluations as Filon-type and Levin-type methods. They are also applicable to multivariate integrals, even to those which do not satisfy the nonresonance condition.

Finally, we give a brief overview of other existing methods in Section 6, for both Fourier oscillators and irregular oscillators. The methods for Fourier oscillators are generally based on standard quadrature methods, so the asymptotic order achievable is limited to either $\mathcal{O}(\omega^{-1})$ or, with the use of endpoint data, $\mathcal{O}(\omega^{-2})$.

2 Asymptotics

Whereas standard quadrature schemes are inefficient, a straightforward alternative exists in the form of asymptotic expansions. Asymptotic expansions actually improve with accuracy as the frequency increases, and—assuming sufficient differentiability of f and g—to arbitrarily high order. Furthermore the number of operations required to produce such an expansion is independent of the frequency, and extraordinarily small. Even more surprising is that this is all obtained by only requiring knowledge of the function at very few critical points within the interval—the endpoints and stationary points—as well as its derivatives at these points if higher asymptotic orders are required. There is, however, one critical flaw which impedes their use as quadrature formulæ: asymptotic expansions do not in general converge when the frequency is fixed, hence their accuracy is limited.

2.1 Asymptotic expansion for integrals without stationary points

Whenever g is free of stationary points—i.e., $g'(x) \neq 0$ within the interval of integration—we can derive an *asymptotic expansion* in a very straightforward manner by repeatedly applying integration by parts. The first term of the expansion is determined as follows:

$$\begin{aligned} I[f] &= \int_a^b f(x) e^{i\omega g(x)} \, dx = \frac{1}{i\omega} \int_a^b \frac{f(x)}{g'(x)} \frac{d}{dx} e^{i\omega g(x)} \, dx \\ &= \frac{1}{i\omega} \left[\frac{f(b)}{g'(b)} e^{i\omega g(b)} - \frac{f(a)}{g'(a)} e^{i\omega g(a)} \right] - \frac{1}{i\omega} \int_a^b \frac{d}{dx} \left[\frac{f(x)}{g'(x)} \right] e^{i\omega g(x)} \, dx. \end{aligned}$$

The term

$$\frac{1}{i\omega} \left[\frac{f(b)}{g'(b)} e^{i\omega g(b)} - \frac{f(a)}{g'(a)} e^{i\omega g(a)} \right] \tag{2.1}$$

approximates the integral $I[f]$ with an error

$$-\frac{1}{i\omega} I\left[\frac{d}{dx} \left[\frac{f(x)}{g'(x)} \right] \right] = \mathcal{O}(\omega^{-2}),$$

using the fact that the integral decays like $\mathcal{O}(\omega^{-1})$ [Ste93]. Thus the more oscillatory the integrand, the more accurately (2.1) can approximate the integral, with a relative accuracy $\mathcal{O}(\omega^{-1})$. Moreover the error term is itself an oscillatory integral, thus we can integrate by parts again to obtain an approximation with an error $\mathcal{O}(\omega^{-2})$. Iterating this procedure results in an asymptotic expansion:

Theorem 1 *Suppose that $g' \neq 0$ in $[a, b]$. Then*

$$I[f] \sim -\sum_{k=1}^{\infty} \frac{1}{(-i\omega)^k} \left\{ \sigma_k(b) e^{i\omega g(b)} - \sigma_k(a) e^{i\omega g(a)} \right\},$$

where

$$\sigma_1 = \frac{f}{g'}, \qquad \sigma_{k+1} = \frac{\sigma_k'}{g'}, \qquad k \geq 1.$$

We can find the error term for approximating $I[f]$ by the first s terms of this expansion:

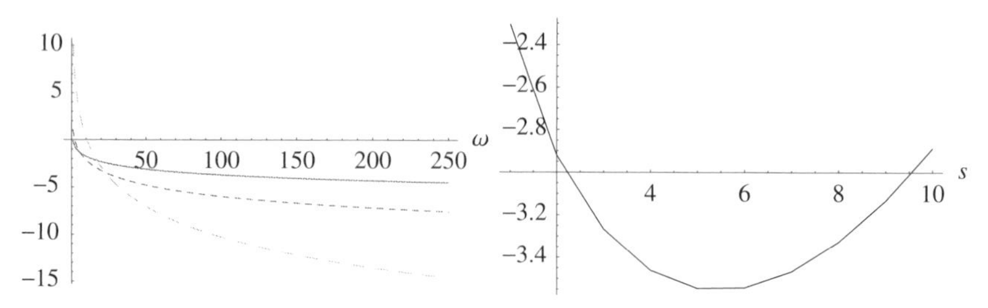

Fig. 2. The base-10 logarithm of the error in approximating the integral $\int_0^1 \cos x \, \mathrm{e}^{\mathrm{i}\omega\left(x^2+x\right)} \, \mathrm{d}x$. The left graph compares the one-term (solid line), three-term (dotted line) and ten-term (dashed line) asymptotic expansions. The right graph shows the error in the s-term asymptotic expansion for $\omega = 20$.

$$I[f] = -\sum_{k=1}^{s} \frac{1}{(-\mathrm{i}\omega)^k} \left\{ \sigma_k(b) \mathrm{e}^{\mathrm{i}\omega g(b)} - \sigma_k(a) \mathrm{e}^{\mathrm{i}\omega g(a)} \right\} + \frac{1}{(-\mathrm{i}\omega)^s} I[\sigma_s'] .$$

In Fig. 2 we use the partial sums of the asymptotic expansion to approximate the integral

$$\int_0^1 \cos x \, \mathrm{e}^{\mathrm{i}\omega\left(x^2+x\right)} \, \mathrm{d}x.$$

We compare three partial sums of the asymptotic expansion in the left graph: s equal to one, three and ten. This graph demonstrates that increasing the number of terms used in the expansion does indeed increase the rate that the error in approximation goes to zero for increasing ω. However, for any given frequency the expansion reaches an optimal error, after which adding terms to the expansion actually increases the error. This is shown in the right graph for ω fixed to be 20, in which case the optimal expansion consists of five terms.

2.2 *Asymptotic expansions in the presence of stationary points*

Consider the following integral with a single stationary point:

$$\int_a^b f(x) \mathrm{e}^{\mathrm{i}\omega g(x)} \, \mathrm{d}x,$$

where, for $\xi \in (a, b)$, $0 = g(\xi) = g'(\xi) = \cdots = g^{(r-1)}(\xi)$, $g^{(r)}(\xi) > 0$ and $g'(x) \neq 0$ for $x \in [a, b] \backslash \xi$. If $g(\xi) \neq 0$, the integral can easily by written in this form by replacing g by $g - g(\xi)$ and multiplying the resulting integral by $\mathrm{e}^{\mathrm{i}\omega g(\xi)}$. If the integral has multiple stationary points, we can write it as multiple integrals of this form.

The method used in the previous section does not work as the stationary point becomes a singularity when we attempt to integrate by parts. Fortunately, it is possible to remove the singularity (for simplicity we assume that $r = 2$):

$$\begin{aligned} I[f] &= I[f - f(\xi)] + f(\xi)I[1] \\ &= \frac{1}{\mathrm{i}\omega}\int_a^b \frac{f(x) - f(\xi)}{g'(x)} \frac{\mathrm{d}}{\mathrm{d}x}\mathrm{e}^{\mathrm{i}\omega g(x)}\,\mathrm{d}x + f(\xi)I[1] \\ &= \frac{1}{\mathrm{i}\omega}\left[\frac{f(b) - f(\xi)}{g'(b)}\mathrm{e}^{\mathrm{i}\omega g(b)} - \frac{f(a) - f(\xi)}{g'(a)}\mathrm{e}^{\mathrm{i}\omega g(a)}\right] \\ &\quad - \frac{1}{\mathrm{i}\omega} I\left[\frac{\mathrm{d}}{\mathrm{d}x}\left[\frac{f(x) - f(\xi)}{g'(x)}\right]\right] + f(\xi)I[1]\,. \end{aligned}$$

Again the error term $I\left[\frac{\mathrm{d}}{\mathrm{d}x}\left[\frac{f(x)-f(\xi)}{g'(x)}\right]\right]$ is an oscillatory integral, and hence we can use induction to obtain a full asymptotic expansion. If there are higher order stationary points, we can subtract out a polynomial to ensure both the function value and necessary derivatives of the integrand vanish in order to make the singularity removable. We thus obtain the following theorem. To simplify subsequent developments, we introduce the Levin differential operator $\mathcal{L}[v] = v' + \mathrm{i}\omega g' v$, which has the property $I[\mathcal{L}[v]] = v(b)\mathrm{e}^{\mathrm{i}\omega g(b)} - v(a)\mathrm{e}^{\mathrm{i}\omega g(a)}$, as explained in Section 4:

Theorem 2 *[IN05] Define* $\mu[f] = \sum_{k=0}^{r-2} \frac{f^k(\xi)}{k!}\mu_k(x)$, *where* $\mathcal{L}[\mu_k](x) = x^k$ *for* $\mathcal{L}[v] = v' + \mathrm{i}\omega g' v$. *Furthermore, let*

$$\sigma_0(x) = f(x), \qquad \sigma_{k+1}(x) = \frac{\mathrm{d}}{\mathrm{d}x}\frac{\sigma_k(x) - \mathcal{L}[\mu[\sigma_k]](x)}{g'(x)}.$$

Then

$$\begin{aligned} I[f] \sim \sum_{k=0}^{\infty} \frac{1}{(-\mathrm{i}\omega)^k}\left\{\mu[\sigma_k](b)\mathrm{e}^{\mathrm{i}\omega g(b)} - \mu[\sigma_k](a)\mathrm{e}^{\mathrm{i}\omega g(a)}\right\} \\ - \sum_{k=0}^{\infty} \frac{1}{(-\mathrm{i}\omega)^{k+1}}\left\{\frac{\sigma_k(b) - \mathcal{L}[\mu[\sigma_k]](b)}{g'(b)}\mathrm{e}^{\mathrm{i}\omega g(b)}\right. \\ \left. - \frac{\sigma_k(a) - \mathcal{L}[\mu[\sigma_k]](a)}{g'(a)}\mathrm{e}^{\mathrm{i}\omega g(a)}\right\}. \end{aligned}$$

This asymptotic expansion depends on knowledge of the moments

$$I\left[x^k\right] = I[\mathcal{L}[\mu_k]] = \mu_k(b)\mathrm{e}^{\mathrm{i}\omega g(b)} - \mu_k(a)\mathrm{e}^{\mathrm{i}\omega g(a)},$$

which is less than ideal. The required functions μ_k are known in closed form only when $g(x) = x^r$. But we need not use the polynomials x^k in the preceding construction: any basis that can interpolate f and its derivatives at the stationary point will do. Thus we can construct a basis so that the moments are known in closed form for any g. A good choice was presented in [Olv07]. Define

$$\phi_{r,k}(x) = D_{r,k}(\operatorname{sgn}(x-\xi))\frac{\omega^{-\frac{k+1}{r}}}{r}\mathrm{e}^{-\mathrm{i}\omega g(x)+\frac{1+k}{2r}\mathrm{i}\pi} \times \left[\Gamma\left(\frac{1+k}{r}, -\mathrm{i}\omega g(x)\right) - \Gamma\left(\frac{1+k}{r}, 0\right)\right],$$

where

$$D_{r,k}(s) = \begin{cases} (-1)^k & s < 0 \text{ and } r \text{ even}, \\ (-1)^k \mathrm{e}^{-\frac{1+k}{r}\mathrm{i}\pi} & s < 0 \text{ and } r \text{ odd}, \\ -1 & \text{otherwise}. \end{cases}$$

Then

$$\mathcal{L}[\phi_{r,k}](x) = \operatorname{sgn}(x)^{r+k+1}\frac{|g(x)|^{\frac{k+1}{r}-1}g'(x)}{r},$$

which is a C^∞ Chebyshev set, hence can interpolate any function at a sequence of nodes and multiplicities. Furthermore, we can compute the moments with respect to this basis:

$$I[\mathcal{L}[\phi_{r,k}]] = \phi_{r,k}(b)\mathrm{e}^{\mathrm{i}\omega g(b)} - \phi_{r,k}(a)\mathrm{e}^{\mathrm{i}\omega g(a)}.$$

We obtain an equivalent theorem to Theorem 2 using this basis:

Theorem 3 *[Olv07] Define* $\mu[f] = \sum_{k=0}^{r-2} c_k\phi_{r,k}$ *so that*

$$\mathcal{L}[\mu[f]](\xi) = f(\xi), \ldots, \mathcal{L}[\mu[f]]^{(r-2)}(\xi) = f^{(r-2)}(\xi).$$

Furthermore, let

$$\sigma_0(x) = f(x), \qquad \sigma_{k+1}(x) = \frac{\mathrm{d}}{\mathrm{d}x}\frac{\sigma_k(x) - \mathcal{L}[\mu[\sigma_k]](x)}{g'(x)}.$$

Then

$$I[f] \sim \sum_{k=0}^{\infty} \frac{1}{(-\mathrm{i}\omega)^k} \left\{ \mu[\sigma_k](b) \mathrm{e}^{\mathrm{i}\omega g(b)} - \mu[\sigma_k](a) \mathrm{e}^{\mathrm{i}\omega g(a)} \right\}$$
$$- \sum_{k=0}^{\infty} \frac{1}{(-\mathrm{i}\omega)^{k+1}} \left\{ \frac{\sigma_k(b) - \mathcal{L}[\mu[\sigma_k]](b)}{g'(b)} \mathrm{e}^{\mathrm{i}\omega g(b)} - \frac{\sigma_k(a) - \mathcal{L}[\mu[\sigma_k]](a)}{g'(a)} \mathrm{e}^{\mathrm{i}\omega g(a)} \right\}.$$

The first term in this expansion is equivalent to the method of stationary phase [Olv74], though derived in a different manner.

2.3 Multivariate asymptotic expansion

A multivariate asymptotic expansion can be derived in a very similar manner to the univariate asymptotic expansion, where a partial integration formula based on generalized Stokes' theorem is used. The expansion no longer has a simple explicit form, but the key point of the theorem is that the asymptotics depends on the vertices of the domain. There is an additional criteria known as the *nonresonance condition*, which is satisfied whenever ∇g does not vanish and is not orthogonal to the boundary. This is similar to requiring the absence of stationary points in one dimension.

Theorem 4 *[IN06b] Suppose that the nonresonance condition is satisfied. Then, for $\omega \to \infty$,*

$$I_g[f, \Omega] \sim \sum_{k=0}^{\infty} \frac{1}{(-\mathrm{i}\omega)^{k+d}} \Theta_k[f],$$

where $\Theta_k[f]$ depends on $f^{(\boldsymbol{m})}$ for $|\boldsymbol{m}| \le k$, where $|\boldsymbol{m}|$ is the sum of the components of $\boldsymbol{m}$, evaluated at the vertices of Ω.

3 Filon method

Though the importance of asymptotic methods cannot be overstated, the lack of convergence forces us to look for alternative numerical schemes. In practice the frequency of oscillations is fixed, and the fact that an approximation method is more accurate for higher frequency is irrelevant; all that matters is that the error for the given integral is small. Thus, though asymptotic expansions lie at the heart of oscillatory quadrature, they are not useful in and of themselves unless the frequency is extremely large. In a nutshell, the basic goal of oscillatory quadrature, then, is to find and investigate methods which preserve the asymptotic properties of an asymptotic expansion, whilst allowing for arbitrarily high accuracy for a fixed frequency. Having been spoilt by the pleasures of asymptotic expansions, we also want methods such that the order of operations is independent of ω, and comparable in cost to the evaluation of the expansion. Fortunately, methods have been developed with these properties, in particular the Filon method and Levin collocation method.

The first known numerical quadrature scheme for oscillatory integrals was developed in 1928 by Louis Napoleon George Filon [Fil28]. Filon presented a method for efficiently computing the Fourier integral

$$\int_a^b f(x) \sin \omega x \, \mathrm{d}x.$$

As originally constructed, the method consists of dividing the interval into $2n$ panels of size h, and applying a modified Simpson's rule on each panel. In other words, f is interpolated at the endpoints and midpoint of each panel by a quadratic. In each panel the integral becomes a polynomial multiplied by the oscillatory kernel $\sin \omega x$, which can be integrated in closed form. This method was generalized in [Luk54] by using higher degree polynomials in each panel, again with evenly spaced nodes. We refer to any method where f is approximated by a polynomial $v = \sum c_k x^k$ which is integrated exactly as a Filon-type method. Thus we define

$$Q^F[f] = I[v] = \sum c_k I\left[x^k\right].$$

We also can obtain a bound via the Cauchy-Shwarz inequality which shows that a Filon-type method converges whenever the polynomial approximation converges in the L_2 norm:

$$\left|I[f] - Q^F[f]\right| = |I[f-v]| \leq \sqrt{b-a}\, \|f-v\|_2 .$$

3.1 Univariate Filon-type methods

Theorem 1 has an interesting consequence for oscillatory integrals without stationary points: if f and its first $s-1$ derivatives vanish at the endpoints of the interval, then the oscillatory integral itself decays like $\mathcal{O}(\omega^{-s-1})$. The error of a Filon-type method is itself a highly oscillatory integral. Thus if v interpolates the first $s-1$ derivatives of f at the

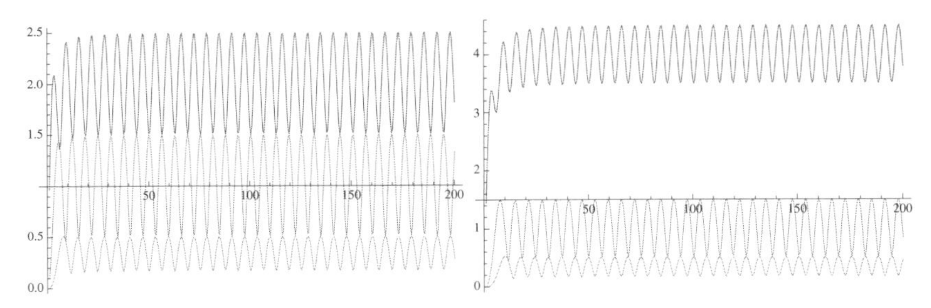

Fig. 3. The error in approximating $\int_0^1 \frac{x+3}{x+1} e^{i\omega x}\, dx$. In the left graph, we scale the error by ω^2 and compare the one-term asymptotic expansion (top) and Filon-type methods with nodes $\{0, 1\}$ and multiplicities both one (middle) and nodes $\{0, \frac{1}{2}, 1\}$ and multiplicities all one (bottom). In the right graph, we scale the error by ω^3 and we compare the two-term asymptotic expansion (top) and Filon-type methods with nodes $\{0, 1\}$ and multiplicities both two (middle) and nodes $\{0, \frac{1}{2}, 1\}$ and multiplicities $\{2, 1, 2\}$ (bottom).

endpoints we know immediately from the asymptotic expansion that the error decays like

$$I[f] - Q^F[f] = I[f - v] \sim \mathcal{O}\left(\omega^{-s-1}\right).$$

Thus we can use Hermite interpolation to achieve higher asymptotic orders. Choose a sequence of nodes $\{x_1, \ldots, x_\nu\}$, where $x_1 = a$ and $x_\nu = b$, and associate a sequence of multiplicities m_k. We then determine $v = \sum c_k x^k$ by solving the following system:

$$v(x_k) = f(x_k), \ldots, v^{(m_k-1)}(x_k) = f^{(m_k-1)}(x_k), \qquad k = 1, 2, \ldots, \nu.$$

In this case, $s = \min\{m_1, m_\nu\}$. This method was proposed in [IN05]. Without delving into details, we note that evaluation of derivatives can be replaced by function evaluation at points that coalesce at the vertices as ω increases [IN04].

If the integral has stationary points, interpolation at the endpoints is not sufficient to accelerate the asymptotic decay. This follows since the asymptotic expansion now depends on f at the stationary point as well. In order to achieve a decay rate of $\mathcal{O}\left(\omega^{-s-1/r}\right)$ we thus need to interpolate f and its first $rs - 2$ derivatives at the stationary point [Olv07], which follows from Theorem 2, L'Hôpital's rule, and the fact that the integral itself decays like $\mathcal{O}\left(\omega^{-1/r}\right)$ [Ste93].

As a simple example, consider the integral

$$I[f] = \int_0^1 \frac{x+3}{x+1} \mathrm{e}^{\mathrm{i}\omega x}\, \mathrm{d}x.$$

In Fig. 3 we compare the asymptotic expansion to four Filon-type methods: two of order $\mathcal{O}(\omega^{-2})$ and two of order $\mathcal{O}(\omega^{-3})$. As can be seen, Filon-type methods allow us to considerably decrease the error compared to the asymptotic expansion, even when only a few nodes are used.

3.2 Moment-free Filon-type methods

The computation of the Filon approximation rests on the ability to compute the moments

$$\int_a^b x^k \mathrm{e}^{\mathrm{i}\omega x}\, \mathrm{d}x.$$

For this particular oscillator the moments are computable in closed form, either through integration by parts or by the identity

$$\int_a^b x^k \mathrm{e}^{\mathrm{i}\omega x}\, \mathrm{d}x = \frac{1}{(-\mathrm{i}\omega)^{k+1}} \left[\Gamma(1+k, -\mathrm{i}\omega a) - \Gamma(1+k, -\mathrm{i}\omega b)\right],$$

where Γ is the incomplete Gamma function [AS65]. But often in applications we have irregular oscillators, giving us integrals of the form

$$\int_a^b f(x) \mathrm{e}^{\mathrm{i}\omega g(x)}\, \mathrm{d}x.$$

In this case knowledge of moments depends on the oscillator g. If we are fortunate, the moments are still known, and Filon-type methods are applicable. This is true if g is a polynomial of degree at most two or if $g(x) = x^r$. But we need not step too far outside the realm of these simple examples before explicit moment calculation falls apart: moments are not even known for $g(x) = x^3 - x$ nor $g(x) = \cos x$. Even when moments are known, they are typically known in terms of special functions, such as the incomplete Gamma function or more generally the hypergeometric function [AS65]. The former of these is efficiently computable [vLT84]. The latter, on the other hand, are significantly harder to compute for the invariably large parameters needed, though some computational schemes exist [For97, Mul01, LO94]. Thus it is necessary that we find an alternative to the Filon method.

We are not required to use polynomial interpolation, however, and

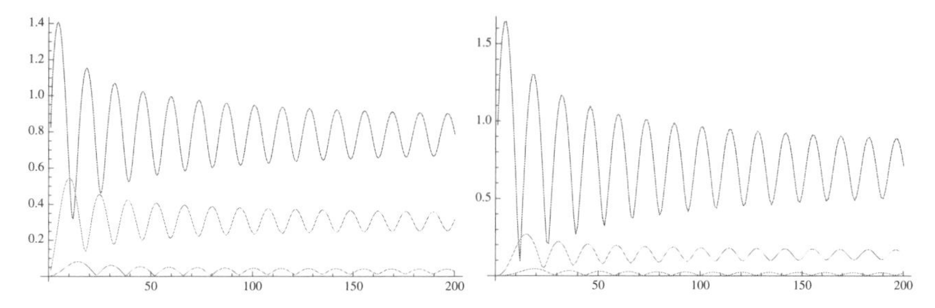

Fig. 4. The error in approximating $\int_{-1}^{1} \frac{x+3}{x+2} \mathrm{e}^{\mathrm{i}\omega(1-\cos x)}\,\mathrm{d}x$. In the left graph, we scale the error by $\omega^{3/2}$ and compare the one-term moment-free asymptotic expansion (top) and Moment-free Filon-type methods with nodes $\{-1,0,1\}$ and multiplicities all one (middle) and nodes $\left\{-1,-\frac{1}{2},0,\frac{1}{2},1\right\}$ and multiplicities all one (bottom). In the right graph, we scale the error by $\omega^{5/2}$ and we compare the two-term moment-free asymptotic expansion (top) and Moment-free Filon-type methods with nodes $\{-1,0,1\}$ and multiplicities $\{2,3,2\}$ (middle) and nodes $\left\{-1,-\frac{1}{2},0,\frac{1}{2},1\right\}$ and multiplicities $\{2,1,3,1,2\}$ (bottom).

results from Subsection 2.2 present an intriguing alternative proposed in [Olv07]: use the basis

$$\mathcal{L}[\phi_{r,k}] = \operatorname{sgn}(x)^{r+k+1} \frac{|g(x)|^{\frac{k+1}{r}-1}\, g'(x)}{r}$$

in place of the polynomial basis x^k. We know this basis has two important properties: its moments are computable and it forms a Chebyshev set. Thus given a sequence of nodes $\{x_1,\ldots,x_\nu\}$ and multiplicities $\{m_1,\ldots,m_\nu\}$ we can successfully solve the system

$$\mathcal{L}[v]\,(x_k) = f(x_k),\ldots,\mathcal{L}[v]^{(m_k-1)}\,(x_k) = f^{(m_k-1)}(x_k), \qquad k=1,\ldots,\nu,$$

for the unknowns c_k, where $v = \sum c_k \phi_{r,k}$. As in the standard Filon-type method, the asymptotic order depends on the number of derivatives interpolated at the endpoints and stationary point.

We now approximate the integral

$$I[f] = \int_0^1 \frac{x+3}{x+2} \mathrm{e}^{\mathrm{i}\omega(1-\cos x)}\,\mathrm{d}x.$$

In Fig. 4 we compare the moment-free asymptotic expansion to four Moment-free Filon-type methods: two of order $\mathcal{O}\big(\omega^{-3/2}\big)$ and two of order $\mathcal{O}\big(\omega^{-5/2}\big)$. Again, we can reduce the error of Filon-type methods by adding interior interpolation points.

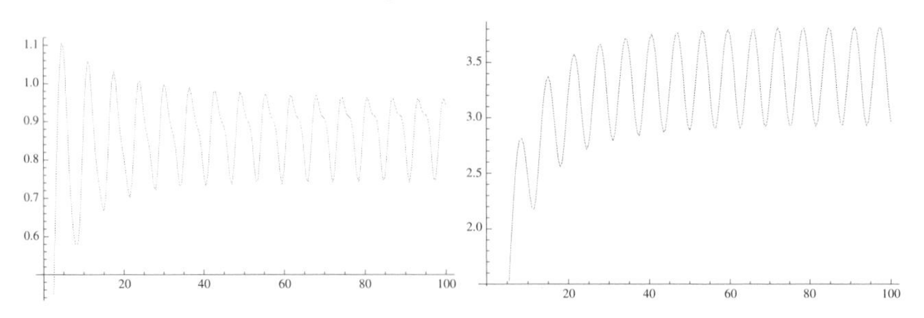

Fig. 5. Scaled error of two Filon-type methods in approximating the integral $\iint_S \frac{e^{i\omega(x-y)}}{2x+y+1}\,dV$. In the left graph, we scale the error by ω^3 where we interpolate at the vertices with multiplicities all one. In the right graph, we scale by ω^4 for interpolating at $\{(0,0),(1,0),(0,1),(\frac{1}{3},\frac{1}{3})\}$ with multiplicities $\{2,2,2,1\}$.

3.3 Multivariate Filon-type methods

We can readily derive a Filon-type method for multivariate oscillatory integrals as well [IN06b]. As before, we interpolate a multivariate function f by a polynomial v at a sequence of points $\{\boldsymbol{x}_1,\ldots,\boldsymbol{x}_\nu\}$, again with multiplicities $\{m_1,\ldots,m_\nu\}$. Assuming such an interpolation is possible—which, unlike univariate polynomial interpolation, is not ensured—we obtain the approximation:

$$Q^F[f] = I[v] = \int_\Omega v(\boldsymbol{x})e^{i\omega g(\boldsymbol{x})}\,dV.$$

Again, this requires knowledge of the moments $I\left[x_1^{i_1}\cdots x_d^{i_2}\right]$, whose evaluation depends not only on the oscillator g but also the domain Ω. There are few cases where these moments are known, but they can be found in closed form if Ω is a simplex and $g(\boldsymbol{x}) = \boldsymbol{\kappa}\cdot\boldsymbol{x}$ is linear. If g is linear, a composite Filon-type rule can be constructed by triangulating Ω, and applying a Filon-type method to each triangle in the mesh.

The asymptotic order of a multivariate Filon-type method follows immediately from Theorem 4, as long as $\nabla g = \boldsymbol{\kappa}$ is not orthogonal to any of the faces of the boundary. Thus if we interpolate at the vertices with multiplicities at least s, we achieve an order

$$Q^F[f] - I[f] \sim \mathcal{O}\left(\omega^{-s-d}\right).$$

Let $S = \{(x,y) : x \geq 0, y \geq 0 \text{ and } x+y \leq 1\}$ be the two-dimensional simplex. We now approximate the integral

$$\iint_S \frac{e^{i\omega(x-y)}}{2x+y+1}\,dV = \int_0^1 \int_0^{1-x} \frac{e^{i\omega(x-y)}}{2x+y+1}\,dy\,dx.$$

Simply by interpolating at the vertices of the domain, we obtain an asymptotic order $\mathcal{O}(\omega^{-3})$, as seen in the left graph of Fig. 5. The right graph demonstrates that increasing multiplicities at the vertices does indeed increase the asymptotic order to $\mathcal{O}(\omega^{-4})$.

4 Levin collocation method

In 1982, David Levin developed the *Levin collocation method* [Lev82], which approximates oscillatory integrals free of stationary points without using moments. A function F such that $\frac{d}{dx}\left[F e^{i\omega g}\right] = f e^{i\omega g}$ satisfies

$$I[f] = \int_a^b f e^{i\omega g}\,dx = \int_a^b \frac{d}{dx}\left[F e^{i\omega g}\right]dx = F(b)e^{i\omega g(b)} - F(a)e^{i\omega g(a)}.$$

By expanding out the derivatives, we can rewrite this condition as $\mathcal{L}[F] = f$ for the operator

$$\mathcal{L}[F] = F' + i\omega g' F.$$

If we can approximate the function F, then we can approximate $I[f]$ easily. In order to do so, we use collocation with the operator $\mathcal{L}$. Let $v = \sum_{k=1}^{\nu} c_k \psi_k$ for some *basis* $\{\psi_1, \ldots, \psi_\nu\}$. Given a sequence of collocation *nodes* $\{x_1, \ldots, x_\nu\}$, we determine the coefficents c_k by solving the collocation system

$$\mathcal{L}[v](x_1) = f(x_1), \ldots, \mathcal{L}[v](x_\nu) = f(x_\nu).$$

We can then define the approximation $Q^L[f]$ to be

$$Q^L[f] = \int_a^b \mathcal{L}[v]\, e^{i\omega g}\,dx = \int_a^b \frac{d}{dx}\left[v e^{i\omega g}\right]dx = v(b)e^{i\omega g(b)} - v(a)e^{i\omega g(a)}.$$

Levin was the first to note the asymptotic properties of oscillatory quadrature schemes, as well as the importance of endpoints in the collocation system. This method has an error $I[f] - Q^L[f] = \mathcal{O}(\omega^{-1})$ when the endpoints of the interval are not included in the collocation nodes. When the endpoints are included, on the other hand, the asymptotic order increases to $I[f] - Q^L[f] = \mathcal{O}(\omega^{-2})$.

A Levin collocation method was also constructed for oscillatory integrals over a square. In this case a Levin differential operator was constructed by iterating the method for each dimension. Though we do investigate multivariate Levin-type methods in Subsection 4.3, we will not use this construction as it is limited to hypercubes.

4.1 Univariate Levin-type methods

Motivated by the asymptotic expansion and the above results for Filon-type methods, we can now develop Levin-type methods, as proposed in [Olv06a]. In addition to collocating function values, we also collocate derivatives at the collocation points $\{x_1, \ldots, x_\nu\}$, up to given multiplicities $\{m_1, \ldots, m_\nu\}$. In other words, for $v = \sum c_k \psi_k$, we determine the unknown coefficients by solving the system

$$\mathcal{L}[v]\,(x_k) = f(x_k), \ldots, \mathcal{L}[v]^{(m_k-1)}\,(x_k) = f^{(m_k-1)}(x_k), \quad k = 1, \ldots, \nu. \tag{4.1}$$

If the basis $\{\psi_1, \psi_2, \ldots\}$ itself can interpolate at the points and multiplicities, then large ω ensures a solution to this system exists, and each of the coefficients is $\mathcal{O}(\omega^{-1})$. This can be seen via the following logic: we can write (4.1) as $(P + \mathrm{i}\omega G)\boldsymbol{c} = \boldsymbol{f}$, where G is the Vandermonde-like matrix obtained from evaluating $g'\psi_k$ at the nodes and multiplicities. Thus G is nonsingular, and

$$(P + \mathrm{i}\omega G)^{-1} = G^{-1}(\mathrm{i}\omega)^{-1}(I + o(1))^{-1}.$$

It is clear that $(I + o(1))^{-1} \to I$, thus this expression is $\mathcal{O}(\omega^{-1})$, and hence $\boldsymbol{c} = (P + \mathrm{i}\omega G)^{-1}\boldsymbol{f} = \mathcal{O}(\omega^{-1})$. It follows that $\mathcal{L}[v]$ and its derivatives are all bounded as $\omega \to \infty$, thus again we can utilize the asymptotic expansion to show that

$$I[f] - Q^L[f] = I[f - \mathcal{L}[v]] \sim \mathcal{O}(\omega^{-s-1}),$$

for $s = \min\{m_1, m_\nu\}$.

An example of an integral which cannot be computed via Filon-type methods is

$$I[f] = \int_0^1 \sinh x\, \mathrm{e}^{\mathrm{i}\omega\left(x^3+x^2+x\right)}\, \mathrm{d}x.$$

We compare the asymptotic expansion to two Levin-type methods in Fig. 6. We use the same nodes and multiplicities as in the Filon-type methods of Fig. 3, and see the same increase of asymptotic order with the use of multiplicities.

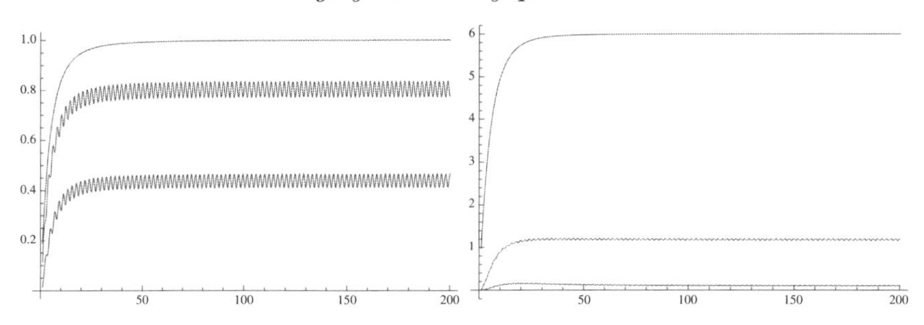

Fig. 6. The error in approximating $\int_0^1 \sinh x \, \mathrm{e}^{\mathrm{i}\omega\left(x^3+x^2+x\right)} \, \mathrm{d}x$. In the left graph, we scale the error by ω^2 and compare the one-term asymptotic expansion (top) and Levin-type methods with nodes $\{0, 1\}$ and multiplicities both one (middle) and nodes $\left\{0, \frac{1}{2}, 1\right\}$ and multiplicities all one (bottom). In the right graph, we scale the error by ω^3 and we compare the two-term asymptotic expansion (top) and Levin-type methods with nodes $\{0, 1\}$ and multiplicities both two (middle) and nodes $\left\{0, \frac{1}{2}, 1\right\}$ and multiplicities $\{2, 1, 2\}$ (bottom).

4.2 Chung, Evans & Webster methods

Often methods represented as weighted sums are preferred. Though Filon and Levin collocation methods are extremely powerful, they do not fall into this framework. In [EW97], Evans and Webster construct such a method for irregular exponential oscillators, based on the Levin collocation method. We want to choose weights w_k and nodes x_k such that

$$\int_{-1}^{1} \phi_k(x) \mathrm{e}^{\mathrm{i}\omega g(x)} \, \mathrm{d}x = \sum_{k=0}^{n} w_k \phi_k(x_k)$$

for some suitable basis ϕ_k. Unlike Gaussian quadrature, we do not choose ϕ_k to be polynomials. Instead, we choose them based on the Levin differential equation:

$$\phi_k = \mathcal{L}[T_k] = T_k' + \mathrm{i}\omega g' T_k,$$

where T_k are the Chebyshev polynomials. The moments with respect to ϕ_k are computable in closed form:

$$\int_{-1}^{1} \phi_k(x) \mathrm{e}^{\mathrm{i}\omega g(x)} \, \mathrm{d}x = T_k(1) \mathrm{e}^{\mathrm{i}\omega g(1)} - T_k(-1) \mathrm{e}^{\mathrm{i}\omega g(-1)}.$$

We can thus determine suitable weights and nodes. Numerical results suggest that this preserves the asymptotic niceties of the original Levin collocation method.

4.3 Multivariate Levin-type methods

We can generalize Levin-type methods to multivariate domains [Olv06b]. The construction of the collocation operator $\mathcal{L}$ followed from the fundamental theorem of calculus, which allowed us to express the value of the integral in terms of a function evaluated at the endpoints of the interval. The construction of a multivariate version will proceed in the same manner, where we use the generalized Stokes theorem to express the value of the integral as an integral over the boundary of the domain.

Define

$$\mathcal{L}[\boldsymbol{v}] = \nabla \cdot \boldsymbol{v} + \mathrm{i}\omega \nabla g \cdot \boldsymbol{v}.$$

The generalized Stokes theorem informs us that

$$\int_{\partial\Omega} \mathrm{e}^{\mathrm{i}\omega g(\boldsymbol{x})} \boldsymbol{v} \cdot \mathrm{d}\boldsymbol{s} = \int_{\Omega} \nabla \cdot (\mathrm{e}^{\mathrm{i}\omega g(\boldsymbol{x})} \boldsymbol{v}) \, \mathrm{d}V = \int_{\Omega} \mathcal{L}[\boldsymbol{v}] \, \mathrm{e}^{\mathrm{i}\omega g(\boldsymbol{x})} \, \mathrm{d}V = I[\mathcal{L}[\boldsymbol{v}]],$$

where $\mathrm{d}\boldsymbol{s} = (\,\mathrm{d}y, -\,\mathrm{d}x)^{\top}$. Thus we use collocation at $\{\boldsymbol{x}_1, \ldots, \boldsymbol{x}_\nu\}$ with multiplicities $\{m_1, \ldots, m_\nu\}$ to determine the coefficients of $\boldsymbol{v} = \sum c_k \boldsymbol{\psi}_k$, i.e., we solve the system

$$\mathcal{L}[\boldsymbol{v}]^{(\boldsymbol{m})}(\boldsymbol{x}_k) = f^{(\boldsymbol{m})}(\boldsymbol{x}_k), \qquad \text{for} \qquad |\boldsymbol{m}| < m_k, \qquad k = 1, \ldots, \nu,$$

where $|\boldsymbol{m}|$ expresses the sum of the individual components. Similar to the univariate case, this system is guaranteed to have a solution for large ω whenever the basis $\nabla g \cdot \boldsymbol{\psi}_k$ can interpolate at the nodes and multiplicities. Hence we obtain

$$I[f] \approx I[\mathcal{L}[\boldsymbol{v}]] = \int_{\partial\Omega} \mathrm{e}^{\mathrm{i}\omega g} \boldsymbol{v} \cdot \mathrm{d}\boldsymbol{s}.$$

We have approximated the oscillatory integral by another oscillatory integral of one dimension less. We can thus iterate the procedure on each dimension, eventually arriving at univariate integrals, which we know how to approximate. It is possible to prove that like multivariate Filon-type methods, an error of order $\mathcal{O}(\omega^{-s-d})$ is achieved when the function value and derivatives of order $s-1$ of f are collocated at each vertex. This does require that $s-1$ derivatives are used for each of the vertices of the boundary integrals as well.

In our example, we define $\boldsymbol{\psi}_k$ as a constant vector $(1,-1)^{\top}$ times the standard multivariate polynomial basis. Thus each $\boldsymbol{\psi}_k$ is of the form $(1,-1)^{\top} x^i y^j$. Letting $H = \{(x,y) : x \geq 0, y \geq 0 \text{ and } x^2 + y^2 \leq 1\}$

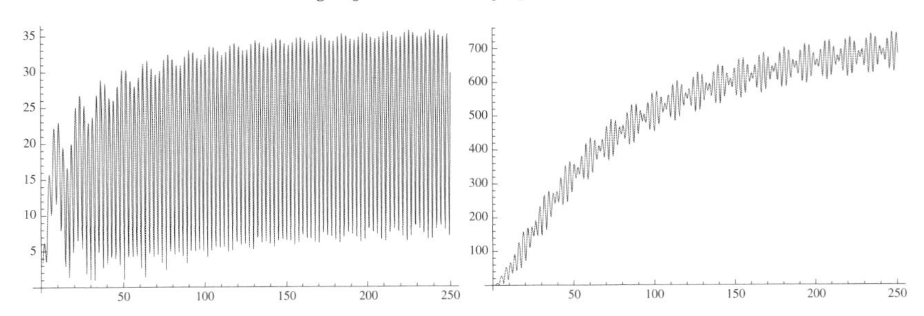

Fig. 7. The error of two Levin-type methods in approximating $\iint_H (\mathrm{e}^{xy} + x^2 + 1)\mathrm{e}^{\mathrm{i}\omega\left(\cos\left(\frac{x}{2}+1\right)+y^2+y\right)}\,\mathrm{d}V$. In the left graph, we scale the error by ω^3 where we interpolate at the vertices with multiplicities all one. In the right graph, we scale by ω^4 for interpolating at $\left\{(0,0),(1,0),(0,1),\left(\frac{1}{4},\frac{1}{4}\right)\right\}$ with multiplicities $\{2,2,2,1\}$.

denote a quarter-disc, in Fig. 7 we approximate the integral

$$\iint_H (\mathrm{e}^{xy} + x^2 + 1)\mathrm{e}^{\mathrm{i}\omega\left(\cos\left(\frac{x}{2}+1\right)+y^2+y\right)}\,\mathrm{d}V$$
$$= \int_0^1 \int_0^{\sqrt{1-x^2}} (\mathrm{e}^{xy} + x^2 + 1)\mathrm{e}^{\mathrm{i}\omega\left(\cos\left(\frac{x}{2}+1\right)+y^2+y\right)}\,\mathrm{d}y\,\mathrm{d}x.$$

Once we push the integral to the boundary we employ three univariate Levin-type methods. In the left graph of this figure we use the endpoints with multiplicities one for each of the univariate methods, in the right graph we use the endpoints with multiplicities two.

5 Steepest descent methods

An entirely different approach becomes applicable once we assume that both f and g are analytic functions. Then, by deforming the path of integration into the complex plane, efficient numerical schemes can again be devised for the evaluation of $I[f]$ with low computational cost and high asymptotic accuracy. The methods require no moments, but they require computations in the complex plane.

5.1 The path of steepest descent

For complex valued functions $g(x)$, we observe that the complex exponential function $\mathrm{e}^{\mathrm{i}\omega g(x)}$ is oscillatory only as a function of the real part of g. The so-called *path of steepest descent* originating at the point $x = a$

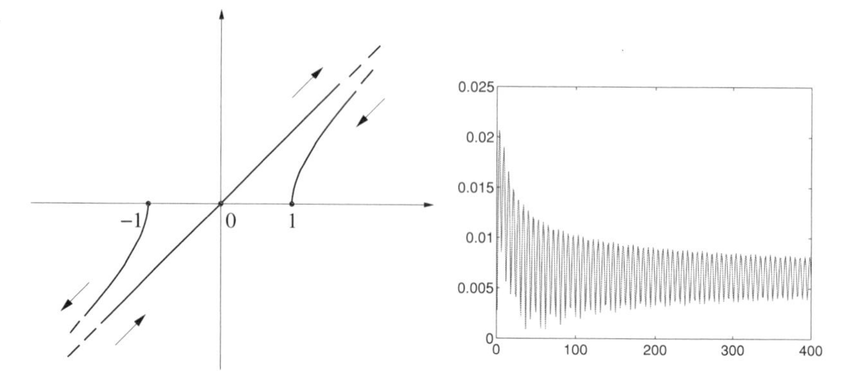

Fig. 8. The absolute error in approximating $\int_{-1}^{1} \frac{1}{(x+3)^2} e^{i\omega x^2} \, dx$ as a function of ω by a numerical steepest descent scheme using 4 quadrature points, 1 near each endpoint and 2 near the stationary point. The error is scaled by $\omega^{5/2}$.

is such that the real part of g is kept fixed along the path. Assuming a parameterization $h_a(p)$ of the path, $p \in [0, P]$, this is achieved by solving

$$g(h_a(p)) = g(a) + ip, \tag{5.1}$$

subject to the condition $h_a(0) = a$. Along such a path, the integrand is nonoscillatory and exponentially decaying, taking the form

$$e^{i\omega g(a)} \int_0^P f(h_a(p)) h_a'(p) e^{-\omega p} \, dp. \tag{5.2}$$

At a stationary point $x = \xi$, equation (5.1) has several solutions since the inverse of g is a multivalued function. Two solutions can easily be identified such that deformation onto a collection of steepest descent contours is justified by Cauchy's integral theorem. This is shown in the left panel of Fig. 8 for the case $g(x) = x^2$.

5.2 Univarate method of steepest descent

The method of steepest descent is traditionally used to obtain the asymptotic expansion of $I[f]$ [Won01]. However, the typical divergence of asymptotic expansions is avoided by evaluating integrals of the form (5.2) numerically. An interesting choice is the use of a Gaussian quadrature rule that incorporates the exponential decay into its weight function [HV06].

For the endpoints $x = a$ or $x = b$, integral (5.2) is approximated by

$$\frac{e^{i\omega g(x)}}{\omega} \sum_{k=1}^{n} w_k f\left(h_x\left(\frac{x_k}{\omega}\right)\right) h'_x\left(\frac{x_k}{\omega}\right),$$

where x_k and w_k are the quadrature points and weights of a classical Gauss–Laguerre rule. It was shown in [HV06] that, using n evaluations of f near a and near b, the error in the approximation behaves as $\mathcal{O}(\omega^{-2n-1})$. The asymptotic order thus obtained is roughly twice that of Filon-type methods or Levin-type methods using n derivatives at each endpoint. This comes at a cost of having to compute $h_x(p)$.

At a simple stationary point, it is beneficial to use a Gauss–Hermite rule, with the weight function e^{-q^2}. For more degenerate stationary points, the optimal Gaussian quadrature rules are no longer classical [DH08]. In all cases however, the asymptotic order is approximately twice that of Filon-type methods and Levin-type methods, when using the same number of function evaluations of f or its derivatives.

An example is shown in the right panel of Fig. 8. Using four function evaluations, the method achieves $\omega^{-5/2}$ asymptotic error for the integral $\int_{-1}^{1} \frac{1}{(x+3)^2} e^{i\omega x^2}\, dx$. A Filon-type method would require at least seven function values to achieve the same order.

5.3 *Multivariate method of steepest descent*

The univariate method of steepest descent can be extended to multivariate oscillatory integrals by recursion. For example, when f and g depend on two variables x_1 and x_2, integration with respect to the variable x_1 may lead to a steepest descent integral of the form

$$e^{i\omega g(a,x_2)} \int_0^P f(h_1(p,x_2),x_2) \frac{\partial h_1}{\partial p}(p,x_2) e^{-\omega p}\, dp,$$

which compares to (5.2). The steepest descent path h_1 now is such that

$$g(h_1(p,x_2),x_2) = g(a,x_2) + ip.$$

The path depends on the starting point $x_1 = a$, but also on the second variable x_2. The recursive process is started by noting that the line integral above is an oscillatory function in x_2, with the known oscillator $g(a,x_2)$. Thus, the integral of this function with respect to x_2 may again be evaluated by deforming onto a path of steepest descent $h_2(q)$. One ends up integrating along a *manifold of steepest descent*, which yields a doubly-infinite, non-oscillatory and exponentially decaying integral

$$e^{i\omega g(a,b)} \int_0^Q \int_0^P f(h_1(p,h_2(q)),h_2(q))h_2'(q)\frac{\partial h_1}{\partial p}(p,h_2(q))e^{-\omega(p+q)}\,\mathrm{d}p\mathrm{d}q.$$

This manifold of steepest descent is not unique, as the lack of symmetry in the expression above indicates. The integrals, as in the univariate case, can be evaluated numerically to high asymptotic order [HV07a].

Much like the result of using generalized Stokes' formula, one can see that each recursive step pushes the integral onto the boundary. The new oscillators that appear in the process may in general have stationary points, which we call *resonance points*. Such resonance points appear on the boundary of the domain as points where $\nabla g \perp \partial\Omega$. Other contributions come from vertices of $\partial\Omega$ and from critical points where $\nabla g = 0$.

The method of steepest descent continues to hold for oscillators g with critical points and resonance points. Integrals with very degenerate oscillators however, for example with a vanishing Hessian, are computable in this manner only subject to certain additional conditions [HV07a].

6 Other methods

There are assorted other numerical methods developed for approximating oscillatory integrals, typically specializing on particular oscillators. We will not investigate these methods in detail, but they are mentioned here for completeness.

6.1 Fourier oscillator methods

Many methods exist for the Fourier oscillator, which were reviewed in [EW99]. They all are based on the fact that moments are computable, and hence are Filon-type methods. The *Bakhvalov and Vasil'eva method* [BV68] interpolates f by Legendre polynomials P_k, and uses the fact that the moments of such polynomials are known explicitly:

$$\int_{-1}^{1} P_k(x)e^{i\omega x}\,\mathrm{d}x = i^k\left(\frac{2\pi}{\omega}\right)^{\frac{1}{2}} J_{k+\frac{1}{2}}(\omega), \tag{6.1}$$

where J_k is a Bessel function [AS65].

A method based on Clenshaw–Curtis quadrature was also devised, where f is interpolated by Chebyshev polynomials T_k. We do not have simple formulæ for the resulting moments, so the polynomials T_k are then expanded into Legendre polynomials and (6.1) is applied [Pat76,

LZ03]. An alternative from [AN02] is to express the moments in terms of the hypergeometric function ${}_0F_1$ [AS65]. Special functions can be avoided in both these methods by expanding the Legendre or Chebyshev polynomials into the standard polynomial basis x^k, whose moments can be found via partial integration [AEH76]. This is not effective for large k due to significant cancellation in the expansions [EW99].

Though it was not observed in any of these papers, all of these Filon-type methods—that is, methods based on interpolating f—have the same asymptotic behaviour as the Filon-type methods developed in Section 3. If the endpoints of the interval are included in the interpolation nodes, then error decays like $\mathcal{O}(\omega^{-2})$; otherwise the error decays at the same rate as the integral $\mathcal{O}(\omega^{-1})$. It means that the number of interpolation points required should actually decrease as the frequency of oscillations increases. Thus at high frequencies we never need to utilize large order polynomials in order to obtain accurate results.

Piessens developed a Gaussian quadrature formula with respect to the weight function $\sin x$ over the interval $[-\pi, \pi]$ [PP71]. It however relies on considering each period separately, thus still requires a large number of function evaluations to obtain accurate approximations. A similar method based on Gaussian quadrature was developed by Zamfirescu [Zam63], and described in [Ise04] (the original paper is in Romanian). We can rewrite the sine Fourier integral as

$$\int_0^1 f(x) \sin \omega x \, \mathrm{d}x = \int_0^1 f(x)(1 + \sin \omega x) \, \mathrm{d}x - \int_0^1 f(x) \, \mathrm{d}x.$$

The second of these integrals is nonoscillatory, so standard quadrature methods can be used to approximate its value. The first integral now has a nonnegative weight function, hence we can approximate it by a weighted sum. Since the moments with respect to the weight function are known, we can successfully compute the quadrature weights needed.

A very effective quadrature scheme is developed in [KCI02] for the standard Fourier oscillator. The paper uses a weighted sum of the value of the functon f and its first derivative at evenly spaced nodes, determining the weights so as to maximize the degree of polynomials integrated exactly. It is noted that, for a method of N points, the error behaves like $\mathcal{O}(\omega^{-N})$. This is generalized to use higher order derivatives of f in [KCI03], resulting in a significant decrease in error.

6.2 Irregular oscillator methods

There is also a comparison of methods for irregular oscillators in [EW99]. In addition to the Levin collocation method and the Chung, Evans and Webster method already discussed, there is method developed by Evans in [Eva94] where the transformation $y = g(x)$ is used to convert the irregular oscillator to a standard Fourier oscillator:

$$I[f] = \int_{g(a)}^{g(b)} \frac{f(g^{-1}(y))}{g'(g^{-1}(y))} \mathrm{e}^{\mathrm{i}\omega y} \, \mathrm{d}y. \tag{6.2}$$

Once the integral is in this form, a Filon-type method can be employed, in particular the method based on Clenshaw–Curtis quadrature. This technique is successful whenever the interval does not contain stationary points. Unfortunately it requires the computation of the inverse of g, albeit only at the interpolation points.

Another method for irregular oscillators is proposed by Evans [Eva97]. Instead of interpolating f by polynomials, we can interpolate $\frac{f(x)}{g'(x)}$ using a basis of the form

$$\sum c_k \psi_k(g(x)).$$

Then making the transformation $y = g(x)$, as in (6.2), does not require the computation of inverses of g. We must, however, be careful in the choice of the basis ψ_k.

In [Bru03], the problem of solving the acoustic equation was tackled. This method required the computation of oscillatory integrals, for which a new quadrature scheme was derived. At high frequencies, univariate oscillatory integrals are dominated by the contribution from the stationary points and endpoints (multivariate integrals are dominated by contributions from stationary points, resonance points and vertices). Thus we can obtain a high accuracy approximation by numerically integrating near these important points, and throwing away the contributions from the more oscillatory regions. This is accomplished by utilizing smooth windowing functions that focus on ϵ neighbourhoods of these important points. The part of the integral which we throw away decays exponentially fast as the frequency increases; though the error in the approximation only decays at the same rate as the integral due to quadrature error in each ϵ neighbourhood.

Bibliography

[AS65] M. Abramowitz, & I. A. Stegun (1965). *Handbook of Mathematical Functions with Formulas, Graphs, and Mathematical Tables* (Dover Publications, New York).

[AN02] G, Adam, & A. Nobile (2002). Product Integration Rules at Clenshaw–Curtis and Related Points: A Robust Implementation, *IMA Journal of Numerical Analysis* **11**(2), 271–296.

[AEH76] A. Alaylioglu, G. Evans, & J. Hyslop (1976). The use of Chebyshev series for the evaluation of oscillatory integrals, *The Computer Journal* **19**(3), 258–267.

[BV68] N. Bakhvalov, & L. Vasilčeva (1968). Evaluation of the integrals of oscillating functions by interpolation at nodes of Gaussian quadratures, *USSR Comp. Math. Phys* **8**, 241–249.

[Bru03] O. P. Bruno (2003). Fast, high-order, high-frequency integral methods for computational acoustics and electromagnetics, in: *Lecture Notes in Computational Science and Engineering 31: Topics in Computational Wave Propagation* (Springer, Berlin).

[DH08] A. Deaño, & D. Huybrechs (2008). Complex Gaussian quadrature of oscillatory integrals. Technical Report 2008/NA04, DAMTP, University of Cambridge.

[Eva94] G. A. Evans (1994). An alternative method for irregular oscillatory integrals over a finite range, *International Journal of Computer Mathematics* **52**(3), 185–193.

[Eva97] G. A. Evans (1997). An expansion method for irregular oscillatory integrals, *Int. J. Comput. Math.* **63**(1), 137–148.

[EW97] G. A. Evans & J. R. Webster (1997). A high order, progressive method for the evaluation of irregular oscillatory integrals, *Appl. Numer. Math.* **23**(2), 205–218.

[EW99] G. A. Evans & J. R. Webster (1999). A comparison of some methods for the evaluation of highly oscillatory integrals, *J. Comput. Appl. Math.* **112**(1), 55–69.

[Fil28] L. N. G. Filon (1928). On a quadrature formula for trigonometric integrals, *Proc. Roy. Soc. Edinburgh* **49**, 38–47.

[For97] R. Forrey (1997). Computing the Hypergeometric function, *J. Comput. Phys.* **137**(1), 79–100.

[HO08] D. Huybrechs & S. Olver (2008). Rapid function approximation by modified Fourier series, This volume.

[HV06] D. Huybrechs & S. Vandewalle (2006). On the evaluation of highly oscillatory integrals by analytic continuation, *SIAM J. Numer. Anal.* **44**(3), 1026–1048.

[HV07a] D. Huybrechs & S. Vandewalle (2007). The construction of cubature rules for multivariate highly oscillatory integrals, *Math. Comp.* **76**(260), 1955–1980.

[HV07b] D. Huybrechs & S. Vandewalle (2007). A sparse discretisation for integral equation formulations of high frequency scattering problems, *SIAM J. Sci. Comput.* **29**(6), 2305–2328.

[Ise04] A. Iserles (2004). On the numerical quadrature of highly-oscillating integrals I: Fourier transforms, *IMA J. Num. Anal.* **24**(3), 365–391.

[IN04] A. Iserles & S. P. Nørsett (2004). On quadrature methods for highly oscillatory integrals and their implementation, *BIT* **44**(4), 755–772.

[IN05] A. Iserles & S. P. Nørsett (2005). Efficient quadrature of highly oscillatory integrals using derivatives, *Proc. R. Soc. Lond. A* **461**, 1383–1399.

[IN06b] A. Iserles & S. P. Nørsett (2006). Quadrature methods for multivariate highly oscillatory integrals using derivatives, *Math. Comp.* **75**, 1233–1258.

[IN06a] A. Iserles & S. P. Nørsett (2008). From high oscillation to rapid approximation I: Modified Fourier expansions, *IMA J. Num. Anal.* **28**, 862-887.

[KCI02] K. J. Kim,, R. Cools & L. G. Ixaru (2002). Quadrature rules using first derivatives for oscillatory integrands, *J. Comput. Appl. Math.* **140**(1-2), 479–497.

[KCI03] K. J. Kim, R. Cools & L. G. Ixaru (2003). Extended quadrature rules for oscillatory integrands, *Appl. Numer. Math.* **46**(1), 59–73.

[Lev82] D. Levin (1982). Procedure for computing one- and two-dimensional integrals of functions with rapid irregular oscillations, *Math. Comp.* **38**(158), 531–538.

[LZ03] R. Littlewood & V. Zakian (2003). Numerical Evaluation of Fourier Integrals, *IMA Journal of Applied Mathematics* **18**(3), 331–339.

[LO94] D. Lozier & F. W. J. Olver (1994). Numerical evaluation of special functions, *Mathematics of Computation, 1943–1993: A Half-century of Computational Mathematics*, ed. W. Gautschi (AMS, Providence).

[Luk54] Y. L. Luke (1954). On the computation of oscillatory integrals, *Proc. Cambridge Phil. Soc.* **50**, 269–277.

[Mul01] K. Muller (2001). Computing the confluent hypergeometric function, $M(a, b, x)$, *Numerische Mathematik* **90**(1), 179–196.

[Olv74] F. W. J. Olver (1974). *Asymptotics and Special Functions* (Academic Press Inc, New York).

[Olv06a] S. Olver (2006). Moment-free numerical integration of highly oscillatory functions, *IMA J. Num. Anal.* **26**(2), 213–227.

[Olv06b] S. Olver (2006). On the quadrature of multivariate highly oscillatory integrals over non-polytope domains, *Numer. Math.* **103**(4), 643–665.

[Olv07] S. Olver (2007). Moment-free numerical approximation of highly oscillatory integrals with stationary points, *Euro. J. Appl. Maths* **18**, 435-447.

[Pat76] T. Patterson (1976). On high precision methods for the evaluation of fourier integrals with finite and infinite limits, *Numerische Mathematik* **27**(1), 41–52.

[PP71] R. Piessens, & F. Poleunis (1971). A numerical method for the integration of oscillatory functions, *BIT Numerical Mathematics* **11**(3), 317–327.

[Ste93] E. M. Stein (1993). *Harmonic Analysis: Real-Variable Methods, Orthogonality and Oscillatory Integrals* (Princeton University Press, Princeton).

[vLT84] C. van der Laan & N. Temme (1984). *Calculation of Special Functions: The Gamma Function, the Exponential Integrals and Error-Like Functions*, (Centrum voor Wiskunde en Informatica Amsterdam, The Netherlands).

[Won01] R. Wong (2001). *Asymptotic Approximation of Integrals* (SIAM, Philadelphia).

[Zam63] I. Zamfirescu (1963). An extension of Gauss's method for the calculation of improper integrals, *Acad. RP Romıne Stud. Cerc. Mat* **14**, 615–631.

3

Rapid function approximation by modified Fourier series

Daan Huybrechs
Department of Computer Science
K.U. Leuven
Celestijnenlaan 200A
BE-3001 Leuven
Belgium
Email: daan.huybrechs@cs.kuleuven.be

Sheehan Olver
St John's College
Oxford University
St Giles'
Oxford OX1 3JP
United Kingdom
Email: Sheehan.Olver@sjc.ox.ac.uk

Abstract

We review a set of algorithms and techniques to approximate smooth functions on a domain $\Omega \subset \mathbf{R}^d$ by an expansion in eigenfunctions of the Laplacian. We refer to such expansions as modified Fourier series. These series converge pointwise everywhere in the domain of approximation, including on the boundary, at an algebraic rate that is essentially arbitrary. The computational complexity of the transformation is only linear in the number of terms of the expansion. Moreover, additional terms can be computed adaptively and efficiently.

1 Introduction

The subject of this review paper is the approximation of smooth functions. The method of approximation is quite simple: a smooth function f on a domain Ω is expanded into eigenfunctions u_n of the Laplace operator, subject to Neumann boundary conditions:

$$\begin{aligned} -\Delta u_n(x) &= \lambda_n u_n(x), & x \in \Omega, \\ \frac{\partial u_n}{\partial \nu}(x) &= 0, & x \in \partial\Omega. \end{aligned}$$

On the face of it, this may not appear to be a recent research topic. It has long been known that such eigenfunctions are orthogonal and dense in $L_2[\Omega]$, making them ideally suitable for series expansions of quite arbitrary functions on domains with great generality. The Laplace operator has probably received the widest study among all partial differential operators. Yet, for the purpose of numerical approximation, it is usually neglected in favour of alternative approaches. Two well-known and successful examples for univariate functions are the approximation by Chebyshev polynomials, for smooth functions on an interval, and the FFT algorithm, for smooth functions that are in addition periodic. The recent research in modified Fourier series brings two properties that are indispensable for any approximation method: fast algorithms and rapid convergence, thereby adding competitiveness to generality.

One of the main results is an $\mathcal{O}(m)$ algorithm for the computation of the first m coefficients in the series. This result holds for periodic and non-periodic functions alike. Needless to say, this computational complexity compares favourably to the alternatives mentioned above. The result has been made possible primarily through the advent of efficient computational schemes for highly oscillatory integrals. Such schemes are the subject of a separate review paper in this same volume, to which the interested reader is referred [HO08]. The connection to modified Fourier series becomes obvious once we observe that Laplace–Neumann eigenfunctions become increasingly oscillatory. Hence, the coefficients of the series are given by increasingly oscillatory integrals. The most important development in the evaluation of highly oscillatory integrals is, arguably, the Filon-type method [IN05]. We will show how Filon-type methods can be extended and adapted to the setting of modified Fourier series. The $\mathcal{O}(m)$ algorithm then follows immediately.

The second property is rapid convergence. In unaltered form, the algorithm briefly described above yields $\mathcal{O}(m^{-2})$ convergence, pointwise in the interior of Ω, and $\mathcal{O}(m^{-1})$ convergence on the boundary $\partial\Omega$. Note that the FFT-algorithm on an interval $[a, b]$, though ideally suited for periodic functions, yields only $\mathcal{O}(m^{-1})$ pointwise convergence for non-periodic functions in the interior, and no convergence at all at the boundary points. The convergence of modified Fourier series can be improved in several ways. In this paper we discuss two approaches: faster initial convergence through the use of eigenfunctions of polyharmonic operators and accelerated convergence through the use of polynomial subtraction. A possible third approach is based on extrapolation [Olv07].

The main results in the approximation by modified Fourier series have been established in a series of papers [IN06a, IN06b, IN07, Olv07, HIN07]. The current paper mostly follows the same pattern of developments. We start by reviewing one-dimensional approximation in §2. The generalization in the direction of polyharmonic operators is discussed in §3. A generalization in a different direction, to multivariate approximation, is reviewed in §4. Finally, we treat the acceleration of convergence in §5. We end with some concluding remarks in §6.

2 Univariate approximation

The simplest setting is that of a function f defined on the interval $\Omega := [-1, 1]$. This univariate setting is well suited to motivate the use of Laplace–Neumann expansions and to appreciate their basic properties.

2.1 Laplace–Neumann expansions

The standard Fourier series of f on the interval $[-1, 1]$ is given by

$$\frac{1}{2}\hat{f}_0^C + \sum_{n=1}^{\infty} \hat{f}_n^C \cos \pi n x + \hat{f}_n^D \sin \pi n x,$$

where

$$\hat{f}_n^C = \int_{-1}^{1} f(x) \cos \pi n x \, \mathrm{d}x, \qquad \hat{f}_n^D = \int_{-1}^{1} f(x) \sin \pi n x \, \mathrm{d}x.$$

Let us assume that the function f is non-periodic. In that case, the Fourier coefficients as defined above behave like $\hat{f}_n^C = \mathcal{O}(n^{-2})$ and $\hat{f}_n^D = \mathcal{O}(n^{-1})$ for $n \gg 1$. We note that the sine coefficients are primarily responsible for the slow convergence rate of the Fourier series.

The expansion in eigenfunctions of the Laplace operator, as defined by (1.1), leads to a very similar series:

$$\frac{1}{2}\hat{f}_0^C + \sum_{n=1}^{\infty} \hat{f}_n^C \cos \pi n x + \hat{f}_n^S \sin \pi (n - \tfrac{1}{2}) x, \tag{2.1}$$

where

$$\hat{f}_n^S = \int_{-1}^{1} f(x) \sin \pi (n - \tfrac{1}{2}) x \, \mathrm{d}x. \tag{2.2}$$

The only difference compared to classical Fourier is a shift by $\frac{1}{2}$ in the argument of the sine functions, hence the name *modified Fourier series*. This small change suffices to yield $\mathcal{O}(n^{-2})$ behaviour of the corresponding coefficients in the series when f is differentiable and its derivative has bounded variation. Moreover, both $\hat{f}_0^C$ and $\hat{f}_0^S$ have an even expansion if f is sufficiently smooth. This follows from integration by parts and establishes a pattern for the upcoming generalizations.

Theorem 1 *If $f \in C^\infty[-1,1]$ then for $n \gg 1$ we have*

$$\hat{f}_n^C \sim (-1)^n \sum_{k=0}^\infty \tfrac{(-1)^k}{(n\pi)^{2k+2}}[f^{(2k+1)}(1) - f^{(2k+1)}(-1)],$$
$$\hat{f}_n^S \sim (-1)^{n-1} \sum_{k=0}^\infty \tfrac{(-1)^k}{[(n-\frac{1}{2})\pi]^{2k+2}}[f^{(2k+1)}(1) + f^{(2k+1)}(-1)].$$

Proof Integrating expression (2.2) by parts once yields

$$\hat{f}_n^S = -\left.\frac{f(x)\cos\pi(n-\frac{1}{2})x}{(n-\frac{1}{2})\pi}\right|_{-1}^{1} + \frac{1}{(n-\frac{1}{2})\pi}\int_{-1}^{1} f'(x)\cos\pi(n-\frac{1}{2})x\,\mathrm{d}x.$$

The first term vanishes due to the homogeneous Neumann boundary conditions. Integrating by parts once more leads to

$$\hat{f}_n^S = \left.\frac{f'(x)\sin\pi(n-\frac{1}{2})x}{(n-\frac{1}{2})^2\pi^2}\right|_{-1}^{1} - \frac{1}{(n-\frac{1}{2})^2\pi^2}\int_{-1}^{1} f^{(2)}(x)\sin\pi(n-\frac{1}{2})x\,\mathrm{d}x.$$

Repeated invocation of integration by parts on the remainder integral yields the result. The proof for the coefficients $\hat{f}_n^C$ is analogous. □

It is thus established that the coefficients of the modified Fourier series (2.1) decay faster than those of a classic Fourier series. We emphasize the fact that this property essentially follows from the homogeneous Neumann boundary conditions satisfied by the basis functions.

The asymptotic expansions in Theorem 1 also indicate that the coefficients asymptotically depend only on odd derivatives of f, evaluated at the two boundary points of the interval $[-1,1]$. This observation will be used later on for efficiently accelerating the decay even further.

The size of the coefficients and the approximation error are illustrated in Fig. 1 (top curve). The figure also shows improved decay rates through polynomial subtraction. This will be discussed later in §5.1.

2.2 Convergence of the modified Fourier series

The Laplace–Neumann eigenfunctions in the univariate case are given by the set of basis functions

$$\{\cos\pi nx : n \geq 0\} \cup \{\sin\pi(n-\frac{1}{2})x : n \geq 1\}. \tag{2.3}$$

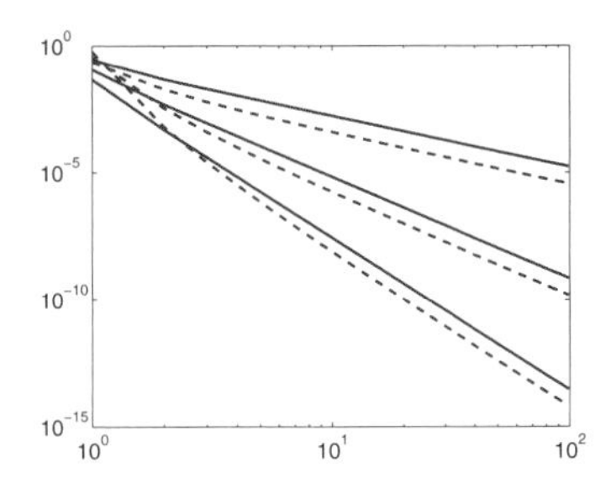

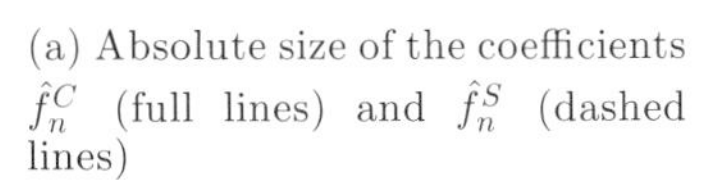

(a) Absolute size of the coefficients $\hat{f}_n^C$ (full lines) and $\hat{f}_n^S$ (dashed lines)

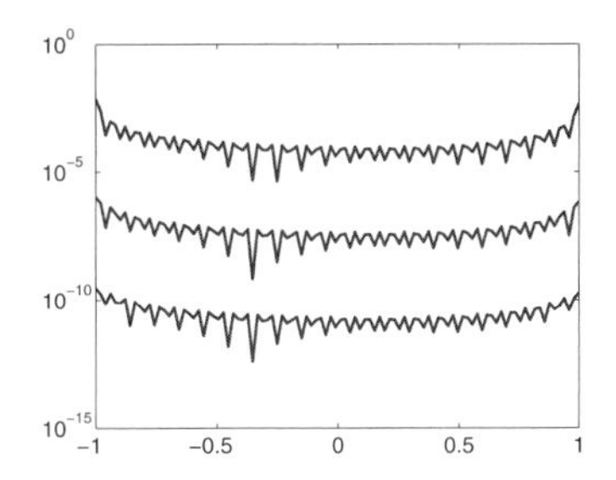

(b) Approximation error for $m =$ 30 terms on the interval $(-1, 1)$

Fig. 1. Approximation of $f(x) = \cos(x+1)\sin(x+1)$ by a modified Fourier series. Polynomial subtraction was used with 0, 1 and 2 steps (from top to bottom), leading to convergence rates of $\mathcal{O}(m^{-2})$, $\mathcal{O}(m^{-4})$ and $\mathcal{O}(m^{-6})$.

These functions form an orthonormal set in $L_2[-1,1]$. It follows that convergence of the modified Fourier series

$$f_m(x) := \sum_{n=0}^{m} \hat{f}_n^C \cos \pi n x + \sum_{n=1}^{m} \hat{f}_n^S \sin \pi (n - \frac{1}{2}) x \tag{2.4}$$

to f is guaranteed in the L_2-norm. However, a more interesting notion in function approximation is pointwise convergence. In this section we review the known results on pointwise convergence of the series (2.4) to the value $f(x)$.

Theorem 2 ([IN06a]) *Suppose that f is Riemann integrable in $[-1,1]$ and that*

$$\hat{f}_n^C, \hat{f}_n^S = \mathcal{O}(n^{-1}), \qquad n \to \infty.$$

If f is Lipschitz at $x \in (-1,1)$ then

$$f_m(x) \to f(x).$$

The result is proved by amending the classical theorems of Féjer and of de la Vallée Poussin. The rate of convergence of the series was later established in [Olv07]. We quote the following theorem.

Theorem 3 ([Olv07]) *Suppose that $f \in \mathcal{C}^2[-1,1]$ and f'' has bounded variation. If $-1 < x < 1$ then*

$$f(x) - f_m(x) = \mathcal{O}(m^{-2}).$$

Otherwise,

$$f(\pm 1) - f_m(\pm 1) = \mathcal{O}(m^{-1}).$$

The series converges everywhere, though at a slower rate in the endpoints $x = \pm 1$. The result was obtained by expanding the difference $f(x) - f_m(x)$ asymptotically in terms of the special Lerch function. An asymptotic expansion for the approximation error was thus constructed that is uniform for $x \in [-1,1]$, i.e., including the endpoints. This opens the possibility of increasing the convergence rate even further, by explicitly adding the first term of the error expansion. We do not delve into this topic further in this paper, but refer the interested reader to [Olv07].

2.3 A geometric interpretation

Theorem 1 showed that the coefficients $\hat{f}_n^C$ and $\hat{f}_n^S$ asymptotically depend only on odd derivatives of f at the boundary points ± 1. This may seem odd at first sight. In this section, we attempt to give a meaning to this result with a geometric argument. In the process, we obtain a close relation to classical Fourier series in this univariate case.

Consider the function $g(x)$, defined on the interval $[-1,3]$ and constructed from $f(x)$ by reflecting evenly around the point $x = 1$,

$$g(x) = \begin{cases} f(x), & \text{if } x \in [-1,1], \\ f(2-x), & \text{if } x \in (1,3]. \end{cases}$$

Next, we extend the function g periodically as illustrated in Fig. 2. Note that reflecting evenly around the point $x = -1$ and then periodically extending would result in the same function.

Now consider the classical Fourier series of g on the interval $[-1,3]$,

$$g_m(x) = \sum_{n=0}^{m} \tilde{g}_n^C \cos\left(\frac{2\pi}{4} n(x+1)\right) + \sum_{n=1}^{m} \tilde{g}_n^S \sin\left(\frac{2\pi}{4} n(x+1)\right),$$

where

$$\begin{aligned} \tilde{g}_n^C &:= \int_{-1}^{3} g(x) \cos\left(\tfrac{2\pi}{4} n(x+1)\right) \mathrm{d}x, \qquad && n = 0, 1, \ldots, \\ \tilde{g}_n^S &:= \int_{-1}^{3} g(x) \sin\left(\tfrac{2\pi}{4} n(x+1)\right) \mathrm{d}x, \qquad && n = 1, 2, \ldots. \end{aligned}$$

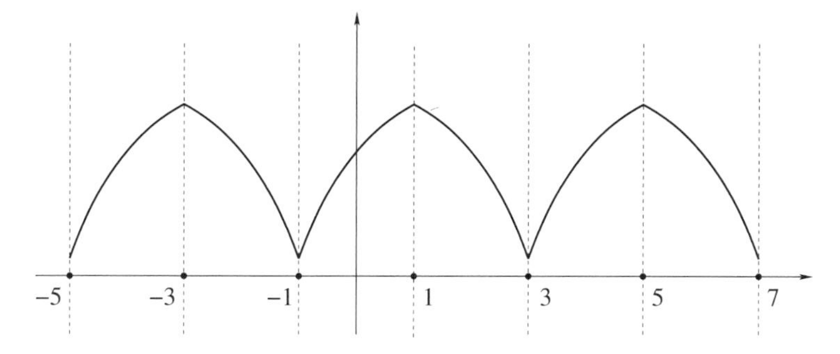

Fig. 2. The function $f(x)$ on $[-1,1]$ is evenly reflected around the points ± 1 and extended periodically.

Since $g(x)$ is even on $[-1,3]$ around the center point $x = 1$, we have $\tilde{g}_n^S = 0$. One can also easily verify that

$$\hat{f}_n^C = (-1)^n \frac{\tilde{g}_{2n}^C}{2}, \qquad \hat{f}_n^S = (-1)^n \frac{\tilde{g}_{2n-1}^C}{2}.$$

In other words, the modified Fourier series of f is equivalent to the classical Fourier series of g. Fig. 2 now explains the importance of odd derivatives at ± 1. If $f'(1) = f'(-1) = 0$, then one sees that the differentiability of g across the reflection points is increased. Thus, its Fourier series converges faster. The values of the even derivatives of f at ± 1 are irrelevant, as they do not limit the continuity of g.

We observe that the modified Fourier series of f converges faster if f can be extended as a smooth and even function across the boundary of the domain Ω. We will see that this observation generalizes to different domains Ω. The analogy to Fourier series, unfortunately, does not.

Due to this analogy with Fourier series, one may be tempted to compute the coefficients $\hat{f}_n^C$ and $\hat{f}_n^S$ using FFT. The alternative approaches discussed in the next section are far more accurate however, because the function g in general will not be a smooth function.

2.4 Computation of the coefficients

The coefficients $\hat{f}_n^C$ and $\hat{f}_n^S$ are given by integrals that become increasingly oscillatory as $n \gg 1$. An account of recent research in efficient methods for the evaluation of oscillatory integrals is given in [HO08]. From the results in that review, it becomes apparent that the coefficients can be computed at a cost that is asymptotically independent of

n. That is, each coefficient $\hat{f}_n^C$ and $\hat{f}_n^S$ can be computed with $O(1)$ evaluations of f as $n \gg 1$. An immediate result is that the first m coefficients can be computed in $O(m)$ operations.

An additional desire in the setting of modified Fourier series is to reuse function evaluations of f for the computation of coefficients with varying values of n. With this goal in mind, we will revisit Filon-type quadrature, though noting that Levin-type methods are equally applicable [Olv07].

The purpose of so-called exotic quadrature is again to maximize reuse of function evaluations, this time for the computation of coefficients with smaller n, in a further attempt to bridge the gap between oscillatory and non-oscillatory quadrature.

2.4.1 Filon-type quadrature

The simplest approximation method for $\hat{f}_n^C$ and $\hat{f}_n^S$ is, arguably, a truncation of the asymptotic expansions in Theorem 1. For example, we may define

$$\hat{A}_{s,n}^C[f] \sim (-1)^n \sum_{k=0}^{s-1} \frac{(-1)^k}{(n\pi)^{2k+2}} [f^{(2k+1)}(1) - f^{(2k+1)}(-1)],$$

which corresponds to truncating the expansion for $\hat{f}_n^C$ after s terms. This approximation carries an error of asymptotic order

$$\hat{f}_n^C - \hat{A}_{s,n}^C[f] \sim \mathcal{O}(n^{-2s-2}), \qquad n \gg 1. \tag{2.5}$$

The error has the same size, asymptotically, as the first discarded term in the expansion. It decreases rapidly as n becomes large.

This asymptotic approximation is not very useful however for small values of n. The idea of Filon-type quadrature is to maintain high asymptotic order, in the sense of (2.5), while also maintaining the classical notion of polynomial exactness. That is, Filon-type quadrature rules are exact when f is a polynomial of a certain degree. For general oscillatory integrals, this leads to quadrature rules involving derivatives. Let

$$-1 = c_1 < c_2 < \ldots < c_\nu = 1$$

be a set of ν quadrature points with associated multiplicities $m_k \in \mathbf{N}$. We construct a polynomial p of degree $\sum_{k=1}^{\nu} m_k - 1$ such that

$$p^{(i)}(c_k) = f^{(i)}(c_k), \qquad i = 0, 1, \ldots, m_k - 1, \quad k = 1, 2, \ldots, \nu.$$

Then a Filon-type method may be defined by

$$\hat{f}_n^C \approx \hat{Q}_n^C[f] := \hat{p}_n^C = \int_{-1}^{1} p(x) \cos \pi n x \, \mathrm{d}x.$$

This can be expressed in closed form since the moments are known, which are computable either via integration by parts or the formula (for $k > 0$):

$$\int_{-1}^{1} x^k \cos \pi n x \, \mathrm{d}x$$
$$= \Re \left\{ (-in\pi)^{-k-1} [\Gamma(1+k, in\pi) - \Gamma(1+k, -in\pi)] \right\},$$

where Γ is the incomplete Gamma function.

In the context of modified Fourier series, this general setting changes in the following way. The key to obtain high asymptotic order is to interpolate the odd derivatives of f at the endpoints. In other words, it is sufficient to interpolate precisely the data on which the early terms in the asymptotic expansion depend. This leads to a set of interpolation conditions of the form

$$p^{(2i+1)}(c_k) = f^{(2i+1)}(c_k), \qquad i = 0, 1, \ldots, m_k - 1, \quad k = 1, 2, \ldots, \nu.$$

Augmented by the condition $p(0) = f(0)$, this interpolation problem has a unique solution. The corresponding quadrature rules take the form

$$\hat{Q}_n^C[f] := \sum_{k=1}^{\nu} \sum_{j=0}^{m_k - 1} \theta_{k,j}^C(n) f^{(2j+1)}(c_k). \tag{2.6}$$

This rule has asymptotic error $\mathcal{O}(n^{-2s-2})$ if $m_1, m_\nu \geq s$. Similar rules can of course be constructed for the sine coefficients $\hat{f}_n^S$.

The weights $\theta_{k,j}^C(n)$ typically depend on n in an explicit manner. Substantial insights in the design of Filon-type quadrature methods are developed in [IN06a, IN06b, IN07], and we refer the reader to those papers for explicit examples of suitable quadrature rules.

2.4.2 Exotic quadrature

For small values of n, the integrals $\hat{f}_n^C$ and $\hat{f}_n^S$ to compute are non-oscillatory in nature. One can resort to any of the known quadrature schemes for smooth and non-oscillatory functions, such as composite Gaussian quadrature [DR84]. Note that for achieving an $O(m)$ algorithm for the computation of the first m coefficients, it is actually irrelevant how the first (finitely many) elements are computed. Nevertheless, one naturally seeks for optimal methods that reduce computation time.

Given that the computation of coefficients for large n requires derivatives at the endpoints, one can reuse this information in the computation of coefficients for small n. This leads to quadrature rules involving derivatives that, for lack of an established name, were dubbed *exotic quadrature* in [IN06b]. They have the general form

$$\int_{-1}^{1} g(x)\,\mathrm{d}x \approx \hat{P}[f] := \sum_{k=1}^{\nu} \sum_{j \in N_{m_k}} \delta_{k,j} g^{(j)}(c_k),$$

and they are typically applied to the function $g(x) = f(x)u_n(x)$, where $u_n(x)$ is one of the Laplace-Neumann eigenfunctions. We have introduced the sets N_m to illustrate that the information about derivatives in exotic quadrature may be lacunary. For example, in the case of univariate modified Fourier series, we may define

$$N_m := \{2j+1\}_{j=0}^{m-1}.$$

The main message embodied in the theory of Filon-type quadrature and exotic quadrature is that the onset of asymptotic behaviour for increasing n is quite rapid. It appears that asymptotic accuracy kicks in for very moderate values of n. It is only reasonable to exploit this.

3 Polyharmonic approximation

The importance of imposing homogeneous Neumann boundary conditions in the general setting (1.1) became visible in Theorem 1. The boundary conditions rendered the first term in the asymptotic expansion of the coefficients zero, thus generating faster decay. This observation leads in a natural way to the first generalization of the theory, where faster convergence is achieved by imposing higher-order Neumann boundary conditions. In particular, in this section we are interested in eigenfunctions of the *polyharmonic operator*

$$u^{(2q)} + (-1)^{q+1}\alpha^{2q}u = 0, \qquad -1 \le x \le 1, \tag{3.1}$$

subject to the Neumann boundary conditions

$$u^{(i)}(-1) = u^{(i)}(1) = 0, \qquad i = q, q+1, \ldots, 2q-1. \tag{3.2}$$

Here q is a fixed parameter determining the order of the polyharmonic operator.

Denote the nth eigenvalue as

$$\kappa_n = (-1)^q \alpha^{2q},$$

with the corresponding nth eigenfunction denoted as u_n. Like the modified Fourier series, the basis of eigenfunctions $\{u_1, u_2, \ldots\}$ form an orthogonal series with respect to the L_2 inner product. Furthermore, they are dense in $L_2[-1, 1]$ [IN06b]. Thus we can successfully utilize them for function approximation, giving us the expansion

$$f(x) \sim \sum_{n=1}^{\infty} \hat{f}_n u_n, \qquad \text{for} \qquad \hat{f}_n = \int_{-1}^{1} f(x) u_n(x) \, \mathrm{d}x.$$

By repeatedly utilizing integration by parts and assuming that f is $(q+1)$-times differentiable, we immediately find that

$$\begin{aligned} \hat{f}_n &= \int_{-1}^{1} f(x) u_n(x) \, \mathrm{d}x = \frac{(-1)^q}{\alpha_n^{2q}} \int_{-1}^{1} f(x) u_n^{(2q)}(x) \, \mathrm{d}x \\ &= \frac{1}{\alpha_n^{2q}} \int_{-1}^{1} f^{(q)}(x) u_n^{(q)}(x) \, \mathrm{d}x \\ &= \frac{1}{\alpha_n^{2q}} [f^{(q)}(1) u_n^{(q-1)}(1) - f^{(q)}(-1) u_n^{(q-1)}(-1)] \\ &\quad - \frac{1}{\alpha_n^{2q}} \int_{-1}^{1} f^{(q+1)}(x) u_n^{(q-1)}(x) \, \mathrm{d}x. \end{aligned}$$

From [IN06b], we know that $u_n^{(i)}(x) = \mathcal{O}(\alpha_n^i)$. Furthermore, we also know that $\alpha_n \sim \mathcal{O}(n)$ [PT87]. It follows immediately that

$$\hat{f}_n = \mathcal{O}(n^{-q-1}).$$

Thus we can obtain any algebraic convergence rate by choosing q large enough.

3.1 The basis of eigenfunctions

In this section we will demonstrate how α_n and the eigenfunctions u_n can be found. For simplicity we focus on the case where $q = 2$, referring the interested reader to [IN06b] for the derivation for other values of q. The first two eigenfunctions are trivial (which we will not include in the enumeration $u_1, u_2, \ldots$): 1 and x. It follows immediately from (3.1) that the other eigenfunctions can be expressed as a sum of exponentials. In particular, for $q = 2$ we obtain

$$u(x) = c_1 \cos \alpha x + c_2 \sin \alpha x + c_3 \cosh \alpha x + c_4 \sinh \alpha x.$$

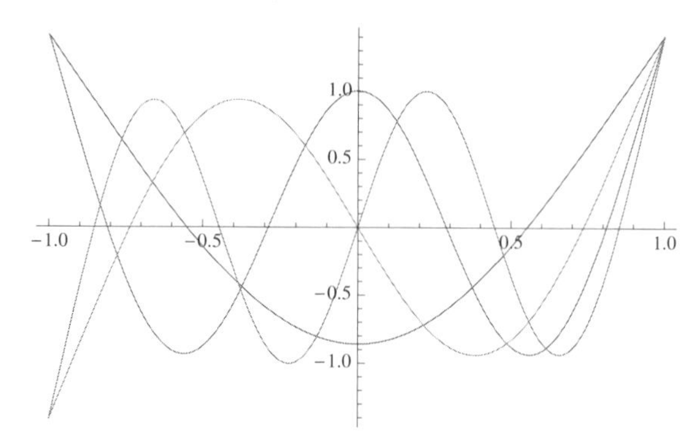

Fig. 3. The functions $u_1, \ldots, u_4$ on $[-1, 1]$ for $q = 2$.

Our goal, then, is to find which values of c_k and α satisfy the boundary conditions. Thus we want to ensure that $0 = u''(\pm 1) = u'''(\pm 1)$. Sparing the algebraic details, this is achieved when

$$
\begin{aligned}
u(x) &= \frac{\sqrt{2}}{2}\left(\frac{\cos \alpha x}{\cos \alpha} + \frac{\cosh \alpha x}{\cosh \alpha}\right) \qquad \text{if} \qquad \tan \alpha + \tanh \alpha = 0, \\
u(x) &= \frac{\sqrt{2}}{2}\left(\frac{\sin \alpha x}{\sin \alpha} + \frac{\sinh \alpha x}{\sinh \alpha}\right) \qquad \text{if} \qquad \tan \alpha - \tanh \alpha = 0.
\end{aligned}
$$

We are thus left with the problem of computing the roots of two transcendental equations. Trigonometric manipulations and calculus inform us that $\tan \alpha + \tanh \alpha$ has a root in each exponentially small window $[(n - \frac{1}{4})\pi, (n - \frac{1}{4})\pi + e^{-2(n-\frac{1}{4})\pi}]$ for $n = 1, 2, \ldots$. Similarly, $\tan \alpha - \tanh \alpha$ has a root in $[(n + \frac{1}{4})\pi - e^{-2(n+\frac{1}{4})\pi}, (n + \frac{1}{4})\pi]$ for $n = 1, 2, \ldots$. These zeros interlace, and we can thus enumerate the zeros of each transcendental equation as α_n, where a sufficiently small neighbourhood is known where the zero exists as to make its computation straightforward. Thus we can successfully compute the eigenfunction u_n associated to α_n, and in Fig. 3 we graph the first four such eigenfunctions.

4 Multivariate approximation

The generalization of univariate modified Fourier expansions to a higher-dimensional domain $\Omega \cup \mathbf{R}^d$ is, at least in principle, straightforward: the expansion is simply defined in terms of eigenfunctions of the Laplace operator on Ω, subject to Neumann boundary conditions. We will show below that this indeed leads to rapid convergence in very general circumstances. Unfortunately, the set of domains for which the eigenfunctions are known with suitably explicit expressions is small. In practice, multivariate modified Fourier expansions are therefor limited to a few domains. In two dimensions, the list includes ellipses, rectangles, annuli and a number of different triangles.

4.1 Laplace–Neumann eigenfunctions

We are looking for eigenfunctions u of the Laplace–Neumann problem in a simply-connected and bounded domain $\Omega \subset \mathbf{R}^d$,

$$-\Delta u = \lambda u, \quad \mathbf{x} \in \Omega, \qquad \frac{\partial u}{\partial \nu} = 0, \quad \mathbf{x} \in \partial\Omega. \tag{4.1}$$

We can in general denote the countable set of eigenvalues and eigenfunctions by λ_n and u_n, $n \geq 0$, with $\lambda_k \leq \lambda_n$ if $k < n$. It is always true that $\lambda_0 = 0$ and $u_0 \equiv 1$, and all subsequent eigenvalues are strictly positive. With this ordering, the Weyl theorem holds:

$$\lambda_n \sim \text{meas}(\Omega) n^{\frac{2}{d}}, \qquad n \gg 1,$$

where $\text{meas}(\Omega)$ denotes the measure of Ω [CH62]. A function f can be expanded in the series

$$f \sim \sum_{n=0}^{\infty} \frac{\hat{f}_n}{\hat{u}_n} u_n, \tag{4.2}$$

where $\hat{u}_n = \langle u_n, u_n \rangle$ and

$$\hat{f}_n = \langle f, u_n \rangle = \int_\Omega f(\mathbf{x})\, u_n(\mathbf{x})\, \mathrm{d}V. \tag{4.3}$$

In practice, other notation is often more convenient, as explicitly known eigenfunctions in more than one dimension typically depend on more than one parameter. This of course depends on the case at hand. We will review the case of the d-dimensional cube in §4.3. First however, we continue the general setting to discuss convergence.

4.2 Convergence of the approximation

Convergence of the series (4.2) implies decay of the coefficients. In one dimension, the rate of decay of the coefficients was established in Theorem 1 via integration by parts. The multivariate counterpart of integration by part is the Stokes theorem. This, in connection with the Weyl theorem, will again establish rapid decay of Laplace–Neumann coefficients in the general setting treated so far.

Given $f \in C^\infty[\Omega]$, we replace u_n with $\lambda^{-1}\Delta u_n$ in (4.3). Applying Stokes theorem twice and substituting homogeneous Neumann boundary conditions [IN07], this leads to

$$\langle f, u_n \rangle = \frac{1}{\lambda_n} \int_{\partial\Omega} \frac{\partial f(\mathbf{x})}{\partial \nu} u_n(\mathbf{x})\,\mathrm{d}S - \frac{1}{\lambda_n} \langle \Delta f, u_n \rangle.$$

Iterating the approach, we obtain the expansion

$$\langle f, u_n \rangle \sim - \sum_{m=0}^{\infty} \frac{1}{(-\lambda_n)^{m+1}} \int_{\partial\Omega} \frac{\partial \Delta^m f(\mathbf{x})}{\partial \nu} u_n(\mathbf{x})\,\mathrm{d}S, \qquad \lambda_n \gg 1. \quad (4.4)$$

We make the following remarks:

1. (4.4) converges only in an asymptotic sense, that is for $\lambda_n \gg 1$ (or, equivalently, $n \gg 1$).
2. the size of the coefficient is given, asymptotically, by the first term in the expansion. Assuming orthonormal u_n, the integral remains bounded as $n \gg 1$. It follows that $\hat{f}_n = \mathcal{O}(\lambda_n^{-1}) = \mathcal{O}(n^{-\frac{2}{d}})$ from the Weyl theorem.
3. the coefficients decay faster if f can be extended as a smooth and even function across $\partial\Omega$ (recall §2.3). This follows because, for such functions, the normal derivatives $\frac{\partial^{2j+1} f}{\nu^{2j+1}}$ of odd order vanish on $\partial\Omega$, $j = 0, 1, \ldots$. Consequently, one can show that the normal derivatives $\frac{\partial \Delta^m f}{\partial \nu}$ also vanish, $m = 0, 1, \ldots$.

It appears that expansion (4.4) is very informative. Yet, at the same time it may also be misleading. Most coefficients are in fact much smaller than the $\frac{1}{\lambda_n}$ estimate obtained above. In particular, for most values of n, the boundary integrals in (4.4) are themselves oscillatory integrals. This means they are much smaller than $\mathcal{O}(1)$ as was assumed in (ii). They can typically be expanded further asymptotically, using either the Stokes theorem or partial integration on the piecewise smooth parts of $\partial\Omega$. The results depend on the particular shape of Ω.

Such further expansion is possible only if u_n is oscillatory along the boundary $\partial\Omega$ of the domain. This is not always the case, even if $n \gg 1$. This phenomenon gives rise to the so-called *hyperbolic cross* [Bab60], to which we will return later in §5.

4.3 d-dimensional cubes

The case of the d-dimensional cube is the most obvious setting for a generalization of modified Fourier series to higher dimension. In this case, the Laplace–Neumann eigenfunctions are given by a tensor-product of the univariate eigenfunctions. We illustrate the points raised in §4.2 above with the two-dimensional square $\Omega = [-1, 1]^2$.

As indicated earlier, it is convenient to switch notation and have eigenfunctions depend on two parameters m and n, rather than just n. These parameters correspond to frequency in x and y direction respectively. There are four kinds of eigenfunctions,

$$\begin{aligned} u_{m,n}^{[0,0]}(x,y) &= \cos(\pi m x)\,\cos(\pi n y), \\ u_{m,n}^{[0,0]}(x,y) &= \cos(\pi m x)\,\sin[\pi(n-\tfrac{1}{2})y], \\ u_{m,n}^{[0,0]}(x,y) &= \sin[\pi(m-\tfrac{1}{2})x]\,\cos(\pi n y), \\ u_{m,n}^{[0,0]}(x,y) &= \sin[\pi(m-\tfrac{1}{2})x]\,\sin[\pi(n-\tfrac{1}{2})y]. \end{aligned}$$

The eigenvalues corresponding to $u_{m,n}^{[0,0]}$, for example, are $\lambda_{m,n}^{[0,0]} = \pi^2(m^2 + n^2)$. This means that the coefficients decay as

$$\langle f, u_{m,n}^{[0,0]} \rangle \sim \frac{1}{\lambda_{m,n}^{[0,0]}} \sim \mathcal{O}\left(\frac{1}{m^2+n^2}\right), \qquad m^2+n^2 \gg 1.$$

This estimate is valid as soon as either $m \gg 1$ or $n \gg 1$. However, if both m and n are large, we actually have

$$\langle f, u_{m,n}^{[0,0]} \rangle \sim \mathcal{O}\left(\frac{1}{m^2 n^2}\right), \qquad m, n \gg 1.$$

This follows from a further integration by parts on the boundary integrals in (4.4). The same procedure shows that such coefficients asymptotically depend only on certain partial derivatives of f at the vertices.

In the (m, n) plane, we conclude that coefficients near the edge $m = 0$ behave as $\mathcal{O}(n^{-2})$ and coefficients near the edge $n = 0$ behave as $\mathcal{O}(m^{-2})$. Elsewhere the coefficients are much smaller and behave as $\mathcal{O}(m^{-2}n^{-2})$. Such setting leads to the typical shape of a hyperbolic cross, that is illustrated further on in Fig. 4. Note that the decay of the coefficients compares to $\mathcal{O}(m^{-1}n^{-1})$ had we considered a classical tensor-product Fourier series instead.

4.4 Computation of the coefficients

The computation of coefficients in the multivariate case is more involved than constructing a Cartesian product generalization of the univariate quadrature. This is true even in the case of a d-dimensional cube. We focus briefly on two of the issues surrounding multivariate quadrature (cubature) in the setting of modified Fourier series.

Cubature is based on polynomial interpolation, but not all sets of points are suitable for such interpolation [Coo97]. The interpolation problem can be elegantly circumvented in the design of Filon-type methods by considering a Filon-type method as a correction to the asymptotic method. We formally write

$$\hat{Q}^F_{s,n}[f] = \hat{A}_{s,n}[f] + \hat{E}_{s,n}[f],$$

where $\hat{A}_{s,n}[f]$ is the asymptotic expansion up to an order defined by s. The correction term $\hat{E}_{s,n}$ is found as the best approximation to the next terms of the expansion, based on the available function evaluations of f. This can be achieved using finite differences. All known Filon-type methods that are exact for polynomials fit this formalism.

The second issues arises in so-called *edge coefficients*. These are coefficients given by integrals that are non-oscillatory in some variables, but oscillatory in the others. For example, in the case of the square as defined above, edge coefficients are those coefficients $\hat{f}^{[0,0]}_{m,n}$ where $m \gg 1$ and $n \approx 1$, or $m \approx 1$ and $n \gg 1$. The issue can be resolved by combining oscillatory and non-oscillatory quadrature, for example combining Filon-type quadrature in x with exotic quadrature in y. We refer the reader to [IN07] for an in-depth discussion.

5 Convergence acceleration

Modified Fourier series converge quite rapidly, at least faster than classical Fourier series, while yielding convergence everywhere. Yet, the convergence rate can also be improved to essentially arbitrarily high algebraic rates. The first method reviewed in §3, by considering polyharmonic operators, yields faster initial convergence of $O(m^{-q-1})$ but does not scale well to high order. An alternative is acceleration of the standard Laplace–Neumann case. In the section we focus on acceleration through the techniques of polynomial subtraction.

5.1 Polynomial subtraction

Asymptotic expansions of the coefficients, such as those given in Theorem 1 and in expression (4.4), are very revealing. Let us consider first the univariate case. The idea of polynomial subtraction is simple: we subtract from f a polynomial such that the first few terms of the expansion vanish. Thus, the coefficients decay faster and the convergence rate improves likewise.

We set

$$g(x) = f(x) - p_s(x), \qquad x \in [-1, 1],$$

where $p_s(x)$ is a polynomial that satisfies

$$p_s^{(2j+1)}(\pm 1) = f^{(2j+1)}(\pm 1), \qquad j = 0, \ldots, s-1.$$

Lemma 4 *Consider the modified Fourier series $g_m(x)$ for g. We have*

$$g(x) - g_m(x) = \mathcal{O}(m^{-2s-2}), \qquad m \gg 1.$$

Proof Follows immediately from Theorem 1. □

We approximate $f(x)$ by $p_s(x) + g_m(x)$ with the same error. Note that $p_s(x)$ itself is not a good approximation to $f(x)$. The convergence rate is $\mathcal{O}(m^{-2s-2})$ for $x \in (-1, 1)$ and $\mathcal{O}(m^{-2s-1})$ for $x = \pm 1$. The Gibbs phenomenon is still present, but clearly it is much less severe: convergence at the endpoints is only slightly slower than in the interior of the domain. The improved decay of the coefficients and accelerated convergence of the series was already illustrated in Fig. 1 in §2. There, we constructed p_1 and p_2 for the function $f(x) = \cos(x + 1)\sin(x + 1)$.

Of course, one is not forced to use a polynomial basis for $p_s(x)$. In any case, all that is required for acceleration is a suitable estimate of the odd derivatives of $f(x)$ at the endpoints.

5.2 Subtraction in a multivariate setting

The above generalizes to multivariate integration in the following way. Based on the asymptotic expansion (4.4), we again set

$$g(\mathbf{x}) = f(\mathbf{x}) - p_s(\mathbf{x}),$$

where now $p_s(\mathbf{x})$ is such that

$$\frac{\partial \Delta^j p}{\partial \nu}(\mathbf{x}) = \frac{\partial \Delta^j f}{\partial \nu}(\mathbf{x}), \qquad \mathbf{x} \in \partial\Omega, \qquad j = 0, \ldots, s-1.$$

Lemma 5 *Consider the modified Fourier series $g_m(\mathbf{x})$ for g. We have*

$$g(\mathbf{x}) - g_m(\mathbf{x}) = \mathcal{O}(\lambda_m^{-s-1})$$

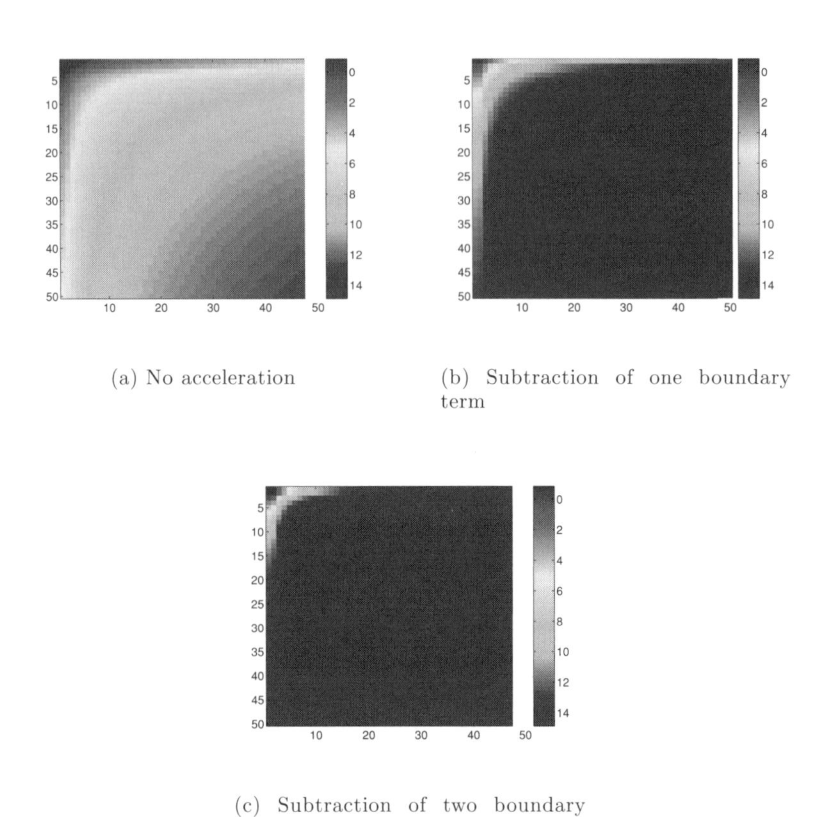

(a) No acceleration

(b) Subtraction of one boundary term

(c) Subtraction of two boundary terms

Fig. 4. The absolute value of coefficients $\hat{f}^{[0,0]}_{m,n}$ for $f(x) = e^{\pi(x-y)}\cos(x)$ with and without acceleration. The values are shown in base-10 logarithmic scale.

Proof Follows immediately from (4.4). □

Though the function $p_s(\mathbf{x})$ can in general not be a polynomial, the interpolation of a function or its derivatives along the boundary of a domain is a topic well studied in the field of Computer Aided Geometric Design (CAGD). A general construction for any s in the case of d-dimensional cubes is given in [HIN07].

Alternatively, it is easier to interpolate only partial derivatives at the vertices of the boundary $\partial\Omega$. This achieves accelerated decay of all coefficients except for edge coefficients. The minimal set of partial derivatives to interpolate depends on the shape of Ω.

We remark however that in boundary value problems, Neumann data is precisely the sort of information that may be available. The use of modified Fourier series in spectral methods for boundary value problems was recently explored in [Adc07]. It was found that modified Fourier series are better behaved in this setting than Chebyshev expansions. In particular, the condition number of the discretization matrix behaves as $\mathcal{O}(m^2)$ for modified Fourier series, but as $\mathcal{O}(m^4)$ for Chebyshev expansions.

Fig. 4 illustrates the size of the modified Fourier coefficients for the case of the square $\Omega := [-1, 1]$ and the smooth function

$$f(x) = e^{\pi(x-y)} \cos(x).$$

Level curves roughly correspond to constant $m^2 n^2$, due to the $\mathcal{O}((mn)^{-2})$ behaviour of the coefficients. The *hyperbolic cross* derives from the hyperbolic shape of these curves. The hyperbolic cross is particularly interesting in higher dimension: one can show that the number of coefficients larger than a given threshold grows only logarithmically with dimension, rather than exponentially [HIN07]. The acceleration procedure in this two-dimensional example is very effective indeed. For example, the number of elements larger than 10^{-5} in panel (a) is 2,113, in panel (b) it is 184 and in panel (c) only 58. In the latter case, the coefficients behave as $O((mn)^{-6})$, for $m, n \gg 1$.

6 Concluding remarks

At the beginning of this paper, we motivated the use of modified Fourier series by comparing to classical Fourier series and Chebyshev expansions. To summarize, we note the following advantages:

1. The first m coefficients can be computed in $\mathcal{O}(m)$ operations.
2. Additional coefficients can be computed adaptively, with an $\mathcal{O}(1)$ cost associated to each term.
3. The method of approximation generalizes in different directions: to eigenfunctions of polyharmonic operators and to general domains Ω including ellipses, cubes and triangles. (And, by extension, to any Ω that can be tessellated by these basic shapes. For example, any polygon can be divided in a set of triangles, and functions defined on a polygon can be approximated in each triangle separately.)

4. Modified Fourier series are more stable in spectral methods than Chebyshev expansions, since they typically exhibit an $\mathcal{O}(m^2)$ condition number of the discretization matrix rather than $\mathcal{O}(m^4)$.

Compared to FFT for non-periodic functions, modified Fourier series overcome the Gibbs phenomenon and yield pointwise convergence at the boundary. These facts are even more true when convergence acceleration is used. An alternative, and very effective, procedure for defeating the Gibbs phenomenon in one dimension was also reviewed in [GS97]. The described method achieves exponential accuracy in the reconstruction of a non-periodic function f from its values in equidistant points, much like FFT for periodic functions, by expanding the partial Fourier sum into a set of Gegenbauer polynomials. The computations involved can become unstable however if f has singularities close to the real axis [Boy05]. Modified Fourier series on the other hand, leaving generality aside, are guaranteed to be stable for all domains Ω: all coefficients are easily bounded in terms of the maximum of $|f|$ on Ω. Moreover, the techniques for convergence acceleration are quite flexible, and we anticipate that the degrees of freedom may be used in the future to produce numerically stable algorithms in a variety of applications.

Bibliography

[Adc07] B. Adcock (2007). Spectral methods and modified Fourier series, Technical Report 2007/NA08, University of Cambridge.

[Bab60] K. Babenko (1960). Approximation of periodic functions of many variables by trigonometric polynomials, *Soviet Maths* **1**, 513–516.

[Boy05] J. P. Boyd (2005). Trouble with Gegenbauer reconstruction for defeating Gibbs' phenomenon: Runge phenomenon in the diagonal limit of Gegenbauer polynomial approximations, *J. Comput. Phys.* **204**, 253–264.

[Coo97] R. Cools (1997). Constructing cubature formulae: The science behind the art, *Acta Numer.* **6**, 1–54.

[CH62] R. Courant & D. Hilbert, D. (1962). *Methods of Mathematical Physics* (Wiley Interscience, New York).

[DR84] P. J. Davis & P. Rabinowitz (1984). *Methods of Numerical Integration*, Computer Science and Applied Mathematics (Academic Press, New York).

[GS97] D. Gottlieb & C.-W. Shu (1997). On the Gibbs phenomenon and its resolution, *SIAM Rev.* **39**, 644–668.

[HIN07] D. Huybrechs, A. Iserles & S. P. Nørsett (2007). From high oscillation to rapid approximation IV: Accelerating convergence, Technical Report NA2007/07, University of Cambridge.

[HO08] D. Huybrechs & S. Olver (2008). Highly oscillatory quadrature, This volume.

[IN05] A. Iserles & S. P. Nørsett (2005). Efficient quadrature of highly oscillatory integrals using derivatives, *Proc. R. Soc. Lond. A* **461**, 1383–1399.

[IN06b] A. Iserles & S. P. Nørsett (2006). From high oscillation to rapid approximation II: Expansions in polyharmonic eigenfunctions, Technical Report 2006/NA07, University of Cambridge.

[IN07] A. Iserles & S. P. Nørsett (2007). From high oscillation to rapid approximation III:Multivariate expansions, Technical Report 2007/NA01, University of Cambridge, to appear in *IMA J. Num. Anal.*

[IN06a] A. Iserles & S. P. Nørsett (2008). From high oscillation to rapid approximation I: Modified Fourier expansions, *IMA J. Num. Anal.* **28**, 862-887.

[Olv07] S. Olver (2007). On the convergence rate of a modified Fourier series, Technical Report 2007/NA02, University of Cambridge.

[PT87] J. Poschel & E. Trubowitz (1987). *Inverse Spectral Theory* (Academic Press, Boston).

4

Approximation of high frequency wave propagation problems

Mohammad Motamed
Department of Numerical Analysis, KTH
100 44 Stockholm
Sweden
Email: mohamad@nada.kth.se

Olof Runborg
Department of Numerical Analysis, KTH
100 44 Stockholm
Sweden
Email: olofr@nada.kth.se

Abstract

The numerical approximation of high frequency wave propagation is important in many applications including seismic, acoustic, optical waves and microwaves. For these problems the solution becomes highly oscillatory relative to the overall size of the domain. Direct simulations using the standard wave equations are therefore very expensive, since a large number of grid points is required to resolve the wave oscillations. There are however computationally much less costly models, that are good approximations of many wave equations at high frequencies. In this paper we review such models and related numerical methods that are used for simulations in applications. We focus on the infinite frequency approximation of geometrical optics and the finite frequency corrections given by the geometrical theory of diffraction. We also briefly discuss Gaussian beams.

1 Introduction

Simulation of high-frequency waves is a problem encountered in a great many engineering and science fields. Currently the interest is driven by new applications in wireless communication (cell phones, Bluetooth, WiFi) and photonics (optical fibers, filters, switches). Simulation is also used increasingly in more classical applications. Some examples in electromagnetism are antenna design, radar signature computation and base station coverage for cell phones. In acoustics simulation is used for noise reduction, underwater communication and medical ultrasonography. Finding the location of an earthquake and oil exploration are some applications of seismic wave simulation in geophysics. Nondestructive testing is another example where both electromagnetic and acoustic waves are simulated.

In this review we consider numerical simulation of waves at high frequencies and the underlying mathematical models used. For simplicity we will mainly discuss the linear scalar wave equation,

$$u_{tt} - c(\boldsymbol{x})^2 \Delta u = 0, \qquad (t, \boldsymbol{x}) \in \mathbb{R}^+ \times \Omega, \qquad \Omega \subset \mathbb{R}^d, \tag{1.1}$$

where $c(\boldsymbol{x})$ is the local speed of wave propagation of the medium. We complement (1.1) with initial or boundary data that generate high-frequency solutions. The exact form of the data will not be important here, but a typical example would be $u(0, \boldsymbol{x}) = A(\boldsymbol{x}) \exp(i\omega \boldsymbol{k} \cdot \boldsymbol{x})$ where $|\boldsymbol{k}| = 1$ and the frequency $\omega \gg 1$. With slight modifications, the techniques we describe will also carry over to systems of wave equations, like the Maxwell equations and the elastic wave equation.

When the frequency of the waves is high, (1.1) has highly oscillatory solutions with a wavelength that is much smaller than the overall size of the computational domain. In the direct numerical simulation of (1.1) the accuracy of the solution is determined by the number of grid points or elements per wavelength. The computational cost to maintain constant accuracy grows algebraically with the frequency, and for sufficiently high frequencies, direct numerical simulation is no longer feasible. Numerical methods based on approximations of (1.1) are needed.

Fortunately, there exist good approximations of many wave equations precisely for very high frequency solutions. In this paper we consider variants of *geometrical optics* (GO), which are asymptotic approximations obtained when the frequency tends to infinity. Instead of the oscillating wave field the unknowns in standard geometrical optics are the phase and the amplitude, which typically vary on a much coarser scale than the full solution. Hence, they should in principle be easier to compute numerically.

The main drawbacks of the infinite frequency approximation of GO are that diffraction effects at boundaries are lost, and that the approximation breaks down at caustics, where the predicted amplitude is unbounded. For these situations more detailed models are needed, such as the *geometrical theory of diffraction* (GTD) [Kel62], which adds diffraction phenomena by explicitly taking into account the geometry of Ω and boundary conditions. The solution's asymptotic behavior close to caustics can also be derived, and a correct amplitude for finite frequency can be computed [Kra64, Lud66, Hör83]. Numerically this can for instance be done with Gaussian beams [Pop82, BP73].

The purpose of this paper is to review the mathematical models and numerical methods for high frequency waves based on GO and GTD, with a special focus on creeping rays. For other reviews of this topic, see [Ben03, ER03, LOT06, Run07]. The paper is organized as follows. In Section 2 the equations used in GO are derived and a survey of numerical methods for such high frequency models is given. The techniques to add diffraction effects and to correct the standard geometrical optics approximation at caustics are discussed in Section 3.

2 Geometrical optics

In this section, we first derive the equations that are used in geometrical optics. We then review numerical methods based on different formulations.

2.1 Mathematical formulation

The starting point is the Cauchy problem for the scalar wave equation (1.1),

$$\begin{aligned} u_{tt}(t,\boldsymbol{x}) - c(\boldsymbol{x})^2 \Delta u(t,\boldsymbol{x}) &= 0, && \boldsymbol{x} \in \mathbb{R}^d,\ t > 0, \\ u(0,\boldsymbol{x}) &= u_0(\boldsymbol{x}), && \boldsymbol{x} \in \mathbb{R}^d, \\ u_t(0,\boldsymbol{x}) &= u_1(\boldsymbol{x}), && \boldsymbol{x} \in \mathbb{R}^d, \end{aligned} \tag{2.1}$$

with highly oscillatory initial data u_0 and u_1. Here $c(\boldsymbol{x})$ is the local wave velocity of the medium. We also define the *index of refraction* as $\eta(\boldsymbol{x}) = c_0/c(\boldsymbol{x})$ with the reference velocity c_0 (*e.g.* the speed of light in vacuum). For simplicity we will henceforth let $c_0 = 1$. The wave equation (2.1) admits solutions of the type

$$u(t,\boldsymbol{x}) = A(t,\boldsymbol{x},\omega)e^{i\omega\phi(t,\boldsymbol{x})}, \tag{2.2}$$

with A and ϕ representing the amplitude and phase functions, respectively. The level curves of ϕ correspond to the wave fronts of a propagating wave.

Since (2.1) is linear, the superposition principle is valid and a sum of solutions is itself a solution (provided initial data is adapted accordingly). The generic solution to (2.1) is, at least locally, described by a finite sum of terms like (2.2),

$$u(t, \boldsymbol{x}) = \sum_{n=1}^{N} A_n(t, \boldsymbol{x}, \omega) e^{i\omega\phi_n(t,\boldsymbol{x})}, \tag{2.3}$$

with A_n, ϕ_n being smooth functions and A_n depending only mildly on the frequency ω. Typically this setting only breaks down at a small set of points, namely focus points, caustics and discontinuities in $c(\boldsymbol{x})$.

In this section, we will assume the geometrical optics approximation that $\omega \to \infty$. This means that we accept the loss of diffraction phenomena in the solution and the amplitude's breakdown at caustics, which we will come back to in Section 3. There are three strongly related formulations of geometrical optics, which we will review here.

2.1.1 Eikonal equation

Let us start by deriving Eulerian PDEs for the phase and the amplitude functions that are formally valid in the limit when $\omega \to \infty$. We consider time harmonic waves of type $u(t, \boldsymbol{x}) = v(\boldsymbol{x}) \exp(i\omega t)$ with ω fixed. Inserting it into the time-dependent wave equation (2.1), we get the Helmholtz equation

$$c^2 \Delta v + \omega^2 v = 0. \tag{2.4}$$

We assume that the solution to (2.1) has the form (2.2) and that the amplitude function in (2.2) can be expanded in a power series in $1/i\omega$. We then get the asymptotic WKB expansion, [Hör83],

$$v = e^{i\omega\tilde{\phi}(\boldsymbol{x})} \sum_{k=0}^{\infty} \tilde{a}_k(\boldsymbol{x})(i\omega)^{-k} \tag{2.5}$$

and substitute it into (2.4). Equating coefficients of powers of ω to zero gives us the *eikonal equation* and the the *transport equation* for the phase and the first amplitude term in the frequency domain,

$$|\nabla\tilde{\phi}| = 1/c = \eta, \qquad 2\nabla\tilde{\phi} \cdot \nabla\tilde{a}_0 + \Delta\tilde{\phi}\tilde{a}_0 = 0. \tag{2.6}$$

For the remaining amplitude terms, we get additional transport equations

$$2\nabla\tilde{\phi} \cdot \nabla\tilde{a}_{k+1} + \Delta\tilde{\phi}\tilde{a}_{k+1} + \Delta\tilde{a}_k = 0.$$

When ω is large, only the first term in the expansion (2.5) is significant, and the problem is reduced to computing the phase $\tilde{\phi}$ and the first amplitude term $\tilde{a}_0$. Since the family of curves $\{\boldsymbol{x} \mid \phi(t, \boldsymbol{x}) = \tilde{\phi}(\boldsymbol{x}) - t = 0\}$, parametrized by $t \geq 0$, describe a propagating wave front, we often directly interpret the frequency domain phase $\tilde{\phi}(\boldsymbol{x})$ as the *travel time* of a wave.

In what follows we will drop the tilde for the frequency domain quantities and denote the first amplitude term a_0 by A (or A_1, A_2, etc. when there are multiple crossing waves).

One problem with the eikonal and transport equations is that they do not accept solutions with multiple phases. There is no superposition principle for the nonlinear eikonal equation. In fact, for the case in (2.3), the derivation must be done term wise, and the $\{\phi_n\}$ and $\{A_n\}$ will, locally, satisfy separate eikonal and transport equation pairs. However, the eikonal equation still has a well-defined solution. It is a nonlinear Hamilton–Jacobi type equation with Hamiltonian $H(\boldsymbol{x}, \boldsymbol{p}) = c(\boldsymbol{x})|\boldsymbol{p}|$. Extra conditions are needed for this type of equations to have a unique solution known as the *viscosity solution*, which is the analogue of the entropy solution for hyperbolic conservation laws, [CL83]. As can be deduced from the previous paragraph, at points where the correct solution should have a multivalued phase, *i.e.* be of the type in (2.3), the viscosity solution picks out the phase corresponding to the first arriving wave, *i.e.* the smallest ϕ_n in (2.3).

It is well known that solutions of Hamilton–Jacobi equations can develop kinks, *i.e.* discontinuities in the gradient, just as shocks appear in the solutions of conservation laws. In the case of the eikonal equation, the kinks are located where the physically correct phase solution should become multivalued. We notice that the transport equation has a factor involving $\Delta\phi$, which is not bounded at kinks, and therefore we can expect blow-up of a_0 at these points.

2.1.2 Ray equations

Another formulation of geometrical optics is *ray tracing*, which gives the solution via ODEs. This Lagrangian formulation is closely related to the method of characteristics. Let $(\boldsymbol{x}(t), \boldsymbol{p}(t))$ be a bicharacteristic pair related to the Hamiltonian $H(\boldsymbol{x}, \boldsymbol{p}) = c(\boldsymbol{x})|\boldsymbol{p}|$. We are interested in solutions for which $H \equiv 1$. In this case the projections on physical space, $\boldsymbol{x}(t)$, are usually called *rays*, and we have

$$\frac{d\boldsymbol{x}}{dt} = \nabla_p H(\boldsymbol{x}, \boldsymbol{p}) = \frac{1}{\eta^2}\boldsymbol{p}, \qquad \boldsymbol{x}(0) = \boldsymbol{x}_0, \tag{2.7}$$

$$\frac{d\boldsymbol{p}}{dt} = -\nabla_x H(\boldsymbol{x}, \boldsymbol{p}) = \frac{\nabla\eta}{\eta}, \qquad \boldsymbol{p}(0) = \boldsymbol{p}_0, \quad |\boldsymbol{p}_0| = \eta(\boldsymbol{x}_0). \tag{2.8}$$

Solving (2.7, 2.8) is called *ray tracing*. In d dimensions the bicharacteristics are curves in $2d$-dimensional *phase space*, $(\boldsymbol{x}, \boldsymbol{p}) \in \mathbb{R}^{d \times d}$. It follows that H is constant along them, $H(\boldsymbol{x}(t), \boldsymbol{p}(t)) = H(\boldsymbol{x}_0, \boldsymbol{p}_0)$. Note that $\boldsymbol{p}(t) \equiv \nabla\phi(\boldsymbol{x}(t))$ if $\boldsymbol{p}_0 = \nabla\phi(\boldsymbol{x}_0)$. Hence, with this initialization, the rays are always orthogonal to the level curves of ϕ, since $d\boldsymbol{x}/dt$ is parallel to $\boldsymbol{p} = \nabla\phi$ by (2.7). Moreover, for our particular H,

$$\begin{aligned} \frac{d}{dt}\phi(\boldsymbol{x}(t)) &= \nabla\phi(\boldsymbol{x}(t)) \cdot \frac{d\boldsymbol{x}(t)}{dt} = \boldsymbol{p}(t) \cdot \nabla_p H(\boldsymbol{x}(t), \boldsymbol{p}(t)) \\ &= H(\boldsymbol{x}(t), \boldsymbol{p}(t)) = 1. \end{aligned}$$

Thus, as long as ϕ is smooth, we have

$$\phi(\boldsymbol{x}(t)) = \phi(\boldsymbol{x}_0) + t.$$

Since ϕ corresponds to travel time, this also shows that the parametrization t in (2.7, 2.8) actually corresponds to unscaled time; the ray $\boldsymbol{x}(t)$ traces one point on a propagating wave front at time t. The absolute value of its time derivative $|d\boldsymbol{x}/dt|$ is precisely the local speed of propagation $c(\boldsymbol{x})$ by (2.7), and since $\boldsymbol{p}$ is parallel to $d\boldsymbol{x}/dt$, while $|\boldsymbol{p}| = H(\boldsymbol{x}, \boldsymbol{p})c(\boldsymbol{x})^{-1} = c(\boldsymbol{x})^{-1}$, the vector $\boldsymbol{p}$ is often called the *slowness* vector.

As was discussed before, the solution of the eikonal equation is valid up to the point where discontinuities appear in the gradient of ϕ. This is where the phase should become multivalued but, by the construction, cannot. The bicharacteristics, however, do not have this problem, and we can extend their validity to all t.

In order to compute the amplitude along a ray we also need information about the local shape of the ray's source. Let $(\boldsymbol{x}(t, \boldsymbol{x}_0), \boldsymbol{p}(t, \boldsymbol{x}_0))$ denote the bicharacteristic originating in $\boldsymbol{x}_0$ with $\boldsymbol{p}(0, \boldsymbol{x}_0) = \nabla\phi(\boldsymbol{x}_0)$, hence $\boldsymbol{x}(0, \boldsymbol{x}_0) = \boldsymbol{x}_0$. Let $J(t, \boldsymbol{x}_0)$ be the Jacobian of $\boldsymbol{x}$ with respect to initial data, $J = D_{x_0}\boldsymbol{x}(t, \boldsymbol{x}_0)$, and assume that it is nonsingular. The amplitude is then given by the expression, [Mas65],

$$A(\boldsymbol{x}(t, \boldsymbol{x}_0)) = A(\boldsymbol{x}_0) \frac{\eta(\boldsymbol{x}_0)}{\eta(\boldsymbol{x}(t, \boldsymbol{x}_0))} \sqrt{\left| \frac{q(0, \boldsymbol{x}_0)}{q(t, \boldsymbol{x}_0)} \right|} e^{-im\frac{\pi}{2}}, \tag{2.9}$$

where $q = \det J$ is the *geometrical spreading* measuring the size change of an infinitesimal area transported by the rays, and $m = m(t)$ is a nonnegative integer called the *Keller–Maslov index*. It represents the number of times $q(\cdot, \boldsymbol{x}_0)$ has changed sign in the interval $[0, t]$, *i.e.* the number of times that the ray has touched a caustic which are sets of points where $q = 0$. These are points where rays concentrate, *cf.* Figure 5. We see clearly from (2.9) that the amplitude is unbounded close to caustic points. This formula is therefore valid only before and after these points.

The elements of the Jacobian is given by another ODE system, obtained by differentiating (2.7, 2.8) with respect to $\boldsymbol{x}_0$,

$$\frac{d}{dt}\begin{pmatrix} D_{x_0}\boldsymbol{x} \\ D_{x_0}\boldsymbol{p} \end{pmatrix} = \begin{pmatrix} D^2_{px}H & D^2_{pp}H \\ -D^2_{xx}H & -(D^2_{px}H)^T \end{pmatrix} \begin{pmatrix} D_{x_0}\boldsymbol{x} \\ D_{x_0}\boldsymbol{p} \end{pmatrix}, \tag{2.10}$$

with initial data

$$D_{x_0}\boldsymbol{x}(0,\boldsymbol{x}_0) = I, \qquad D_{x_0}\boldsymbol{p}(0,\boldsymbol{x}_0) = D^2\phi(\boldsymbol{x}_0).$$

Here D^2 represents the Hessian.

Note that since we have the constraint $H(\boldsymbol{x},\boldsymbol{p}) = 1$, or $|\boldsymbol{p}| = \eta(\boldsymbol{x})$, the dimension of the phase space $(\boldsymbol{x},\boldsymbol{p})$ can actually be reduced by one. For example in two dimensions, with $\boldsymbol{x} = (x, y)$, by setting $\boldsymbol{p} = \eta(\cos\theta, \sin\theta)$ and using θ as a dependent variable in (2.7, 2.8) instead of $\boldsymbol{p}$, we get the reduced equations,

$$\frac{dx}{dt} = c(x,y)\cos\theta, \tag{2.11a}$$

$$\frac{dy}{dt} = c(x,y)\sin\theta, \tag{2.11b}$$

$$\frac{d\theta}{dt} = \frac{\partial c}{\partial x}\sin\theta - \frac{\partial c}{\partial y}\cos\theta. \tag{2.11c}$$

2.1.3 Kinetic equations

Finally, we can adopt a purely kinetic viewpoint. This is based on the interpretation that rays are trajectories of particles following the Hamiltonian dynamics of (2.7, 2.8). We introduce the phase space $(t, \boldsymbol{x}, \boldsymbol{p})$, where $\boldsymbol{p}$ is the slowness vector defined above, and we let $f(t, \boldsymbol{x}, \boldsymbol{p})$ be a particle ("photon") density function. It will satisfy the Liouville equation,

$$f_t + \nabla_p H \cdot \nabla_x f - \nabla_x H \cdot \nabla_p f = 0.$$

We are only interested in solutions to (2.7, 2.8) for which $H(\boldsymbol{x},\boldsymbol{p}) = c(\boldsymbol{x})|\boldsymbol{p}| \equiv 1$, meaning that f only has support on the sphere $|\boldsymbol{p}| = \eta(\boldsymbol{x})$ in phase space. We then arrive at the Vlasov-type equation

$$f_t + \frac{1}{\eta^2}\boldsymbol{p} \cdot \nabla_x f + \frac{1}{\eta}\nabla_x \eta \cdot \nabla_p f = 0, \tag{2.12}$$

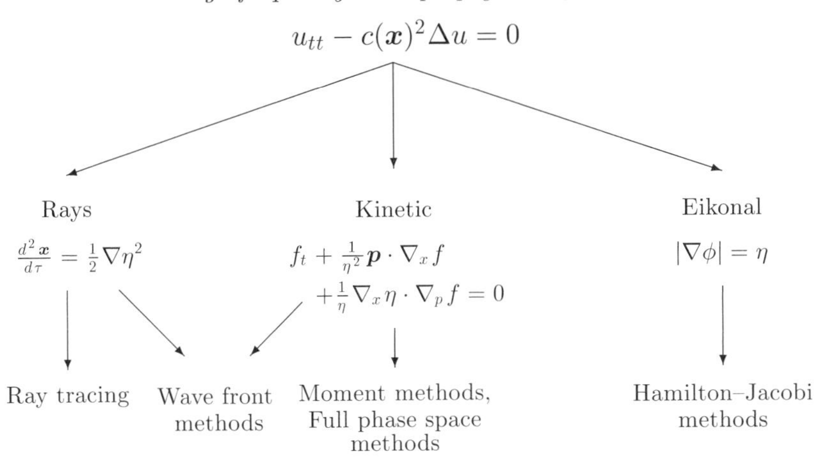

Fig. 1. Mathematical models and numerical methods. Wave front methods use aspects of both the ray and the kinetic model.

with initial data $f_0(\boldsymbol{x}, \boldsymbol{p})$ vanishing whenever $|\boldsymbol{p}| \neq \eta$. We note that, if $\eta \equiv 1$, the equation (2.12) is just a free transport equation with solution $f(t, \boldsymbol{x}, \boldsymbol{p}) = f_0(\boldsymbol{x} - t\boldsymbol{p}, \boldsymbol{p})$ which corresponds to straight line ray solutions of (2.7, 2.8).

By the Wigner transform, one can show that a sum of simple wave solutions (2.3) to the wave equation (2.1) corresponds to a sum of particle solutions to (2.12),

$$f(t, \boldsymbol{x}, \boldsymbol{p}) = \sum_{n=1}^{N} A_n^2(t, \boldsymbol{x})\, \delta(\boldsymbol{p} - \nabla\phi_n(t, \boldsymbol{x})), \tag{2.13}$$

when $\omega \to \infty$. See [JL03, SMM02].

2.2 Numerical methods

There are different classes of computational techniques, based on the different mathematical models for geometrical optics discussed in Section 2.1: see Figure 1. We give a brief introduction here.

2.2.1 Ray tracing

The ray equations derived in Section 2.1.2 are the basis for ray tracing. The ray $\boldsymbol{x}(t)$ and slowness vector $\boldsymbol{p}(t) = \nabla\phi(\boldsymbol{x}(t))$ are governed by the ODE system (2.7, 2.8). This system together with another ODE system for the amplitude, (2.10), are solved with standard ODE solvers giving the phase and amplitude along the ray. The solution at a desired point is then interpolated from the solutions along the rays. This can be rather difficult in the regions where ray tracing produces diverging or crossing rays. Moreover, ray tracing is only of interest for problems involving a few number of source points. For problems with many source points, ray tracing may be computationally expensive. Some general references on ray tracing are [CMP77, LLC85].

2.2.2 Hamilton–Jacobi methods

To avoid the problem of diverging rays, several PDE-based methods have been proposed for the eikonal and transport equations. When the solution is sought in a domain, this is also computationally a more efficient and robust approach. The equations are solved directly, using numerical methods for PDEs, on a uniform Eulerian grid to control the resolution. Different types of numerical techniques have been proposed to compute the unique viscosity solution of the eikonal equation, including upwind methods of ENO or WENO type [vTS91, OS91], fast marching method [Set99], group marching method [Kim00] and sweeping method [TCOZ03]. However, since the eikonal equation is a nonlinear equation for which the superposition principle does not hold, these methods fail to capture multivalued solutions corresponding to crossing rays. There are also Hamilton–Jacobi based methods proposed for computing multivalued solutions. Among those are a domain decomposition based method by detecting kinks [FEO95], big ray tracing [Ben96] and the slowness matching method [SQ03]. The multivalued solutions, in these methods, are constructed by putting together the solutions of several eikonal equations. Nevertheless, finding a robust technique to compute multivalued solutions is still a computational challenge.

2.2.3 Wave front methods

Wave front methods are closely related to standard ray tracing, but instead of computing a sequence of individual rays, the location of many rays coming from one source is computed. At fixed times, those points form a wavefront whose evolution is tracked in the physical or the phase space. This can be based on the ODE formulation (2.7, 2.8) or the PDE formulation (2.12).

Wave front construction, [VIG93], is an ODE-based front tracking method in which Lagrangian markers on the phase space wave front is propagated according to the ray equations (2.7, 2.8). To maintain an accurate description of the front, new markers are adaptively inserted by interpolation when the resolution of the front deteriorates. For the PDE-based wave front methods in phase space, the evolution of the front is given by the Liouville equation (2.12) and the front is represented by some interface propagation technique, such as the segment projection method, [ERT02], and level set techniques, [OS88]. The segment projection method uses an explicit representation of the wave front while the level set method uses an implicit representation.

2.2.4 Moment-based methods

Moment-based methods take as their starting point the kinetic formulation of geometrical optics (2.12). This equation has the advantage of the linear superposition property of the ray equations and like the eikonal equation, the solution is defined by a PDE and can be computed efficiently on a uniform Eulerian grid. Direct numerical approximation of (2.12) is, however, rather costly, because of the large set of independent variables (six in 3D). Instead one can use the classic technique of approximating a kinetic transport equation set in high-dimensional phase space $(t, \boldsymbol{x}, \boldsymbol{p})$, by a finite system of moment equations in the reduced space $(t, \boldsymbol{x})$. The moments m_{ij} are defined as

$$m_{ij}(t, \boldsymbol{x}) = \frac{1}{\eta(\boldsymbol{x})^{i+j}} \int_{\mathbb{R}^2} p_1^i p_2^j f(t, \boldsymbol{x}, \boldsymbol{p}) d\boldsymbol{p}, \quad \boldsymbol{p} = (p_1, p_2)^T. \tag{2.14}$$

Multiplying (2.12) by $\eta^{2-i-j} p_1^i p_2^j$ and integrating over $\mathbb{R}^2$ with respect to $\boldsymbol{p}$ gives us the infinite system of moment equations

$$\begin{aligned}(\eta^2 m_{ij})_t + (\eta m_{i+1,j})_x + (\eta m_{i,j+1})_y& \\ = i\eta_x m_{i-1,j} + j\eta_y m_{i,j-1} - (i+j)(\eta_x m_{i+1,j} + \eta_y m_{i,j+1})&,\end{aligned} \tag{2.15}$$

valid for all $i, j \geq 0$. Since the system is not closed, we have to make the closure assumption that at most N rays cross at any given point in time and space, [Run00]. This amounts to requiring that (2.13) holds. The closed system is then a $2N \times 2N$ system of nonlinear hyperbolic conservation laws with source terms, which can be solved with finite difference methods on fixed grids.

2.2.5 Full phase space methods

As was discussed, solving the full phase space Liouville equation (2.12) is significantly more expensive than solving *e.g.* the eikonal and transport equation. This is, however, only under the assumption that we are interested in the solution for just *one* set of initial data. In many applications, we seek the solution for *many* different initial data (sources), with the same index of refraction $\eta(\boldsymbol{x})$. Examples include the inverse problem in geophysics and the computation of radar cross sections. For these cases, solving a PDE in the full phase space can be an attractive alternative. Full phase space methods include the fast phase space method, [FS02], and the phase flow method, [YC06b].

3 Geometrical theory of diffraction

The main shortcomings of standard geometrical optics are the failure to include diffraction effects and its breakdown at caustics. In this section we give a brief introduction to how diffracted waves can be added to geometrical optics and how to correct the approximation close to caustics.

3.1 Mathematical formulation

The geometrical theory of diffraction (GTD) can be seen as a generalization of geometrical optics. It was pioneered by Keller in the 1960s, [Kel62], and provides a technique for adding diffraction effects to the geometrical optics approximation. GTD is often used in scattering problems in computational electromagnetics, where boundary effects are of major importance, for example in radar cross section calculations and in the optimization of base station locations for cell phones in a city.

3.1.1 Diffraction at nonsmooth boundaries

In general, diffracted rays are induced at discontinuities in the standard geometrical optics solution. By the reflection law, this happens primarily at singular points of the boundary, such as at corners and edges where the normal, and therefore the reflected field, is discontinuous. At these points an infinite set of diffracted rays are produced which obey the usual geometrical optics equations. The amplitude of each diffracted ray is proportional to the amplitude of the ray hitting the corner and a diffraction coefficient D. The coefficient D depends on the directions of the inducing and diffracted rays, the frequency, the local boundary geometry and the shape of the incident wave front.

An example is given in Fig. 2 where the incident plane wave is reflected off a half plane. This divides the space into regions A, B and C according to the number ray families present (two, one and zero respectively). The resulting geometrical optics solution is discontinuous at the region

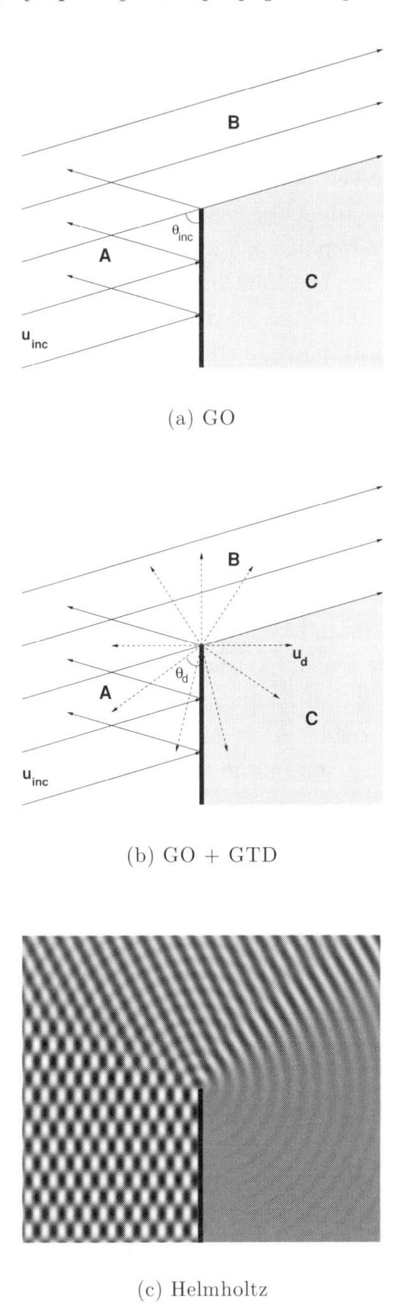

(a) GO

(b) GO + GTD

(c) Helmholtz

Fig. 2. A typical geometrical optics solution in two dimensions and a constant medium ($c \equiv 1$) around a perfectly reflecting halfplane (a), and the same problem augmented with diffracted waves given by GTD (b). In the geometrical optics case, region A contains two phases (incident and reflected), region B one phase (incident), and region C is in shadow, with no phases and hence a zero solution. On the boundaries between the regions the geometrical optics solution is discontinuous. Real part of a solution to the Helmholtz equation for this problem is shown in (c). The diffracted wave is faintly visible as a circular wave centered at the halfplane tip.

interfaces. Infinitely many diffracted rays shoot out in all directions at the singular tip of the half plane, which thus acts as an (anisotropic) point source.

By (2.5) the error in standard geometrical optics solutions is of the order $\mathcal{O}(1/\omega)$. However, the derivation of (2.6) from (2.5) does not take into account the effects of geometry and boundary conditions. In these cases the series expansion (2.5) is not adequate and extra terms must be added to match the solution to the boundary conditions. These terms represent the diffracted waves. They are of the order $\mathcal{O}(1/\omega^\alpha)$ for some $\alpha \in (0, 1)$ and hence normally much larger than the the usual error in standard geometrical optics, but still small for large frequencies. Discarding diffraction phenomena may therefore be too crude an approximation for a scattering problem at moderate frequencies.

One typical improved expansion that includes diffraction terms is

$$u(\boldsymbol{x}) = e^{i\omega\phi(\boldsymbol{x})} \sum_{k=0}^{\infty} A_k(\boldsymbol{x})(i\omega)^{-k} + \frac{1}{\sqrt{\omega}}\, e^{i\omega\phi_{\mathrm{d}}(\boldsymbol{x})} \sum_{k=0}^{\infty} B_k(\boldsymbol{x})(i\omega)^{-k},$$

which is similar to the standard geometrical optics ansatz (2.5), only that a new diffracted wave scaled by $\sqrt{\omega}$ has been added (index d). For high frequencies, the diffraction term B_0 is also retained, together with the geometrical optics term A_0. More elaborate expansions must sometimes be used, such as those given by the *uniform theory of diffraction* (UTD), [KP74].

It is important to note that the diffraction coefficients only depend on the *local* geometry of the boundary. Relatively few types of coefficients are therefore sufficient for a systematic use of GTD. Diffraction coefficients have been computed for many different canonical geometries, such as wedges, slits and apertures, different wave equations, in particular Maxwell equations, and different materials and boundary conditions.

3.1.2 Creeping rays

Another type of diffraction is generated even for smooth scatterers. When an incident field hits a smooth body there will be a shadow zone behind it and the geometrical optics solution will again be discontinuous. There is a curve (point in 2D) dividing the shadow part and the illuminated part of the body. Along this *shadow line* (shadow point in 2D) the incident rays are tangent to the body surface. The shadow line will act as a source for *creeping rays*, that propagate along geodesics on the scatterer surface, if the surrounding medium is homogeneous, $\eta \equiv 1$. The creeping ray carries an amplitude proportional to the amplitude of

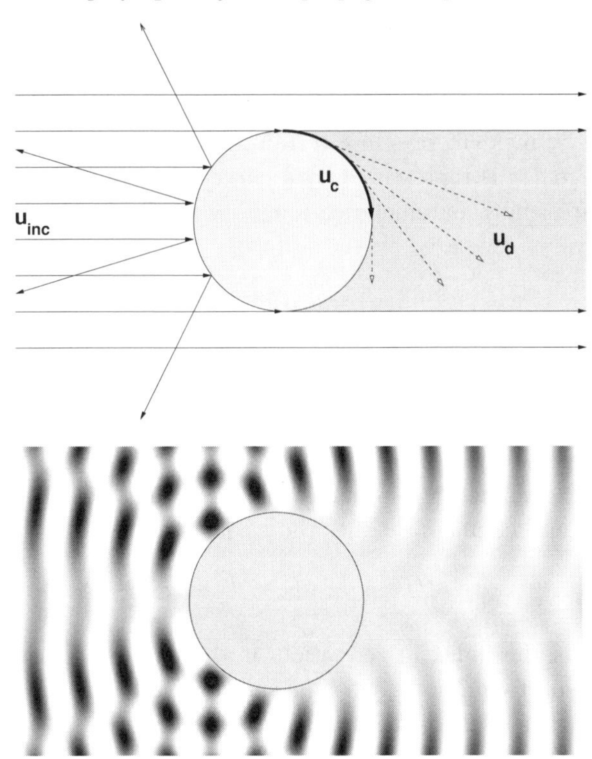

Fig. 3. Diffraction by a smooth cylinder. Top figure shows the solution schematically. The incident field u_{inc} induces a creeping ray u_{c} at the north (and south) pole of the cylinder. As the creeping ray propagates along the surface, it continuously emits surface-diffracted rays u_{d} with exponentially decreasing initial amplitude. Bottom figure shows the real part of a solution to the Helmholtz equation. The surface diffracted waves can be seen behind the cylinder.

the inducing ray. At each point on a convex surface with perfectly conducting material, the creeping ray sheds surface-diffracted rays in the tangential direction, with its current amplitude. The amplitude decays exponentially along the creeping ray's trajectory. In three dimensions, the amplitude also changes through geometrical spreading on the surface. The diffracted rays follow the usual geometrical optics laws. A 2D example is shown in Fig. 3.

We will now derive the equations used for computing the creeping wave field generated on the scatterer surface. Assume that the scatterer surface can be represented by a regular parameterization $\mathbf{x} = \bar{X}(\mathbf{u})$, where $\mathbf{x} = (x, y, z) \in \mathbb{R}^3$ is the coordinate in $3D$ physical space, and the parameters $\mathbf{u} = (u, v)$ belong to a set $\Omega \subset \mathbb{R}^2$. Let the scatterer be illuminated by incident rays in a certain direction. We assume that the shadow line $\mathbf{u}_0(s)$ is represented by a curve in parameter space, with s being the arc length parameterization. A wave field, associated to the creeping rays, is generated on the surface

$$v_s(\mathbf{u}) = a(\mathbf{u}) e^{i\omega\phi(\mathbf{u})}, \tag{3.1}$$

where $\phi(\mathbf{u})$ and $a(\mathbf{u})$ are surface phase and amplitude. The creeping rays are related to (3.1) in the same way as the standard GO rays are related to the leading term of the series (2.5). Like in GO, the surface wave field can be formulated either as PDEs or as a system of ODEs.

Let us first give the PDE formulations. According to Keller and Lewis [KL95], the surface phase satisfies the *surface eikonal equation*,

$$|\tilde{\nabla}\phi| = \eta, \qquad \tilde{\nabla}\phi := JG^{-1}\nabla\phi, \quad G = J^\top J, \quad J = [\bar{X}_u\, \bar{X}_v] \in \mathbb{R}^{3\times 2}, \tag{3.2}$$

where $\eta(\mathbf{u})$ is the index of refraction at the surface. There is also a surface transport equation giving the surface amplitude $a(\mathbf{u})$.

One can write (3.2) as a Hamilton-Jacobi equation $H(\mathbf{u}, \nabla\phi) = 0$, with the Hamiltonian

$$H(\mathbf{u}, \mathbf{p}) \equiv \frac{1}{2}\mathbf{p}^\top G^{-1}(\mathbf{u})\,\mathbf{p} - \frac{\eta^2(\mathbf{u})}{2}. \tag{3.3}$$

Note that in the case $\eta =$ constant, the rays associated with the surface eikonal equation (3.2) are geodesics, or shortest paths between two points on the surface. Henceforth, we will assume $\eta \equiv 1$.

Another formulation of the creeping wave field is based on ODEs. Introducing a parameter τ, the bicharacteristics $\big(\mathbf{u}(\tau), \mathbf{p}(\tau)\big)$ are determined by the solution of the following Hamiltonian equations

$$\dot{\mathbf{u}} = H_{\mathbf{p}} = G^{-1}\mathbf{p}, \tag{3.4a}$$

$$\dot{\mathbf{p}} = -H_{\mathbf{u}}. \tag{3.4b}$$

Here the dot denotes differentiation with respect to the parameter τ. As for standard GO, one can show that if $\mathbf{p}(0) = \nabla\phi(\mathbf{u}(0))$, which we assume, then $\mathbf{p}(\tau) = \nabla\phi(\mathbf{u}(\tau))$ for all $\tau > 0$, as long as ϕ is smooth. As a consequence, from (3.2) and (3.4) we obtain that $|\dot{\bar{X}}| = |J\dot{\mathbf{u}}| = |JG^{-1}\mathbf{p}| = 1$, and therefore we can identify the parameter τ with arc length along the creeping rays $\bar{X}(\mathbf{u}(\tau))$. Setting $\dot{u} = \rho\cos\theta$ and $\dot{v} =$

$\rho \sin\theta$, we can reduce the system (3.4) to

$$\dot{\gamma} = \mathbf{g}(\gamma), \qquad \mathbf{g}(\gamma) := \rho(\gamma)\,(\cos\theta, \sin\theta, \mathcal{V}(\gamma))^\top, \quad \gamma := (u, v, \theta)^\top, \tag{3.5}$$

where

$$\rho(\gamma) = \left|\bar{X}_u \cos\theta + \bar{X}_v \sin\theta\right|^{-1},$$

$$\mathcal{V}(\gamma) = (\Gamma^1_{11}\cos^2\theta + 2\Gamma^1_{12}\cos\theta\sin\theta + \Gamma^1_{22}\sin^2\theta)\sin\theta - \\ (\Gamma^2_{11}\cos^2\theta + 2\Gamma^2_{12}\cos\theta\sin\theta + \Gamma^2_{22}\sin^2\theta)\cos\theta.$$

Here, $\Gamma^k_{ij}(\mathbf{u})$ are Christoffel symbols. Moreover, we have

$$\dot{\phi} = \nabla\phi^\top\dot{\mathbf{u}} = \mathbf{p}^\top G^{-1}\mathbf{p} = 1, \quad \phi(0) = \phi_0(\mathbf{u}_0), \tag{3.6}$$

implying that the phase ϕ corresponds to the length of the ray,

$$\phi(\mathbf{u}(\tau)) = \phi_0(\mathbf{u}_0) + \tau.$$

In order to determine an equation for the amplitude, we apply the optical form of energy conservation principle in a small interval from τ to $\tau + d\tau$, [LK59], and obtain

$$a(\mathbf{u}(\tau)) = a_0\,\mathcal{Q}(s,\tau)^{\frac{-1}{2}}\exp\left(-\omega^{1/3}\beta(\tau)\right),$$

where a_0 is the amplitude at the starting point on the shadow line, s is the arc length parameterization along the shadow line, $\mathcal{Q}(s,\tau)$ is the geometrical spreading at distance τ from the starting point, and $\beta(\tau)$ is a function representing the attenuation given by the ODE

$$\dot{\beta} = \frac{q_0}{\rho_g(\gamma)}\exp\left(i\frac{\pi}{6}\left(\frac{\rho_g(\gamma)}{2}\right)^{1/3}\right) =: \alpha(\gamma), \quad \beta(0) = 0. \tag{3.7}$$

Here $q_0 \approx 2.33811$ is the smallest positive zero of the Airy function, and $\rho_g(\gamma)$ is the radius of curvature of the surface along the ray trajectory.

We now set $\tilde{\mathbf{u}}(s,\tau) := \mathbf{u}(\tau)$, with $\tilde{\mathbf{u}}(s,0) = \mathbf{u}_0(s)$ and let $\tilde{X}(s,\tau) := \bar{X}\left(\tilde{\mathbf{u}}(s,\tau)\right)$ be a point on the geodesic at the distance τ from the starting point $\tilde{X}_0(s) = \tilde{X}(s,0)$ on the shadow line. The geometrical spreading of the creeping ray at $\tilde{X}(s,\tau)$ in the physical space is given by, [MR06],

$$\mathcal{Q}(s,\tau) = \frac{\tilde{X}^\perp_\tau \cdot \tilde{X}_s}{\tilde{X}^\perp_{0\tau}\cdot\tilde{X}_{0s}}.$$

To compute $\tilde{X}_s$, we first note that $\tilde{X}_s = J\tilde{\mathbf{u}}_s$. We then differentiate (3.5) with respect to s and derive the following ODE system

$$\dot{\tilde{\gamma}}_s = D_\gamma\mathbf{g}\,\tilde{\gamma}_s, \qquad \tilde{\gamma}_s(s,0) = \tilde{\gamma}_{0s}(s). \tag{3.8}$$

By solving this system, $\tilde{\mathbf{u}}_s$ and therefore $\tilde{X}_s$ can be computed. All equations (3.5,3.6,3.7,3.8) are referred to as *surface ray equations* by which we can compute the phase and amplitude of the surface wave field (3.1).

There is yet another formulation, which can be seen as an analogue to the kinetic formulation of standard GO, [FS02, MR07a]. We introduce the phase space $\mathbb{P} = \mathbb{R}^2 \times \mathbb{S}$, where $\mathbb{S} = [0, 2\pi]$, and consider the triplet $\gamma = (u, v, \theta)$ as a point in this space. The ray trajectories on the scatterer, given by (3.5), are then confined to a subdomain $\Omega_p = \Omega \times \mathbb{S} \subset \mathbb{P}$ in phase space. We consider a ray $\bar{\gamma}(\tau)$ satisfying (3.5), starting at $\bar{\gamma}(0) = \gamma = (u, v, \theta) \in \Omega_p$ and ending at the boundary $\partial\Omega_p = \partial\Omega \times \mathbb{S}$. We call this end point $(U, V, \Theta) \in \partial\Omega_p$ the *escape point* of the ray. See Fig. 4. We then define three types of unknown *escape functions* for this starting point γ, as follows:

- $F : \mathbb{P} \to \mathbb{P}$, $F(\gamma) = (U, V, \Theta)$ is the escape point.
- $\Phi : \mathbb{P} \to \mathbb{R}$ is the length of the ray starting at γ and ending at $F(\gamma)$. We also refer to this as the travel-time of the ray.
- $B : \mathbb{P} \to \mathbb{R}$ is the difference between the β-values at the escape and starting points, where β satisfies (3.7).

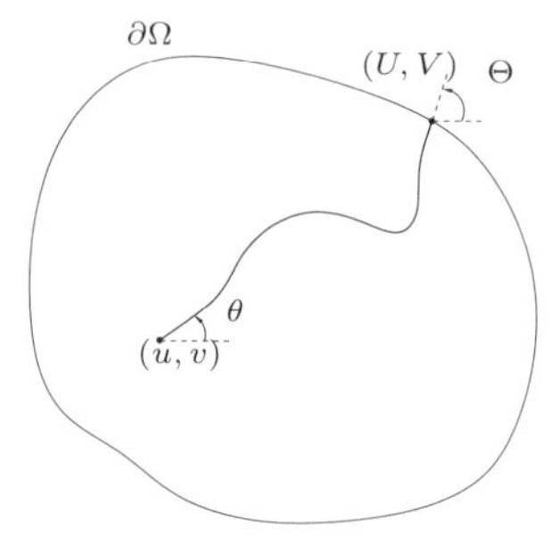

Fig. 4. A ray trajectory in the parameter space, starting at $\gamma = (u, v, \theta) \in \Omega_p$ and ending at the escape point $F(\gamma) = (U, V, \Theta) \in \partial\Omega_p$.

In order to derive equations for these functions, we notice that F is constant for all $\gamma(\tau)$ along the geodesic, and therefore

$$\frac{d}{d\tau} F(\gamma(\tau)) = 0 \Rightarrow g_1\, F_u + g_2\, F_v + g_3\, F_\theta = 0, \quad \gamma \in \Omega_p. \tag{3.9}$$

Here the coefficients $\mathbf{g} = (g_1, g_2, g_3)^\top$ are known and given by (3.5).

Similarly, by (3.6) and (3.7), we can write the equations for Φ and B. Each escape function $f(\gamma)$ of the above types satisfies the *escape* PDE

$$g_1(\gamma)\, f_u + g_2(\gamma)\, f_v + g_3(\gamma)\, f_\theta = h(\gamma), \qquad \gamma \in \Omega_p, \tag{3.10}$$

where the forcing term h is 0, 1 and $\alpha(\gamma)$ for $f = F$, $f = \Phi$ and $f = B$, respectively. The boundary condition at inflow points of $\partial\Omega_p$ are

$$f(\gamma) = b, \qquad \gamma \in \partial\Omega_p^{\text{inflow}}, \qquad \partial\Omega_p^{\text{inflow}} = \left\{\gamma \in \partial\Omega_p \,|\, \hat{n}(\gamma)^\top \mathbf{g}(\gamma) < 0\right\},$$

with $\hat{n}$ being the outward normal vector in the phase space. The boundary value b is γ, 0 and 0 for $f = F$, $f = \Phi$ and $f = B$, respectively. Note that one can also derive escape equations for computing geometrical spreading, using the ODE (3.8).

The escape equation (3.10) is a linear hyperbolic equation, and the variable velocity coefficients $\mathbf{g} = (g_1, g_2, g_3)^\top$ are known and determine the characteristic direction at every point $\gamma \in \Omega_p$. One important property of the solutions to the escape PDEs is that they are in general discontinuous due to discontinuous boundary conditions.

3.1.3 Caustics

Close to caustics the amplitude grows rapidly in the geometrical optics approximation and blows up at the caustic itself. In reality the amplitude remains bounded, but increases with the frequency ω, see Fig. 5. The error in the standard series expansion (2.5) is thus unbounded around caustics. To capture the actual solution behavior there are better expansions that have small errors uniformly in ω, derived *e.g.* by Ludwig [Lud66] and Kravtsov [Kra64]. The expansions are different for different types of caustics. For a fold caustic there are two ray families meeting at the caustic, with phases ϕ^+ and ϕ^-. Letting $\rho = \frac{3}{4}(\phi^+ - \phi^-)$ a more suitable description of the solution u in this case is

$$u(\boldsymbol{x}) = \omega^{1/6}\, e^{i\omega\phi(\boldsymbol{x})} \left(\mathrm{Ai}(-(\omega\rho(\boldsymbol{x}))^{2/3}) \sum_{k=0}^{\infty} A_k(\boldsymbol{x})(i\omega)^{-k} \right.$$
$$\left. + \, i\omega^{-1/3} \mathrm{Ai}'(-(\omega\rho(\boldsymbol{x}))^{2/3}) \sum_{k=0}^{\infty} B_k(\boldsymbol{x})(i\omega)^{-k} \right),$$

where Ai is the Airy function. The dominant term close to the caustic, $|\rho|\omega \ll 1$ is of the order $\mathcal{O}(\omega^{1/6})$ with an error of $\mathcal{O}(\omega^{-1/3})$. Away from the caustic, on the convex side where $\rho > 0$, we can use the fact that $|\mathrm{Ai}(-x)| \sim x^{-1/4}$ and $|\mathrm{Ai}'(-x)| \sim x^{1/4}$ for large x, to conclude that the dominant term is of the order $\mathcal{O}(1)$ with an error of $\mathcal{O}(\omega^{-1})$, *i.e.* the standard situation for geometrical optics.

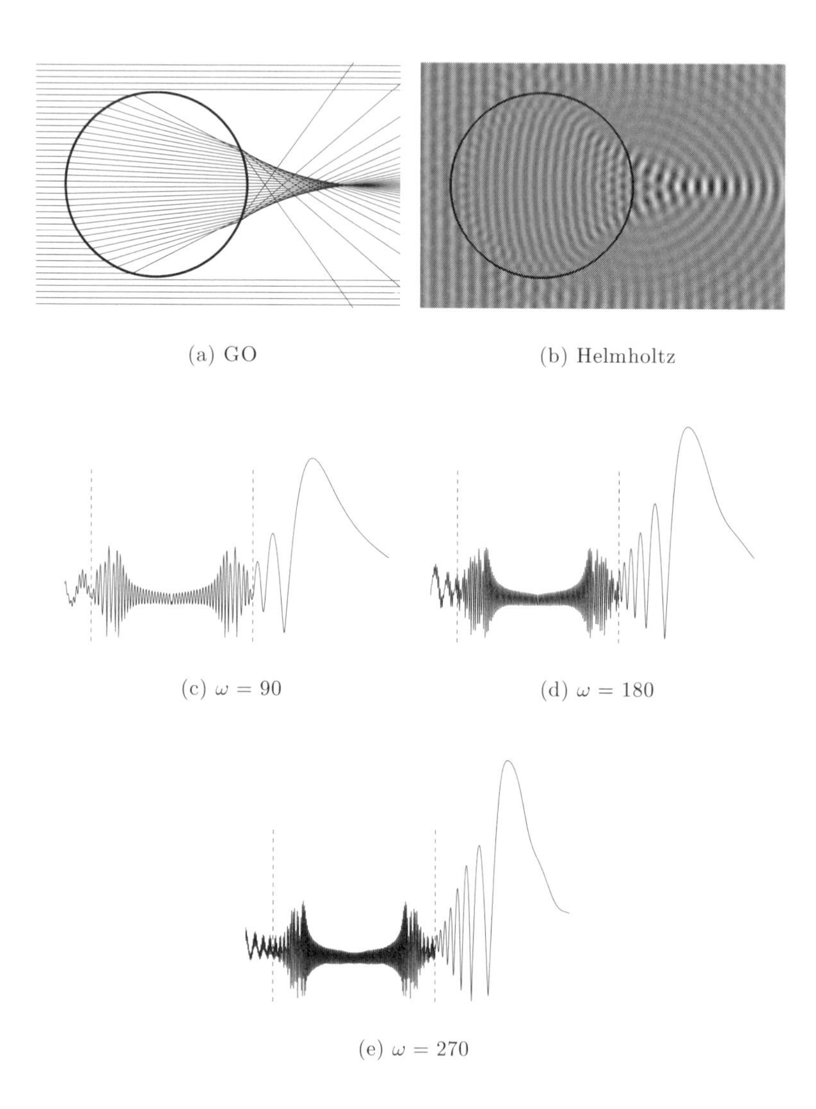

Fig. 5. Caustics generated when a plane wave is refracted by a cylinder of higher refractive index than the surrounding media. The geometrical optics solution by ray tracing is shown in (a), while (b) contains an actual solution of Helmholtz equation (real part). The concentration of rays coincides with the pronounced dark/light pattern (high amplitude) in the solution. The figures in the bottom row (c–e) show the absolute value of the Helmholtz solution in a horizontal cut in the middle of the top figures, for increasing frequencies ω. The cylinder boundaries are indicated by dashed lines. The amplitude away from the caustic is independent of ω but grows slowly with ω at the caustic.

3.2 Numerical methods

The diffracted rays generated by discontinuities and shed by creeping rays obey the usual geometrical optics equations. The main computational task is thus based on the standard GO approximation discussed in Section 2. However, computing creeping ray contribution to the field involves more technicalities, and one needs to find geodesics on the scatterer surface as well. We therefore here discuss computation of creeping rays in more detail. We then review numerical methods for computing the wave field at caustics.

3.2.1 Creeping rays computation

There are different numerical techniques for computing creeping rays. Similar to the numerical methods in standard GO, these techniques have advantages and disadvantages. Here, we will briefly review the methods based on the surface ray equations and surface eikonal/transport equations. We then discuss in more detail a new method based on the escape equations.

Lagrangian techniques are based on surface ray equations. The simplest and most common method is standard ray tracing which solves these ODEs on triangulated surfaces [HORSKP00]. Assuming the geodesic paths are given by piecewise linear curves, it is possible to find the linear ray path on each triangle, analytically. This method gives the surface phase and amplitude solutions along creeping rays. Interpolation must then be applied to obtain the solution everywhere. But, in regions where rays cross or diverge this can be rather difficult. However, the interpolation can be simplified by using wave front methods [VIG93, Hag05] in which, instead of individual rays, an interface representing a wave front is evolved. Nevertheless, for some problems, such as radar cross section (RCS) computations, where creeping rays from all illumination angles must be computed, Lagrangian methods can be computationally expensive.

Eulerian techniques are based on surface eikonal and surface transport equations. These PDEs are discretized on fixed computational grids, and there is no problem with interpolation [KS98]. However, these equations only give the correct solution when it is a single wave. In the case of crossing waves, more elaborate schemes have been devised to capture multivalued solutions, [MR06, YC06a].

We now discuss an adaptation of the fast phase space method, [FS02], for standard geometrical optics to computation of creeping waves, presented in [MR06]. This method is based on the escape equations (3.10). As a first step, the escape equations must be solved numerically. There are different approaches for solving (3.10). One is to discretize the PDEs in the phase space using a finite difference or finite volume approximation and arrive at a system of linear equations $Af = \bar{b}$, where A is a sparse $N^3 \times N^3$ matrix, with N being the number of grid points in each coordinate direction of Ω_p, and $\bar{b}$ represents the boundary conditions. This system can then be solved iteratively, and one can speed up the computations using suitable preconditioners [BH99]. However, in the case that characteristics change direction many times in the phase space domain, it is difficult to find good preconditioners. Another way to solve the escape equations is to write them as

$$f_t + g_1\, f_u + g_2\, f_v + g_3\, f_\theta = h,$$

and solve these time-dependent equations until the steady state $f_t = 0$. However, finding a fast algorithm which is not much restricted by the CFL condition is analogous to finding a good preconditioner in the iterative method. Yet, another method is a version of fast marching algorithm given by Fomel and Sethian [FS02]. The basic idea of the algorithm is to march the solution outwards from the boundary and use the characteristic directions to update grid values by interpolating them from their known neighboring grid points. Note that since the solutions to the escape PDEs can be *discontinuous*, it is important to use suitable interpolation techniques, such as essentially non-oscillatory (ENO) interpolations, to avoid the unphysical Gibbs oscillations. The algorithm is of complexity $\mathcal{O}(N^3 \log N)$.

The escape PDEs solutions contain information about all possible creeping rays in all directions. To extract properties like phase and amplitude for a ray family, post-processing of the solution is needed. We first note that $F(u_1, v_1, \theta_1) = F(u_2, v_2, \theta_2)$ implies that (u_1, v_1, θ_1) and (u_2, v_2, θ_2) lie on the same geodesic. Suppose we want to compute the surface phase at a point on the illuminated scatterer. We assume that the shadow line $\gamma_0(s)$ in known. For each point $\mathbf{u} \in \Omega$ covered by the surface wave, there is at least one creeping ray, starting at the shadow line, which passes through it. We can thus find $s = s^*(\mathbf{u})$ and phase angle $\theta = \theta^*(\mathbf{u})$, as the solution to

$$F(\gamma_0(s)) = F(\mathbf{u}, \theta). \tag{3.11}$$

The phase at $\mathbf{u}$ is then given by

$$\phi(\mathbf{u}) = \phi_0(\mathbf{u}_0(s^*)) + \Phi(\gamma_0(s^*)) - \Phi(\gamma^*), \quad \gamma^* = (\mathbf{u}, \theta^*) \in \mathbb{L}(\gamma_0),$$

where $\mathbb{L}(\gamma_0)$ is a sub-manifold of phase space $\mathbb{P}$ on which the creeping rays generated at $\gamma_0(s)$ lie. Note that there may be multiple solutions (s^*, θ^*) to (3.11), giving multiple phases. Using the same technique, the amplitude can also be computed.

To solve (3.11), we note that since F is a point on the phase space boundary, it can be reduced to a point in $\mathbb{R}^2$. The left and right hand sides of (3.11) are then curves in $\mathbb{R}^2$, and solving (3.11) amounts to finding crossing points of these curves. Discretizing the shadow line in N grid points, we need to find crossing points of two complex lines of N straight line segments, which can be done in $\mathcal{O}(N)$. For all N^2 points on the scatterer surface, the computational cost for solving (3.11) will be $\mathcal{O}(N^3)$. The total complexity, including solving the escape PDEs and solving (3.11), is therefore $\mathcal{O}(N^3 \log N)$. This is expensive for computing the field for only one shadow line. For example by using wave front tracking or solvers based on the surface eikonal equation, the complexity is $\mathcal{O}(N^2)$. However, if the solutions are sought for many shadow lines and only a few points on the scatterer, the phase space method is more efficient. One such example is when computing the monostatic RCS.

The method described above requires one fixed parameterization $\bar{X}(\mathbf{u})$ of the scatterer. It has however been modified in [MR07a] for more complex scatterer surfaces which cannot be represented by a single non-singular explicit parameterization. The surface is split into several simpler surfaces with explicit parameterizations. These multiple patches collectively cover the scatterer surface in a non-singular manner. The escape PDEs are solved in every patch, individually. The creeping rays on the scatterer are then computed by connecting all individual solutions through a fast post-processing. The inter-patch boundaries are treated by the continuity of creeping rays.

3.2.2 Wave field computation at caustics

Here we focus on the Gaussian beam method for computing the wave field at caustics. See [BLSS02] for another method based on the GO approximation.

The Gaussian beam method is an asymptotic method for computing high-frequency wave fields in smoothly varying inhomogeneous media. It was proposed by Popov [Pop82], based on an earlier work of Babic and Pankratova [BP73]. Gaussian beams are closely related to ray tracing, but instead of viewing rays just as characteristics of the eikonal equation, Gaussian beams are fatter rays: They are approximate high frequency solution to the wave equation or the Helmholtz equation which are concentrated on a standard ray. Contrary to standard GO rays, Gaussian beams accept complex valued phase functions. The main advantage of this construction is that Gaussian beams give the correct solution also at caustics where standard geometrical optics breaks down.

We now review the governing equations. We consider a ray in a two-dimensional Cartesian coordinate system x, y given by the ray tracing system (2.11). In orthogonal ray-centered coordinates (t, q), where q is the axis perpendicular to the ray at point t with the origin on the ray, the paraxial Gaussian beam solution closely concentrated about the central ray is given by

$$u(t, q, \omega) = A(t, q) \exp\{i\omega\phi(t, q)\}. \tag{3.12}$$

Here the complex-valued amplitude A and the phase ϕ are given by the eikonal and transport equations with complex initial data for ϕ of the type $\phi(0, q) \sim iq^2$ to give $u(0, q)$ a Gaussian profile. They are approximated by Taylor expansions

$$A \approx A(t, 0) = \sqrt{c(x(t), y(t))/Q(t)}, \tag{3.13}$$

$$\phi \approx \phi(t, 0) + q\phi_q(t, 0) + \frac{q^2}{2}\phi_{qq}(t, 0) = t + \frac{q^2}{2}\frac{P(t)}{Q(t)}. \tag{3.14}$$

The complex-valued scalar functions P and Q satisfy the *dynamic ray tracing* system

$$\begin{aligned} \frac{dQ}{dt} &= c^2 P, \\ \frac{dP}{dt} &= -\frac{1}{c}\,(c_{xx}\sin^2\theta - 2c_{xy}\sin\theta\cos\theta + c_{yy}\cos^2\theta)\, Q. \end{aligned} \tag{3.15}$$

As initial data for (3.15), we may choose

$$Q(0) = Q_0 > 0, \qquad P(0) = i.$$

One can show that this choice will guarantee that two important conditions are satisfied along the ray: $Q(t) \neq 0$ and $Im(P(t)/Q(t)) > 0$. The first condition guarantees the regularity of the Gaussian beam (with finite amplitudes at caustics). The second condition guarantees the concentration of the solution close to the ray. Note that in order to have a good accuracy in the Taylor expansions (3.13,3.14), we should have beams which are as narrow as possible. We therefore choose the Q_0 which gives minimum beam width.

In the Gaussian beam method, the initial/boundary condition for the wave field is decomposed into initial conditions for Gaussian beams. Individual Gaussian beams are computed by solving the ray tracing and dynamic ray tracing systems (2.11,3.15). The contributions of the beams concentrated close to their central rays are determined by the approximations (3.13,3.14) entered in (3.12). The wave field at a receiver is then obtained by a superposition of the Gaussian beams situated close to the receiver, [PCP82]. One can also construct a wave front Gaussian beam method in which a front of Gaussian beams is tracked [MR07b] or level-set based Eulerian methods [LQB07].

Bibliography

[BP73] V. M. Babič & T. F. Pankratova (1973). On discontinuities of Green's function of the wave equation with variable coefficient, *Problemy Matem. Fiziki*, **6**, 9–27 (in Russian).

[Ben96] J.-D. Benamou (1996). Big ray tracing: Multivalued travel time field computation using viscosity solutions of the eikonal equation, *J. Comput. Phys.* **128**, 463–474.

[Ben03] J.-D. Benamou (2003). An introduction to Eulerian geometrical optics (1992-2002), *J. Sci. Comput.* **19**, 63–93.

[BLSS02] J.-D. Benamou, O. Lafitte, R. Sentis & I. Solliec (2002). A geometric optics based numerical method for high frequency fields computations near fold caustics - Part I, Technical report, INRIA, France, No. 4422.

[BH99] H. Brandén & S. Holmgren (1999). Convergence acceleration for hyperbolic systems using semicirculant approximations, *J. Sci. Comput.* **14**, 357–393.

[CMP77] V. Červený, I. A. Molotkov & I. Psencik (1977). *Ray Methods in Seismology* (Univ. Karlova Press).

[CL83] M. G. Crandall & P.-L. Lions (1983). Viscosity solutions of Hamilton–Jacobi equations, *Trans. Amer. Math. Soc.* **277**, 1–42.

[ER03] B. Engquist & O. Runborg (2003). Computational high frequency wave propagation, *Acta Numerica* **12**, 181–266.

[ERT02] B. Engquist, O. Runborg & A.-K. Tornberg (2002). High frequency wave propagation by the segment projection method, *J. Comput. Phys.* **178**, 373–390.

[FEO95] E. Fatemi, B. Engquist & S. J. Osher (1995). Numerical solution of the high frequency asymptotic expansion for the scalar wave equation, *J. Comput. Phys.* **120**, 145–155.

[FS02] S. Fomel & J. A. Sethian (2002). Fast phase space computation of multiple arrivals, *Proc. Natl. Acad. Sci. USA* **99**, 7329–7334 (electronic).

[Hag05] S. Hagdahl (2005). *Hybrid Methods for Computational Electromagnetics in Frequency Domain*, PhD thesis (NADA, KTH, Stockholm).

[Hör83] L. Hörmander (1983–85). *The analysis of linear partial differential operators. I-IV* (Springer-Verlag, Berlin).

[HORSKP00] P. E. Hussar, V. Oliker, H. L. Riggins, E.M. Smith-Rowlan, W.R. Klocko, & L. Prussner (2000). An implementation of the UTD on facetized CAD platform models, *IEEE Antennas Propag.* **42**, 100–106.

[JL03] S. Jin & X. Li (2003). Multi-phase computations of the semiclassical limit of the Schrödinger equation and related problems: Whitham vs. Wigner, *Physica D* **182**, 46–85.

[Kel62] J. B. Keller (1962). Geometrical theory of diffraction, *J. Opt. Soc. Amer.*, **52**, 113–130.

[KL95] J. B. Keller & R. M. Lewis (1995). Asymptotic methods for partial differential equations: the reduced wave equation and Maxwell's equations, *Surveys Appl. Maths* **1**, 1–82.

[Kim00] S. Kim (2000). An $\mathcal{O}(N)$ level set method for eikonal equations, *SIAM J. Sci. Comput.* **22**, 2178–2193.

[KS98] R. Kimmel & J. A. Sethian (1998). Computing geodesic paths on manifolds, *Proc. Natl. Acad. Sci. USA* **95**, 8431–8435 (electronic).

[KP74] R. G. Kouyoumjian & P. H. Pathak (1974). A uniform theory of diffraction for an edge in a perfectly conducting surface, *Proc. IEEE* **62**, 1448–1461.

[Kra64] Yu. A. Kravtsov (1964). On a modification of the geometrical optics method, *Izv. VUZ Radiofiz.* **7**, 664–673.

[LLC85] R. T. Langan, I. Lerche, & R. T. Cutler (1985). Tracing of rays through heterogeneous media: An accurate and efficient procedure, *Geophysics* **50**, 1456–1465.

[LQB07] S. Leung, J. Qian, & R. Burridge (2007). Eulerian gaussian beams for high frequency wave propagation, *Geophysics* **72**, SM61–SM76.

[LK59] B. R. Levy & J. B. Keller (1959). Diffraction by a smooth object, *Comm. Pure Appl. Maths* **12**, 169-209.

[LOT06] H. Liu, S. Osher, & R. Tsai (2006). Multi-valued solution and level set methods in computational high frequency wave propagation, *Commun. Comput. Phys.* **1**, 765–804.

[Lud66] D. Ludwig (1966). Uniform asymptotic expansions at a caustic, *Comm. Pure Appl. Math.* **19**, 215–250.

[Mas65] V. P. Maslov (1965). *Théorie des perturbations et méthodes asymptotiques* (Izd. Moskov. Univ.) (in Russian). French translation: Dunod, Paris, 1972.

[MR06] M. Motamed & O. Runborg (2006). A fast phase space method for computing creeping rays, *J. Comput. Phys.* **219**, 276–295.

[MR07a] M. Motamed & O. Runborg (2007). A multiple-patch phase space method for computing trajectories on manifolds with applications to wave propagation problems, *Commun. Math. Sci.* **5**, 617–648.

[MR07b] M. Motamed & O. Runborg (2007). A wave front Gaussian beam method for high-frequency wave propagation. In *Proceedings of WAVES 2007* (University of Reading, UK).

[OS88] S. J. Osher & J. A. Sethian (1988). Fronts propagating with curvature-dependent speed: algorithms based on Hamilton-Jacobi formulations, *J. Comput. Phys.* **79**, 12–49.

[OS91] S. J. Osher & C.-W. Shu (1991). High-order essentially nonoscillatory schemes for Hamilton-Jacobi equations, *SIAM J. Numer. Anal.* **28**, 907–922.

[Pop82] M. M. Popov (1982). A new method of computation of wave fields using gaussian beams. *Wave Motion* **4**, 85–97.

[PCP82] M. M. Popov V. Červený & I. Psencik (1982). Computation of wave fields in inhomogeneous media - Gaussian beam approach, *Geophys. J. R. Astr. Soc.* **70**, 109–128.

[Run00] O. Runborg (2000). Some new results in multiphase geometrical optics, *M2AN Math. Model. Numer. Anal.* **34**, 1203–1231.

[Run07] O. Runborg (2007). Mathematical models and numerical methods for high frequency waves, *Commun. Comput. Phys.* **2**, 827–880.

[Set99] J. A. Sethian (1999). *Level set methods and fast marching methods* (2nd edition) (Cambridge University Press, Cambridge).

[SMM02] C. Sparber, N. Mauser, & P. A. Markowich (2003). Wigner functions vs. WKB-techniques in multivalued geometrical optics, *J. Asympt. Anal.* **33**, 153–187.

[SQ03] W. W. Symes & J. Qian (2003). A slowness matching Eulerian method for multivalued solutions of eikonal equations, *J. Sci. Comput.* **19**, 501–526.

[TCOZ03] Y. R. Tsai, L. T. Cheng, S. Osher, & H. K. Zhao (2003). Fast sweeping algorithms for a class of Hamilton–Jacobi equations, *SIAM J. Numer. Anal.* **41**, 673–694 (electronic).

[vTS91] J. van Trier & W. W. Symes (1991). Upwind finite-difference calculation of traveltimes, *Geophysics* **56**, 812–821.

[VIG93] V. Vinje, E. Iversen, & H. Gjøystdal, 1993. Traveltime and amplitude estimation using wavefront construction, *Geophysics* **58**, 1157–1166.

[YC06a] L. Ying & E. J. Candès (2006). Fast geodesics computation with the phase flow method, *J. Comput. Phys.* **220**, 6–18.

[YC06b] L. Ying & E. J. Candès (2006). The phase flow method, *J. Comput. Phys.* **220**, 184–215.

5

Wavelet-based numerical homogenization

Björn Engquist
Department of Numerical Analysis, KTH
100 44 Stockholm
Sweden
Email: engquist@nada.kth.se

Olof Runborg
Department of Numerical Analysis, KTH
100 44 Stockholm
Sweden
Email: olofr@nada.kth.se

Abstract

We consider multiscale differential equations in which the operator varies rapidly over fine scales. Direct numerical simulation methods need to resolve the small scales and they therefore become very expensive for such problems when the computational domain is large. Inspired by classical homogenization theory, we describe a numerical procedure for homogenization, which starts from a fine discretization of a multiscale differential equation, and computes a discrete coarse grid operator which incorporates the influence of finer scales. In this procedure the discrete operator is represented in a wavelet space, projected onto a coarser subspace and approximated by a banded or block-banded matrix. This wavelet homogenization applies to a wider class of problems than classical homogenization. The projection procedure is general and we give a presentation of a framework in Hilbert spaces, which also applies to the differential equation directly. We show numerical results when the wavelet based homogenization technique is applied to discretizations of elliptic and hyperbolic equations, using different approximation strategies for the coarse grid operator.

1 Introduction

In the numerical simulation of partial differential equations, the existence of subgrid scale phenomena poses considerable difficulties. With subgrid scale phenomena, we mean those processes which could influence the solution on the computational grid but which have length scales shorter than the grid size. Highly oscillatory initial data may, for example, interact with fine scales in the material properties and produce coarse scale contributions to the solution.

We consider the general problem where L_ε is a linear differential operator for which ε indicates small scales in the coefficients. The solution u_ε of the differential equation

$$L_\varepsilon u_\varepsilon = f_\varepsilon, \tag{1.1}$$

will typically inherit the small scales from the operator L_ε or the data f_ε. A concrete example could be the simple model problem

$$\begin{aligned} L_\varepsilon u_\varepsilon = -\frac{d}{dx}\left(g^\varepsilon(x)\frac{d}{dx}\right)u_\varepsilon(x) &= f(x), \qquad 0 < x < 1, \\ u_\varepsilon(0) = u_\varepsilon(1) &= 0, \end{aligned} \tag{1.2}$$

where the coefficient $g^\varepsilon(x)$ has a fine scale structure; it may for instance be highly oscillatory, or have a localized sharp transition. Numerical difficulties originate from the small scales in L_ε and f_ε. Let

$$L_{\varepsilon h} u_{\varepsilon h} = f_{\varepsilon h}. \tag{1.3}$$

be a discretization of (1.1), with the typical element size or step size h. If ε denotes a typical wave length in u_ε then h must be substantially smaller than ε in order to resolve the ε-scale in the numerical approximation. This can be costly if ε is small compared to the overall size of the computational domain.

There are a number of traditional ways to deal with this multiple scale problem. Several methods are based on physical considerations for a specific application, such as turbulence models in computational fluid dynamics, [Wil93], and analytically derived local subcell models in computational electromagnetics, [Taf95]. Geometrical optics or geometrical theory of diffraction approximations of high frequency wave propagation are other classical techniques to overcome the difficulty of highly oscillatory solutions, [Kel62]. All these techniques result in new sets of approximative equations that do not contain the small scales, but which anyway attempt to take the effect of these scales into account. A more general analytical technique for achieving this goal is classical homogenization, discussed below.

If the small scales are localized, there are some direct numerical procedures which are applicable. Local mesh refinement is quite common but could be costly if the small scales are very small or distributed. There are also problems with artificial reflections in mesh size discontinuities and time step limitations for explicit techniques. Numerical shock tracking or shock fitting can also be seen as subgrid models, [And95]. Material interfaces can be handled by grid adaptation or the immersed interface method, [Li94].

In this paper we present a general procedure for constructing subgrid models to be used on a coarse grid where the smallest scales are not resolved. The objective is to find a finite dimensional approximation of (1.1),

$$\bar{L}_{\varepsilon\bar{h}} \bar{u}_{\varepsilon\bar{h}} = \bar{f}_{\varepsilon\bar{h}},$$

that accurately reproduces the effect of subgrid scales and that in some sense is similar to a discretization of a differential equation. The discrete operator $\bar{L}_{\varepsilon\bar{h}}$ should thus resemble a discretized differential operator, and it should be designed such that $\bar{u}_{\varepsilon\bar{h}}$ is a good approximation of u_ε even if $\bar{h}$ is not small compared to ε. This goal resembles that of classical analytical homogenization, which we will now briefly discuss.

1.1 Classical homogenization

Homogenization is a well established analytical technique to approximate the effect of smaller scales onto larger scales in multiscale differential equations. The problem is often formulated as follows. Consider a set of operators L_ε indexed by the small parameter ε. Find the limit solution $\bar{u}$ and the *homogenized operator* $\bar{L}$ defined by

$$L_\varepsilon u_\varepsilon = f, \qquad \lim_{\varepsilon \to 0} u_\varepsilon = \bar{u}, \qquad \bar{L}\bar{u} = f, \tag{1.4}$$

for all f in some function class. In certain cases the convergence above and existence of the homogenized operator can be proved, [BLP78].

For simple model problems, with coefficients that are periodic on the fine scale, exact closed form solutions can be obtained. For instance, with $g(y)$ positive, 1-periodic and bounded away from zero, we have for the one-dimensional elliptic example (1.2),

$$L_\varepsilon = -\frac{d}{dx}\left(g(x/\varepsilon)\frac{d}{dx}\right), \qquad \bar{L} = -\hat{g}\frac{d^2}{dx^2}, \qquad \hat{g} = \left(\int_0^1 \frac{dy}{g(y)}\right)^{-1}.$$

With the same $\hat{g}$ we get for the hyperbolic operators,

$$L_\varepsilon = \frac{\partial}{\partial t} + g(x/\varepsilon)\frac{\partial}{\partial x}, \qquad \bar{L} = \frac{\partial}{\partial t} + \hat{g}\frac{\partial}{\partial x}. \tag{1.5}$$

These model examples are used in Sect. 4.1 and Sect. 4.2.

In higher dimensions, the solution to (1.4) is more complicated. For (1.5), even the type of the homogenized equation depends strongly on the properties of the coefficients, see [E92, HX92]. For the multidimensional elliptic case (1.2) the structure of the homogenized operator can still be written down, as long as the coefficients are periodic or stochastic. Let $G(y) : \mathbb{R}^d \mapsto \mathbb{R}^{d\times d}$ be uniformly elliptic and 1-periodic in each of its arguments. Let I_d denote the unit square in d dimensions. It can then be shown, [BLP78], that

$$L_\varepsilon = -\nabla\cdot\left(G\left(\frac{x}{\varepsilon}\right)\nabla\right), \quad \bar{L} = -\nabla\cdot(\hat{G}\nabla), \quad \hat{G} = \int_{I_d} G(y) - G(y)\frac{d\chi(y)}{dy}dy, \tag{1.6}$$

where $d\chi/dy$ is the jacobian of the function $\chi(y) : \mathbb{R}^d \mapsto \mathbb{R}^d$, given by solving the so called cell problem,

$$\nabla\cdot G(y)\frac{d\chi(y)}{dy} = \nabla\cdot G(y), \qquad y \in I_d,$$

with periodic boundary conditions for χ. Note that $\hat{G}$ is a constant matrix.

1.2 Numerical homogenization

Classical homogenization is very useful when it is applicable. The original problem with small scales is reduced to a homogenized problem that is much easier to approximate numerically. See the left path in Fig. 1. The final discretization $\bar{L}_{\bar{h}}\bar{u}_{\bar{h}} = \bar{f}_{\bar{h}}$ satisfies the criteria we set up at the end of Sect. 1; since there are no small scales in the homgenized equation (1.4) the size of $\bar{h}$ can be chosen independently of ε and $\bar{u} \approx u_\varepsilon$ for small ε.

If analytical homogenization is not possible, one can instead numerically compute a suitable discrete operator $\bar{L}_{\varepsilon\bar{h}}$ which has the desired properties. We call such a procedure *numerical homogenization*. The numerical homogenization can be done directly as indicated by the middle path in Fig. 1, or by first discretizing the original problem and then compressing the operator $L_{\varepsilon h}$ and the data $f_{\varepsilon h}$ as indicated by the right path. In this paper we follow the latter strategy, and present wavelet based methods in order to achieve numerical homogenization. The great advantage of this procedure in deriving subgrid models is its generality. It can be used on any system of differential equations and does not require separation into distinct $\mathcal{O}(\varepsilon)$ and $\mathcal{O}(1)$ scales or periodic coeffi-

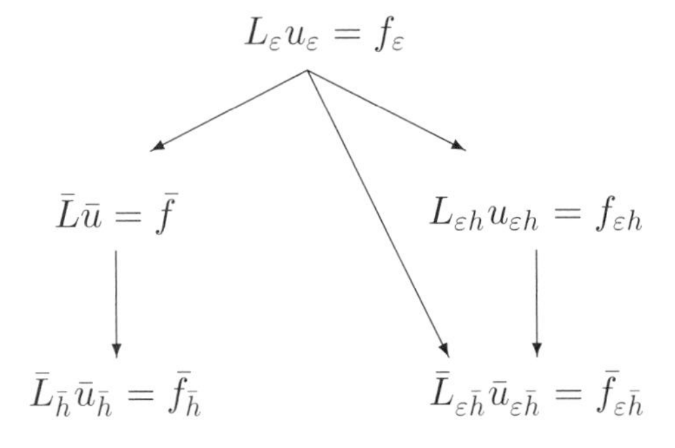

Fig. 1. Schematic steps in homogenization.

cients. It can also be used to test if it is physically reasonable to represent the effect of fine scales on a coarse scale grid with a local operator.

There are other similar methods based on coarsening techniques from algebraic multigrid, see for example, Knapek [Kna99] and Neuss et. al. [NJW98, Neu95]. In the finite element setting the effect of the microstructure can be incorporated in a Galerkin framework, as done in the work of Hughes and collaborators [HFMQ98, Hug95], Hou and collaborators [EHW00, HW97], Matache and Schwab [MS00] and more recently Allaire and Brizzi [AB05]. See also [OZ07] for a general theoretical justification of these methods.

This article is based on [AELR99, ER02a] which is a continuation of the work by Dorobantu and Engquist, [Dor98]. The original ideas are from Beylkin and Brewster, [BB94]. See also [Gil98] for analysis in the one-dimensional case and [CL04] for new strategies to approximate the numerically homogenized operators. The wavelet based numerical homogenization technique has been applied to several problems. We refer the reader to [PR01] for a waveguide filter containing a fine scale structure, with examples of how to use the numerical homogenization technique to construct subgrid models, in particular 1d models from 2d models. In [Leo00, EL97] the numerically homogenized operator was used as a coarse grid operator in multigrid methods. Applications to porous media flow were considered in [Wan05]. Extension of the procedure to nonlinear problems can be found in [BBG99, KRK04]. For a survey of wavelet based numerical homogenization see [ER02b].

2 Projection generated homogenization

In this section we describe the following approach to homogenization. Consider an equation $Lu = f$ where L is a linear operator, f a right hand side and u a solution that contains fine scales. Let P be a projection operator onto a subspace where the fine scales in the original solution do not exist. Our objective is to find the (projection generated) *homogenized* operator $\bar{L}$ such that $\bar{L}Pu = f$ for all f such that $Pf = f$. (When $Pf \neq f$ we also need to find the homogenized right hand side $\bar{f}$.) Here, we confine ourselves to the case of Hilbert spaces.

Let X be a Hilbert space of functions, typically a Sobolev space. Let $X_0 \subset X$ be a closed subspace representing the coarse part of the functions, and denote by P the orthogonal projection operator in X onto X_0. Let the spaces X_0 and $X_0^\perp$ inherit the innerproduct and norm of X, so that $\|u\|_X = \|u\|_{X_0}$ when $u \in X_0$, and similar for $X_0^\perp$. In addition, set $Q = I - P$ where I is the identity operator in X, and introduce the unitary operator $\mathcal{W}$ on X defined by

$$\mathcal{W} : X \mapsto X_0 \times X_0^\perp, \qquad \mathcal{W}u = \begin{pmatrix} Qu \\ Pu \end{pmatrix}. \tag{2.1}$$

Let $\mathcal{L}(X, Y)$ be the set of bounded linear maps from X to Y. In the numerical finite dimensional case $X = Y = \mathbb{R}^n$ and $L \in \mathcal{L}(X, Y)$ is identified by a matrix in $\mathbb{R}^{n\times n}$. In order to simplify the presentation we shall consider $X = Y$ also in infinite dimensions. For $L \in \mathcal{L}(X, X)$ we then have

$$\begin{aligned} \mathcal{W}L\mathcal{W}^* \begin{pmatrix} u \\ v \end{pmatrix} &= \mathcal{W}L(P+Q)(u+v) = \begin{pmatrix} QL(P+Q)(u+v) \\ PL(P+Q)(u+v) \end{pmatrix} \\ &= \begin{pmatrix} QL(Pv+Qu) \\ PL(Pv+Qu) \end{pmatrix} \equiv \begin{pmatrix} A & B \\ C & D \end{pmatrix} \begin{pmatrix} u \\ v \end{pmatrix} \end{aligned} \tag{2.2}$$

where

$$\begin{array}{ll} A = QLQ \in \mathcal{L}(X_0^\perp, X_0^\perp), & B = QLP \in \mathcal{L}(X_0, X_0^\perp), \\ C = PLQ \in \mathcal{L}(X_0^\perp, X_0), & D = PLP \in \mathcal{L}(X_0, X_0). \end{array} \tag{2.3}$$

When A is invertible the following definition can be stated:

Definition 1 *Suppose $L \in \mathcal{L}(X, X)$ and $f \in X$. When A in (2.2, 2.3) is invertible we define the homogenized operator $\bar{L} : X_0 \mapsto X_0$ as the Schur complement with respect to the decomposition in (2.2),*

$$\bar{L} = D - CA^{-1}B,$$

and the homogenized right hand side as

$$\bar{f} = Pf - CA^{-1}Qf.$$

We will write $\bar{L}_{X,X_0}$ and $\bar{f}_{X,X_0}$ when there is a need to show explicitly between which spaces the homogenization step is made. One can then show the following properties of $\bar{L}$.

Proposition 1 *Suppose $Lu = f$, where $L \in \mathcal{L}(X, X)$, $u \in X$ and $f \in X$.*

- *If A^{-1} exists,*

$$\bar{L}Pu = \bar{f}.$$

- *Let $L \in \mathcal{L}(X, X)$ be such that $\bar{L}$ exists. If L is onto, then $\bar{L}$ is onto. If L is self-adjoint, then $\bar{L}$ is self-adjoint. If L is also positive definite, then*

$$0 < \min_{||u||_X=1} \langle Lu, u\rangle \leq \min_{||v||_{X_0}=1} \langle \bar{L}v, v\rangle,$$

$$\max_{||v||_{X_0}=1} \langle \bar{L}v, v\rangle \leq \max_{||v||_{X_0}=1} \langle Dv, v\rangle \leq \max_{||u||_X=1} \langle Lu, u\rangle.$$

- *If $L \in \mathcal{L}(X, X)$ is self-adjoint and positive definite, then A^{-1} exists.*
- *Let X_0 and X_1 be two closed subspaces of X, nested such that $X_0 \subset X_1 \subset X$. Suppose $L \in \mathcal{L}(X, X)$ is onto and that $\bar{L}_{X,X_0}$, $\bar{L}_{X,X_1}$ and $\overline{\bar{L}_{X,X_1}}_{X_1,X_0}$ all exist. If $\bar{L}_{X,X_0}$ is one-to-one, then*

$$\bar{L}_{X,X_0} = \overline{\bar{L}_{X,X_1}}_{X_1,X_0}, \qquad \bar{f}_{X,X_0} = \overline{\bar{f}_{X,X_1}}_{X_1,X_0}.$$

Proofs of these points can be found in [ER02b].

2.1 Relationship between the homogenization approaches

In the elliptic case, there is a striking similarity between the classical homogenized operator in (1.6) and the the Schur complement in Definition 1, repeated here for convenience,

$$PLP - PLQ(QLQ)^{-1}QLP,$$

$$\nabla\left(\int_{I_d} G(y)dy\right)\nabla - \nabla\left(\int_{I_d} G(y)\frac{d\chi(y)}{dy}dy\right)\nabla.$$

Both are written as the average of the original operator minus a correction term, which is computed in a similar way for both operators. For the analytical case, a local elliptic cell problem is solved to get $G\partial_y \chi$, while in the projection case, a positive operator $A = QLQ$ defined on a subspace is inverted to obtain $LQA^{-1}B$. The average over the terms is obtained by integration in the analytical case, and by applying P in the projection case.

The relationship can be made more precise, and we will illustrate it by considering the one-dimensional case, with the coarse space given by the lowest Fourier modes. Similar results in one dimension have been shown for wavelet bases, see [Gil98, SMM00].

Consider (1.2) with $g^\varepsilon = g(x/\varepsilon)$ where $g \in L^\infty(\mathbb{T})$ is 1-periodic. Let X be $H^1(\mathbb{T})$ and $X_0^\varepsilon \subset X$ be the set of functions bandlimited to the lowest $1/\varepsilon$ Fourier modes,

$$X_0^\varepsilon = \left\{ u \in L^2(\mathbb{T}) \;\middle|\; u = \sum_{|k|<1/2\varepsilon} a_k e^{2\pi i k x},\ a_k \in \mathbb{C} \right\}.$$

Then, if P_ε is the projection on X_0^ε, one can show

Theorem 2 *The projection generated homogenized operator $\bar{L}_\varepsilon$ and the classical homogenized operator $\bar{L}$ corresponding to (1.2) satisfy*

$$\bar{L}_\varepsilon u = P_\varepsilon \bar{L} u, \qquad \forall u \in X_0^\varepsilon$$

if $\varepsilon = 1/n$ and $n \in \mathbb{Z}^+$. Moreover, if $\bar{L}_\varepsilon \bar{u}_\varepsilon = \bar{L}\bar{u} = f$, then

$$\lim_{\varepsilon \to 0} ||\bar{u}_\varepsilon - \bar{u}||_{H_1} = 0.$$

See [ER02a] for the proof. We note that the convergence in this theorem is much stronger than that of the solution u_ε of the original problem $L_\varepsilon u_\varepsilon = f$; that solution only converges weakly in H_1 to $\bar{u}$.

3 Wavelet-based numerical homogenization

In this section we show how the projection generated homogenization works when we choose wavelet spaces as our coarse and fine space decomposition. We start by briefly reviewing the theory of multiresolution analysis. We then let $Lu = f$ be a finite-dimensional approximation of a partial differential equation and apply the procedure in Sect. 2 for this case. In particular, we demonstrate how the resulting $\bar{L}$ can be approximated by a sparse matrix and used for computing an approximate solution to the differential equation on a coarse grid.

3.1 Wavelets

A wavelet representation lends itself naturally to analyzing the fine and coarse scales as well as the localization properties of a function. Here we restrict our attention to orthogonal wavelets, to align with the theory in Sect. 2. The theory could easily be modified to allow also biorthogonal wavelets.

We introduce the usual scaling spaces V_j,

$$\ldots \subset V_j \subset V_{j+1} \subset \ldots,$$

where j increases for spaces containing finer details. The corresponding wavelet spaces are the orthogonal complements of V_j in V_{j+1}, denoted by W_j,

$$V_{j+1} = V_j \oplus W_j, \qquad V_j \perp W_j. \tag{3.1}$$

Under some addition standard assumptions we obtain a multiresolution analysis (MRA) where

$$\overline{\bigcup_{j \in \mathbb{Z}} V_j} = \overline{\bigoplus_{j \in \mathbb{Z}} W_j} = L^2(\mathbb{R}), \qquad \bigcap_{j \in \mathbb{Z}} V_j = \{0\}.$$

The scaling and wavelet spaces are spanned by translations of a dilated shape function $\varphi(x)$ and mother wavelet $\psi(x)$ respectively,

$$\{\varphi_{j,k}(x) = 2^{j/2}\varphi(2^j x - k);\ k \in \mathbb{Z}\} \text{ is an orthonormal basis for } V_j, \tag{3.2}$$

$$\{\psi_{j,k}(x) = 2^{j/2}\psi(2^j x - k);\ k \in \mathbb{Z}\} \text{ is an orthonormal basis for } W_j. \tag{3.3}$$

The simplest MRA is the one built on the Haar basis, where the shape function and the mother wavelet are given by

$$\varphi(x) = \begin{cases} 1, & \text{if } 0 \le x \le 1, \\ 0, & \text{otherwise,} \end{cases} \qquad \psi(x) = \begin{cases} 1, & \text{if } 0 \le x \le 1/2, \\ -1, & \text{if } 1/2 < x \le 1, \\ 0, & \text{otherwise.} \end{cases}$$

For a detailed description of MRA we refer the reader to e.g. the book by Daubechies, [Dau91].

In more than one dimension, the MRA is extended by considering tensor product spaces. In two dimensions, we set

$$\boldsymbol{V}_j = V_j \otimes V_j, \qquad \ldots \subset \boldsymbol{V}_j \subset \boldsymbol{V}_{j+1} \subset \ldots, \qquad j \in \mathbb{Z},$$

with V_j being the one-dimensional spaces introduced above. As before, we let the wavelet spaces $\boldsymbol{W}_j$ be the orthogonal complements of $\boldsymbol{V}_j$ in $\boldsymbol{V}_{j+1}$, so that $\boldsymbol{V}_{j+1} = \boldsymbol{V}_j \oplus \boldsymbol{W}_j$ and $\boldsymbol{V}_j \perp \boldsymbol{W}_j$. In this case $\boldsymbol{W}_j$ is composed of three parts,

$$\boldsymbol{W}_j = (W_j \otimes W_j) \oplus (V_j \otimes W_j) \oplus (W_j \otimes V_j),$$

where, the W_j spaces are those of the one-dimensional case, given by (3.1). Similar tensor product extensions can be made also for higher dimensions.

3.2 Wavelet projections of difference operators

We will now discuss homogenization in the finite dimensional wavelet spaces discussed above. The typical operators that we homogenize are discrete approximations of differential operators. In projection generated homogenization, we are given an operator defined on a fine grid and seek an operator defined on a smaller space that extracts only the coarse scales of the solution. For a function in the wavelet space V_{j+1}, the coarse scale is represented by V_j, and we are in the first step thus interested in the homogenized operator given by $\bar{L}_{V_{j+1},V_j}$. We will mostly use the standard Haar basis.

Let us consider the equation

$$L_{j+1}U = F, \qquad U, F \in V_{j+1}, \qquad L_{j+1} \in \mathcal{L}(V_{j+1}, V_{j+1}).$$

This equation may originate from a finite difference, finite element or finite volume discretization of a given differential equation. In the Haar case U can be identified as a piecewise constant approximation of $u(x)$, the solution to the continuous problem.

Let $\mathcal{W}_j : V_{j+1} \mapsto V_j \times W_j$ be the unitary operator corresponding to (2.1). (By (3.1) $V_j^\perp = W_j$.) As in the infinite dimensional case, we apply the transformation $\mathcal{W}_j$ on L_{j+1} from the left and get

$$\mathcal{W}_j L_{j+1} \mathcal{W}_j^* (\mathcal{W}_j U) = \mathcal{W}_j F \tag{3.4}$$

and if we decompose $\mathcal{W}_j L_{j+1} \mathcal{W}_j^*$ according to (2.2),

$$\begin{pmatrix} A_j & B_j \\ C_j & L_j \end{pmatrix} \begin{pmatrix} U_\mathrm{f} \\ U_\mathrm{c} \end{pmatrix} = \begin{pmatrix} F_\mathrm{f} \\ F_\mathrm{c} \end{pmatrix}, \qquad U_\mathrm{f}, F_\mathrm{f} \in W_j, \qquad U_\mathrm{c}, F_\mathrm{c} \in V_j. \tag{3.5}$$

Here the subindex "f" means projection onto the fine scale subspace, W_j in this case, and subindex "c" stands for projection onto the coarse scale subspace, V_j here. This is a convention we will use below. Also note that we changed the notation of the block previously called D_j to L_j, to indicate that $L_j = PL_{j+1}P$ actually corresponds to one type of direct discretization on the coarse scale.

The homogenized "coarse grid operator" is now given by block Gaussian elimination of (3.5) as before,

$$(L_j - C_j A_j^{-1} B_j) U_{\mathrm{c}} = F_{\mathrm{c}} - C_j A_j^{-1} F_{\mathrm{f}}.$$

For simplicity we will use the notation $\overline{L_{j+1}}_{V_{j+1}, V_j} \equiv \bar{L}_j$ and $\bar{F}_{V_{j+1}, V_j} \equiv \bar{F}_j$. Hence,

$$\bar{L}_j = L_j - C_j A_j^{-1} B_j, \qquad \bar{F}_j = F_{\mathrm{c}} - C_j A_j^{-1} F_{\mathrm{f}}. \tag{3.6}$$

Since L_j corresponds to a direct coarse discretization, we can interpret $C_j A_j^{-1} B_j$ as a correction term, which includes subgrid phenomena in $\bar{L}_j$.

In the finite dimensional case, we represent the operators using their matrix representations with respect to the bases given by (3.2) and (3.3). For instance in the Haar basis, we get the following matrix representation of $\mathcal{W}_j$,

$$\mathcal{W}_j = \begin{pmatrix} Q_j \\ P_j \end{pmatrix} = \frac{1}{\sqrt{2}} \begin{pmatrix} 1 & -1 & 0 & \cdots & & \\ 0 & 0 & 1 & -1 & 0 & \cdots \\ \vdots & \vdots & & \ddots & \ddots & \\ 0 & 0 & \cdots & 0 & 1 & -1 \\ 1 & 1 & 0 & \cdots & & \\ 0 & 0 & 1 & 1 & 0 & \cdots \\ \vdots & \vdots & & \ddots & \ddots & \\ 0 & 0 & \cdots & 0 & 1 & 1 \end{pmatrix}.$$

The matrix representation of (3.5) is a block matrix decomposition of the same form as the operator decomposition.

The procedure to obtain $\bar{L}_j$ described above can be applied recursively on $\bar{L}_j$ itself to get $\bar{L}_{j-1}$ and so on. That this is possible can easily be verified when L_{j+1} is symmetric positive definite. Then Proposition 1 shows that A_j is invertible and that $\bar{L}_j$ is again symmetric positive definite. By induction $\bar{L}_k$ exists for $k \leq j$. Also note that Proposition 1 gives $\bar{L}_k = \overline{L_{j+1}}_{V_{j+1}, V_k}$ and $\bar{F}_k = \bar{F}_{V_{j+1}, V_k}$ for $k \leq j$. Moreover, an improvement in the condition number κ can often be estimated. Typically for standard discretizations

$$\max_{||v||=1} \langle L_k v, v \rangle < \max_{||u||=1} \langle L_{j+1} u, u \rangle,$$

when $k \leq j$. Then, from Proposition 1,

$$\kappa(\bar{L}_k) = \frac{\max_{||v||=1}\langle \bar{L}_k v, v\rangle}{\min_{||v||=1}\langle \bar{L}_k v, v\rangle} \leq \frac{\max_{||v||=1}\langle L_k v, v\rangle}{\min_{||v||=1}\langle L_{j+1} v, v\rangle}$$
$$< \frac{\max_{||v||=1}\langle L_{j+1} v, v\rangle}{\min_{||v||=1}\langle L_{j+1} v, v\rangle} = \kappa(L_{j+1}).$$

The two-dimensional wavelet transform

$$\boldsymbol{\mathcal{W}}_j : \boldsymbol{V}_{j+1} \mapsto \boldsymbol{W}_j \times \boldsymbol{V}_j$$

can be written as a tensor product of one-dimensional transforms,

$$\boldsymbol{\mathcal{W}}_j = \mathcal{W}_j \otimes \mathcal{W}_j. \tag{3.7}$$

A linear operator L_{j+1} that acts on the space $\boldsymbol{V}_{j+1}$ can be decomposed in a way similar to the one-dimensional case. The equation

$$L_{j+1}U = F, \qquad U, F \in \boldsymbol{V}_{j+1}, \qquad L_{j+1} \in \mathcal{L}(\boldsymbol{V}_{j+1}, \boldsymbol{V}_{j+1})$$

can then be transformed to

$$\begin{pmatrix} A_j & B_j \\ C_j & L_j \end{pmatrix} \begin{pmatrix} U_{\mathrm{f}} \\ U_{\mathrm{c}} \end{pmatrix} = \begin{pmatrix} F_{\mathrm{f}} \\ F_{\mathrm{c}} \end{pmatrix}, \qquad U_{\mathrm{f}}, F_{\mathrm{f}} \in \boldsymbol{W}_j, \qquad U_{\mathrm{c}}, F_{\mathrm{c}} \in \boldsymbol{V}_j, \tag{3.8}$$

and the coarse grid operator is again the Schur complement,

$$\overline{L}_{\boldsymbol{V}_{j+1}, \boldsymbol{V}_j} \equiv \bar{L}_j = L_j - C_j A_j^{-1} B_j.$$

To get the matrix representation of (3.8), we take the one-dimensional matrix representations of $\mathcal{W}_j$ and, following (3.7), compute its Kronecker tensor product with itself. We must thereafter also apply a suitable permutation matrix to the product in order to get the same block structure of the matrix as in (3.8).

Note that the fine scale part of U can be decomposed as

$$U_{\mathrm{f}} = \begin{pmatrix} U_{\mathrm{ff}} \\ U_{\mathrm{cf}} \\ U_{\mathrm{fc}} \end{pmatrix}, \qquad U_{\mathrm{ff}} \in W_j \otimes W_j, \qquad U_{\mathrm{cf}} \in V_j \otimes W_j, \qquad U_{\mathrm{fc}} \in W_j \otimes V_j.$$

In some cases, the homogenized operator keeps important properties of the original operator. Let the forward and backward undivided differences be defined as

$$\Delta_+ u_i = u_{i+1} - u_i, \qquad \Delta_- u_i = u_i - u_{i-1}.$$

In [Dor98] it was shown that the one-dimensional elliptic model equation $-(gu')' = f$ discretized as

$$L_{j+1}U = -\frac{1}{h^2}\Delta_+ \operatorname{diag}(g)\Delta_- U = F \tag{3.9}$$

will preserve its divergence form during homogenization. That is, we will get

$$\bar{L}_j = -\frac{1}{(2h)^2}\Delta_+ H_j \Delta_-, \tag{3.10}$$

where H_j is a strongly diagonal dominant matrix which can be interpreted as the effective material coefficient related to g. Analogously, for the first order differential operator $g(x)\frac{\partial}{\partial x}$ the discretized form $\operatorname{diag}(g)\Delta_-/h$ is preserved during homogenization,

$$\bar{L}_j = \frac{1}{2h} H_j \Delta_-. \tag{3.11}$$

In two dimensions, the elliptic model equation $-\nabla(g(x,y)\nabla u) = f$ can be discretized as

$$L_{j+1} = -\frac{1}{h^2}\left(\Delta_+^x G \Delta_-^x + \Delta_+^y G \Delta_-^y\right), \qquad L_{j+1}U = F.$$

Then $\bar{L}_j$ is no longer on exactly the same form as L_{j+1}. The cross-derivatives must also be included. We get

$$\bar{L}_j = -\frac{1}{(2h)^2}\left(\Delta_+^x H^{xx} \Delta_-^x + \Delta_+^y H^{yx} \Delta_-^x + \Delta_+^x H^{xy} \Delta_-^y + \Delta_+^y H^{yy} \Delta_-^y\right). \tag{3.12}$$

3.3 Compact representation of projected operators

When the operator L_{j+1} is derived from a finite difference, finite element or finite volume discretization, it is sparse and of a certain structure. In one dimension it might, for instance, be tridiagonal. However, in general $\bar{L}_j$ will not be represented by a sparse matrix even if L_{j+1} is, because A_j^{-1} would typically be dense. Computing all components of $\bar{L}_j$ would be inefficient. Fortunately, $\bar{L}_j$ will be diagonal dominant in many important cases and we can then find a sparse matrix that is a close approximation of $\bar{L}_j$. If this sparse matrix is of banded form, it can be seen as a discretization of a local differential operator acting on the coarse space.

Diagonal dominance of $\bar{L}_j$

We now consider some cases where $\bar{L}_j$ will be strongly diagonal dominant. This is related to the corresponding properties of A_j.

In some simple cases the matrix A_j is in fact diagonal. Examples include integral operators L of the form

$$Lu = \int_0^x a(t)u(t)dt + b(x)u(x). \tag{3.13}$$

Let P_j be the orthogonal projection on V_j. Suppose L is discretized in V_{j+1} as $L_{j+1} = P_{j+1}LP_{j+1}$ with a in (3.13) replaced by $P_{j+1}aP_{j+1}$, and similar for b. When the Haar system is used, A_j is diagonal and $\bar{L}_j$ is of the same form as L_{j+1}. Let $\bar{A}_k$ be related to $\bar{L}_{k+1}$ via (3.4, 3.5) in the same way as A_j relates to L_{j+1}. By induction, $\bar{A}_k$ is also diagonal for $k < j$. The operators in (3.13) turn up for instance in problems with systems of ordinary differential equations and one-dimensional elliptic equations, see [BB94, Gil98]. In these cases, an explicit recurrence relation between scale levels can be established, which permits the computation of $\bar{L}_k$ on any fixed level k as the starting level, $j+1$, tends to infinity.

For more general problems one must instead rely on the rapid decay of elements in A_j and $\bar{L}_j$ off the diagonal, which is a consequence of the wavelet spaces' good approximation properties. The decay rate of A_j for Calderon–Zygmund and pseudo-differential operators were given by Beylkin, Coifman and Rokhlin in [BCR91]. Letting $L_{j+1} = P_{j+1}LP_{j+1}$, $A_j = \{a^j_{k\ell}\}$, $B_j = \{b^j_{k\ell}\}$ and $C_j = \{c^j_{k\ell}\}$, they show that

$$|a^j_{k\ell}| + |b^j_{k\ell}| + |c^j_{k\ell}| \leq \frac{2^{-\lambda j} C_M}{1 + |k-\ell|^{M+1}}, \qquad |k-\ell| \geq \nu, \tag{3.14}$$

when the wavelet system has M vanishing moments. For Calderon–Zygmund operators $\lambda = 0$ and $\nu = 2M$. For a pseudo-differential operator $\nu = 0$ and the symbol $\sigma(x,\xi)$ and its adjoint should both belong to the symbol class $S^\lambda_{1,1}$, i.e. they should satisfy the estimate

$$|\partial^\alpha_\xi \partial^\beta_x \sigma(x,\xi)| \leq C_{\alpha,\beta}(1+|\xi|)^{\lambda-\alpha+\beta},$$

for some constants $C_{\alpha,\beta}$. For instance, in the second order elliptic case $\lambda = 2$. Moreover, Beylkin and Coult, [BC98], showed that if (3.14) holds with $\lambda = 0$ for A_j, B_j and C_j given by L_{j+1} in (3.4, 3.5), then the same estimate also holds for $\bar{A}_{j'}$, $\bar{B}_{j'}$ and $\bar{C}_{j'}$, here given by $P_{j'}\bar{L}_{k+1}P_{j'}$ for $j' \leq k < j$ with $j+1$ being the starting homogenization level. Hence, the decay rate is preserved after homogenization.

The decay estimate in [BC98] for $\bar{A}_{j'}$ is uniform in k and may not be sharp for a fixed k. There is, for example, a general result by Concus, Golub and Meurant, [CGM85], for diagonal dominant, symmetric and tridiagonal matrices. For those cases, which include A_j corresponding to the discretization in (3.9) of the one-dimensional elliptic operator, the inverse has exponential decay,

$$\left|\left(A_j^{-1}\right)_{k\ell}\right| \leq C\varrho^{|k-\ell|}, \qquad 0 < \varrho < 1.$$

This holds also when the elliptic operator has a lower order term of type $b(x)\partial_x$ discretized with upwinding, [Leo00].

Approximating $\bar{L}_j$

We now discuss different approximation strategies for $\bar{L}_j$. A simple approach is the basic thresholding method used in [BCR91], where small elements of $\bar{L}_j$ are set to zero. This is, however, not practical here since the location of the non-zero elements cannot be controlled, and we want to obtain a banded approximation of $\bar{L}_j$, which corresponds to a discretization of a local differential operator.

The first, and simplest, approximation method that we use is instead to set all components outside a prescribed bandwidth ν equal to zero. This is motivated by the decay of elements off the diagonal in $\bar{L}_j$. Let us define

$$\operatorname{trunc}(M,\nu)_{ij} = \begin{cases} M_{ij}, & \text{if } 2\,|i-j| \leq \nu - 1 \\ 0, & \text{otherwise.} \end{cases} \tag{3.15}$$

For $\nu = 1$ the matrix is diagonal. For $\nu = 3$ it is a tridiagonal and so on. We refer to it as *truncation*.

In the second approximation method, the matrix $\bar{L}_j$ is projected onto banded form in a more effective manner. The aim is that the projected matrix should give the same result as the original matrix on a given subspace, e.g. when applied to vectors representing smooth functions. Let $\{\boldsymbol{v}_j\}_{j=1}^{\nu}$ be a set of linearly independent vectors in $\mathbb{R}^N$. Denote by $\mathcal{T}_\nu$ the subspace of $\mathbb{R}^{N\times N}$ with matrices essentially* of bandwidth ν. Moreover, let

$$\mathcal{L}_\nu = \{M \in \mathbb{R}^{N\times N} : \operatorname{span}\{\boldsymbol{v}_1, \boldsymbol{v}_2, \ldots, \boldsymbol{v}_\nu\} \subset \mathsf{N}(M)\},$$

where $\mathsf{N}(M)$ represents the null space of M. Then

$$\mathbb{R}^{N\times N} = \mathcal{T}_\nu \oplus \mathcal{L}_\nu$$

*We must require that each row of the matrices in $\mathcal{T}_\nu$ has the same number of elements. Therefore, the first and last $\nu-1$ rows will have additional elements located immediately to the right and left of the band, respectively.

and we define the *band projection* of a matrix $M \in \mathbb{R}^{N\times N}$ as the projection of M onto $\mathcal{T}_\nu$ along $\mathcal{L}_\nu$, with the notation

$$\text{band}(M,\nu) = \text{Proj}_{\mathcal{T}_\nu} M.$$

As a consequence,

$$M\vec{x} = \text{band}(M,\nu)\vec{x}, \qquad \forall \vec{x} \in \text{span}\{\boldsymbol{v}_1, \boldsymbol{v}_2, \ldots, \boldsymbol{v}_\nu\}.$$

In our setting M will usually operate on vectors representing smooth functions, for instance solutions to elliptic equations, and a natural choice for $\boldsymbol{v}_j$ vectors are thus the first ν polynomials,

$$\boldsymbol{v}_j = \{1^{j-1}, 2^{j-1}, \ldots, N^{j-1}\}^*, \qquad j = 1, \ldots, \nu.$$

Smooth solutions to the homogenized problem should be well approximated by these vectors. For the case $\nu = 1$ we get the standard "mass-lumping" of a matrix, often used in the context of finite element methods.

This technique is similar to the probing technique used by Chan et al., [CM92]. In that case the vectors $\boldsymbol{v}_j$ are sums of unit vectors. Other probing techniques have been suggested by Axelsson, Pohlman and Wittum, see Chapter 8 in [Axe94]. The choice of $\boldsymbol{v}_j$ vectors could be optimized if there is some a priori knowledge of the homogenized solution. In some cases the band projection technique only gives improvements for small values of ν, see Fig. 2. The probing technique of [CM92] could be used for larger ν in our examples. Numerical evidence indicate that for small values of ν, the band projection technique is more efficient.

The two truncation methods described above are even more efficient when applied to H_j, the effective coefficient, instead of directly to the homogenized operator $\bar{L}_j$. For (3.9, 3.10) we could for instance approximate

$$\bar{L}_j \approx -\frac{1}{(2h)^2}\Delta_+ \,\text{trunc}(H,\nu)\Delta_-.$$

The following proposition shows that when the solution to the homogenized problem belongs to the Sobolev space H^1, the accuracy of this approach is one order higher.

Proposition 3 *Suppose $L = \Delta_+ H \Delta_-$ and $LU = h^2 f$. Consider the perturbed problems*

$$(L + \delta L)(U + \delta U_L) = h^2 f, \qquad \Delta_+(H + \delta H)\Delta_-(U + \delta U_H) = h^2 f.$$

For small enough perturbations, and the same constant C,

$$||\delta U_L|| \leq C\frac{||\delta L||}{||L||}||U||, \qquad ||\delta U_H|| \leq \frac{h}{2}C\frac{||\delta H||}{||H||}||u||_{H^1},$$

where $||\cdot||$ denotes the discrete L_2-norm and u is any H^1-function such that $U_j = u(jh)$.

See [AELR99] for a proof.

Another approach proposed by Chertock and Levy [CL04] is to only approximate A_j^{-1}, the high frequency part of the projected operator. In this case we would approximate

$$\bar{L}_j \approx L_j - C_j \operatorname{trunc}(A_j^{-1}, \nu)B_j.$$

Numerical evidence in [CL04] strongly suggests that this gives a better approximation than truncating the full operator. In particular the strategy works better when truncation is done in each homogenization step, instead of only in the final step. It is also simpler than approximating H_j in higher dimensions.

The focus in this paper is on the principle and in the numerical examples the truncation $\operatorname{trunc}(\bar{L}_j, \nu)$ is done after computing the full inverse of A_j, which is expensive. By capitalizing on the nearly sparse structure of the matrices involved, it was however shown in [BC98] that the cost can be reduced to $\mathcal{O}(N)$ operations for N unknowns and fixed accuracy. This method uses a multiresolution based LU decomposition procedure described in [BDG98]. Moreover, the same homogenized operator will typically be reused multiple times, for instance with different right hand sides, or in different places of the geometry as a subgrid model. The computation of $\operatorname{band}(\bar{L}_j, \nu)$ can be based on $\operatorname{trunc}(\bar{L}_j, \mu)$, $\mu > \nu$. The additional computational cost is proportional to $(\nu^3 + \mu\nu)N$. The $\nu^3 N$ term corresponds to solving N $\nu \times \nu$ systems and $\mu\nu N$ to computing the right hand sides, see also [Axe94].

In two dimensions truncation to simple banded form is in general not adequate, since the full operator will typically be block banded. However, both the crude truncation and the band projection generalize easily to treat block banded form instead of just banded. Let M be the tensor product of two $N \times N$ matrices. Then we define truncation as

$$\operatorname{trunc}_2(M, \nu)_{ij} = \begin{cases} M_{ij}, & \text{if } 2\,|i - j - rN| \leq \nu - 1 - |2r|, \\ 0, & \text{otherwise}, \end{cases} \qquad |2r| + 1 \leq \nu. \tag{3.16}$$

This mimics the typical block structure of a discretized differential operator. For the band projection, the space $\mathcal{T}_\nu$ of banded matrices in the one-dimensional definition, is simply replaced by the space of matrices with the block banded sparsity pattern defined in (3.16).

In two dimensions, untangling the various H components of (3.12) from $\bar{L}_j$ is more complicated than finding the H in (3.10) and (3.11) for one-dimensional problems. Although, in principle, this can still be done, it may be easier to truncate A_j^{-1} than the H components for two-dimensional problems, as suggested in [CL04].

4 Numerical examples

In this section we present numerical results for the algorithms described above. Elliptic and hyperbolic model problems are studied. In one of the examples the wavelet based homogenization is applied to the Helmholtz equation for the generation of a coarse grid approximation. For many more numerical examples, also of the approximation strategy involving A_j^{-1}, see [CL04].

4.1 Elliptic problems

We first consider elliptic problems in one dimension and the Helmholtz equation in one and two dimensions. Uniform, central, finite volume discretization are used in the numerical experiments. The computational domain is the unit interval (square in 2D). The grid size is denoted n, and the cell size $h = 1/n$.

The 1D elliptic model equation

We approximate

$$\partial_x g(x) \partial_x u = 1, \qquad u(0) = u'(1) = 0,$$

where the coefficient $g(x)$ has a uniform random distribution in the interval $[0.5, 1]$. We take $n = 256$ grid points and make three homogenization steps. The coarsest level then contains 32 grid points.

In Fig. 2 different truncation strategies are compared. The exact reference solution is given by the numerically homogenized operator at the coarsest level without any truncation. This is equivalent to the projection onto the coarse scale of the solution on the finest scale. In the top two subplots we use crude truncation, (3.15). In the bottom two subplots we use the band projection described in Sect. 3.3. The approximation is

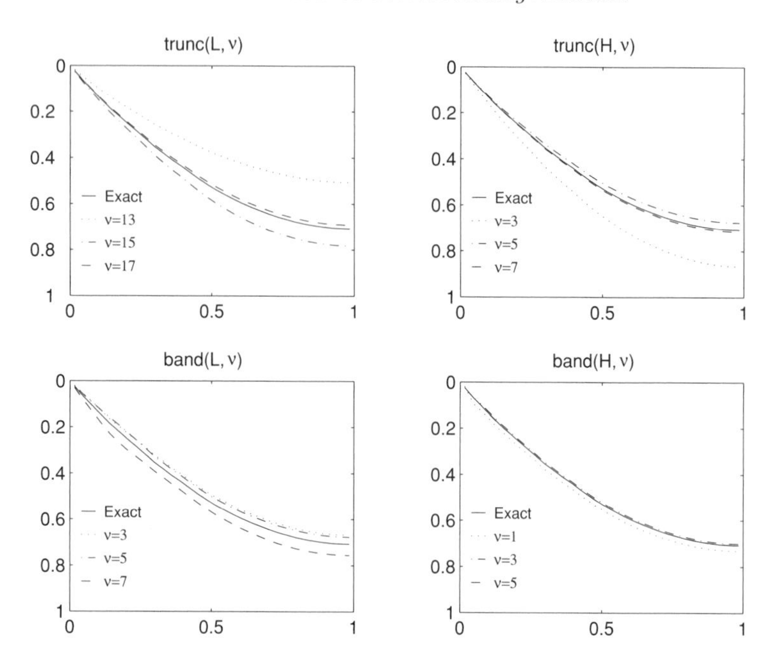

Fig. 2. Result for the elliptic model problem, $g(x)$ random, when the homogenized operator is approximated in different ways. The "exact" solution refers to the solution with the full 32×32 homogenized operator.

performed on H, see (3.10), and on $\bar{L}$ after all three homogenizations. We see that band projection gives a better approximation. We also see that it is more efficient to truncate H than to truncate $\bar{L}$.

Next, the coefficient in the differential equation is changed to

$$g(x) = \begin{cases} 1/6, & 0.45 < x < 0.55, \\ 1, & \text{otherwise.} \end{cases} \tag{4.1}$$

All other characteristics are kept. The result is given in Fig. 3 and it shows that the relative merits of the different methods are more or less the same. The structures of the untruncated $\bar{L}$ and H matrices are shown in Fig. 4. It should be noted that the local inhomogeneity of the full operator has spread out over a larger area, but it is still essentially local.

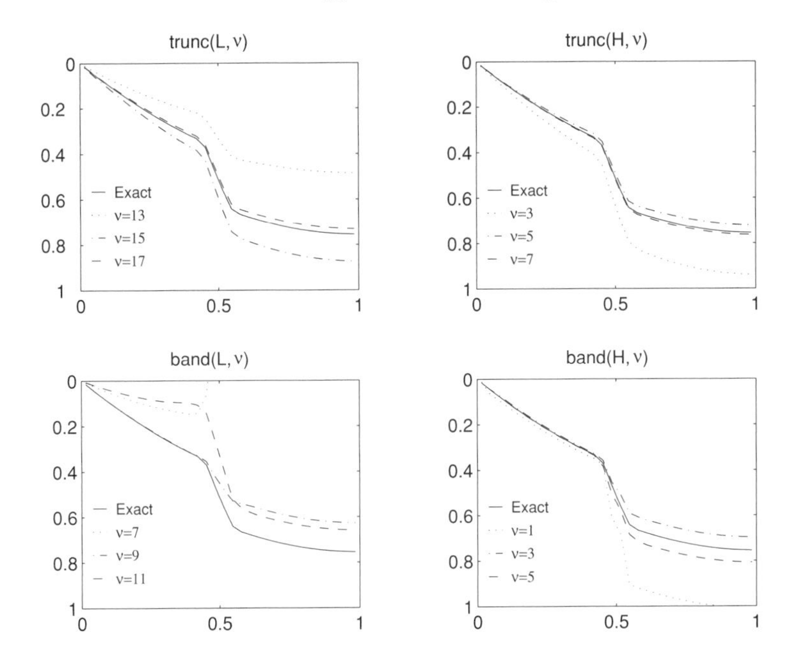

Fig. 3. Result for the elliptic model problem, $g(x)$ a slit, when the homogenized operator is approximated in different ways. The "exact" solution refers to the solution with the full 32×32 homogenized operator.

The 1D Helmholtz equation

In this section we solve the Helmholtz equation

$$\partial_x g(x) \partial_x u + \omega^2 u = 0, \qquad u(0) = 1, \quad u'(1) = 0.$$

We use $\omega = 2\pi$ and the same $g(x)$ as in (4.1) and again we take $n = 256$ and use three homogenizations. We get

$$\bar{L}u = (\tilde{L} - \omega^2 I)u = 0.$$

Truncation is performed on $\tilde{L}$ (or $\tilde{H}$) and not on $\bar{L}$. The result is in Fig. 5. We see that Helmholtz equation gives results similar to those of the model equation. Again band projection is more efficient than truncation and approximating H is more efficient than approximating $\bar{L}$.

The 2D Helmholtz equation

We consider the two-dimensional version of Helmholtz equation,

$$\nabla \cdot g(x, y) \nabla u + \omega^2 u = 0, \qquad (x, y) \in (0, 1)^2,$$

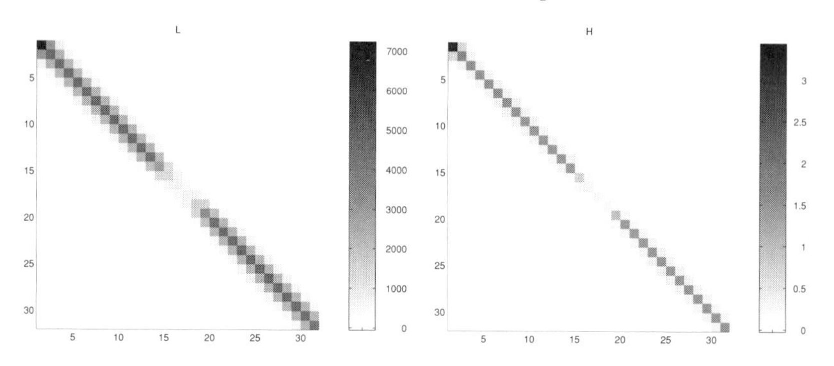

Fig. 4. Structure of the untruncated homogenized operators $\bar{L}$ (left) and H (right) for the elliptic model problem, $g(x)$ a slit. Gray level indicates absolute value of elements.

with periodic boundary conditions in the y-direction, and at the left and right boundaries, $u(0, y) = 1$, $u_x(1, y) = 0$ respectively. This is a simple model of a plane time-harmonic wave of amplitude one entering the computational domain at the line $x = 0$, passing through a medium defined by the coefficent $g(x, y)$ and flowing out at $x = 1$. As an example we choose the $g(x, y)$ shown in Fig. 6, which represents a wall with a small hole where the incoming wave can pass through. With $\omega = 3\pi$ and $n = 48$, we obtained the results presented in Fig. 7.

The operator is homogenized following the theory for two-dimensional problems in Sect. 3.2. After one homogenization step is truncated according to (3.16). We show the results of truncation in Fig. 8, for various values of ν. The case $\nu = 9$ corresponds to a compression to approximately 7% of the original size. The structure of this operator is shown in Fig. 9.

4.2 Hyperbolic problems

In this section we consider the time-dependent hyperbolic model equation in one space dimension,

$$\frac{\partial u}{\partial t} + g(x)\frac{\partial u}{\partial x} = f(x, t), \qquad u(0, t) = 0, \quad u(x, 0) = b(x). \tag{4.2}$$

The variable coefficient $g(x)$ is positive and bounded.

The most straightforward way to homogenize (4.2) is by a semi-discrete

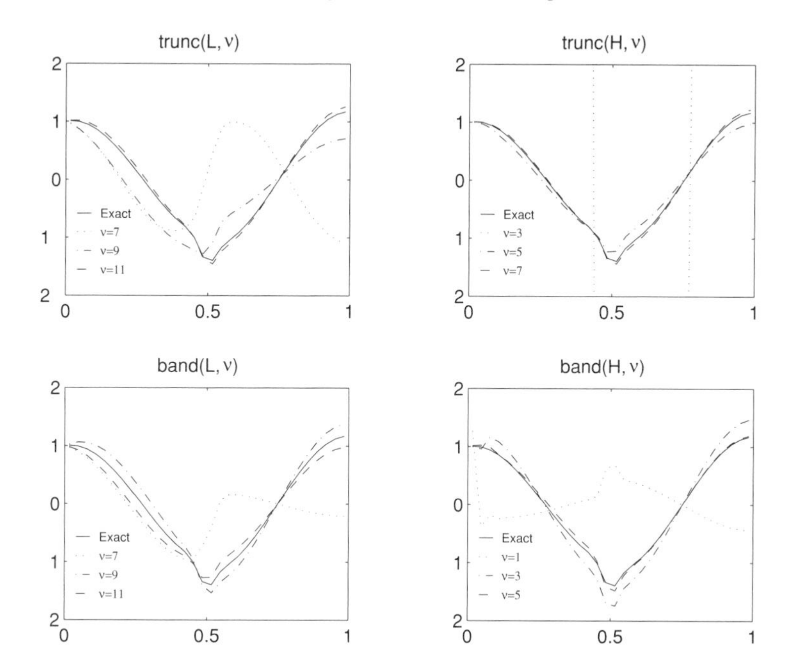

Fig. 5. Result for the Helmholtz equation, $g(x)$ a slit, when the homogenized operator is approximated in different ways. The "exact" solution refers to the solution with the full 32×32 homogenized operator.

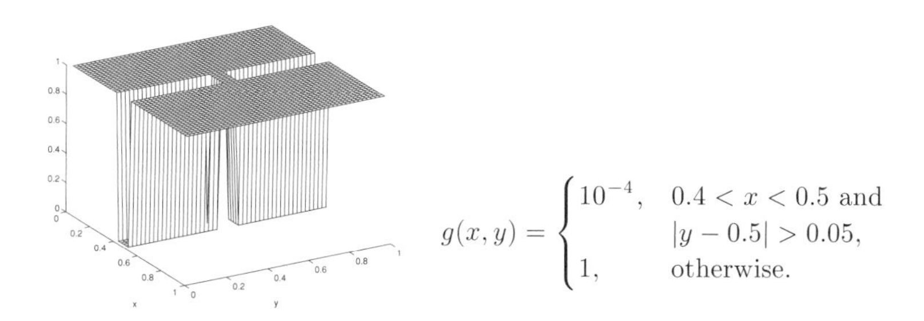

$$g(x,y) = \begin{cases} 10^{-4}, & 0.4 < x < 0.5 \text{ and} \\ & |y - 0.5| > 0.05, \\ 1, & \text{otherwise.} \end{cases}$$

Fig. 6. The variable coefficient $g(x, y)$ used in the 2D Helmholtz example.

approximation,

$$\frac{\partial u_i}{\partial t} + \frac{1}{h} g_i \Delta_- u_i = f_i(t), \qquad i = 1, \dots, n, \qquad u_0 = -u_1, \quad u_i(0) = b_i,$$

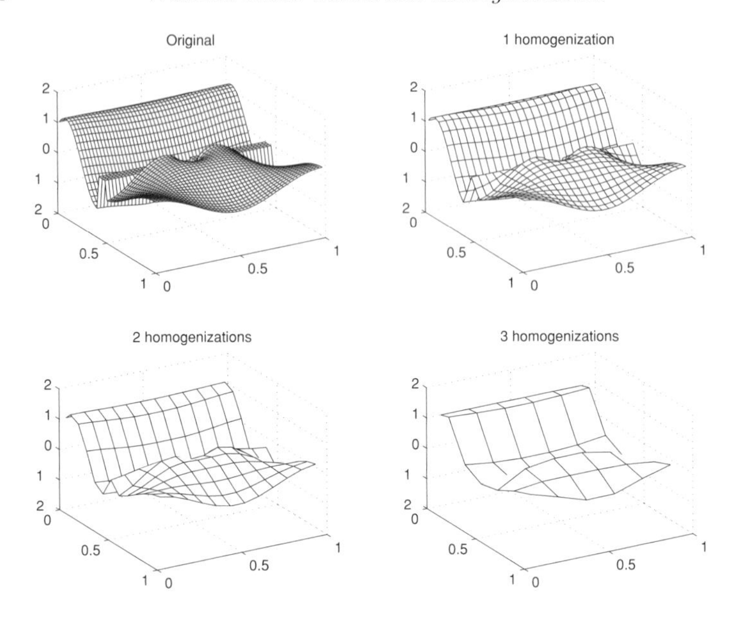

Fig. 7. Result for the 2D Helmholtz example. Solution shown for $0, \ldots, 3$ homogenization steps.

with $g_i = g((i - 1/2)h)$. In matrix form we will have

$$U_t + L_{j+1}U = F, \qquad 2^{j+1} = n,$$

which can be expressed in a wavelet basis as

$$\frac{\partial}{\partial t}\begin{pmatrix} U_{\mathrm{f}} \\ U_{\mathrm{c}} \end{pmatrix} + \begin{pmatrix} A_j & B_j \\ C_j & D_j \end{pmatrix}\begin{pmatrix} U_{\mathrm{f}} \\ U_{\mathrm{c}} \end{pmatrix} = \begin{pmatrix} F_{\mathrm{f}} \\ F_{\mathrm{c}} \end{pmatrix}, \qquad U_{\mathrm{f}}, F_{\mathrm{f}} \in W_j, \qquad U_{\mathrm{c}}, F_{\mathrm{c}} \in V_j. \tag{4.3}$$

In principle, L_{j+1} could be homogenized in the same way as in the elliptic case, using the Schur complement. The motivation for this is (1.5), which shows that when the scales of $g(x/\varepsilon)$ become very small, $\varepsilon \to 0$, the effect of the time-derivative vanishes, $\frac{\partial}{\partial t}U_{\mathrm{f}} \to 0$. As a limiting process in classical homogenization, this is fully justified, compare [BLP78]. This argument implies that in the first set of equations of (4.3),

$$\frac{\partial U_{\mathrm{f}}}{\partial t} + A_j U_{\mathrm{f}} + B_j U_{\mathrm{c}} = F_{\mathrm{f}} \quad \Rightarrow \quad U_{\mathrm{f}} = A_j^{-1}(-B_j U_{\mathrm{c}} + F_{\mathrm{f}} - \frac{\partial U_{\mathrm{f}}}{\partial t}), \tag{4.4}$$

the last term, $-A_j^{-1}\frac{\partial}{\partial t}U_{\mathrm{f}}$ is eliminated. Substitution into the second set

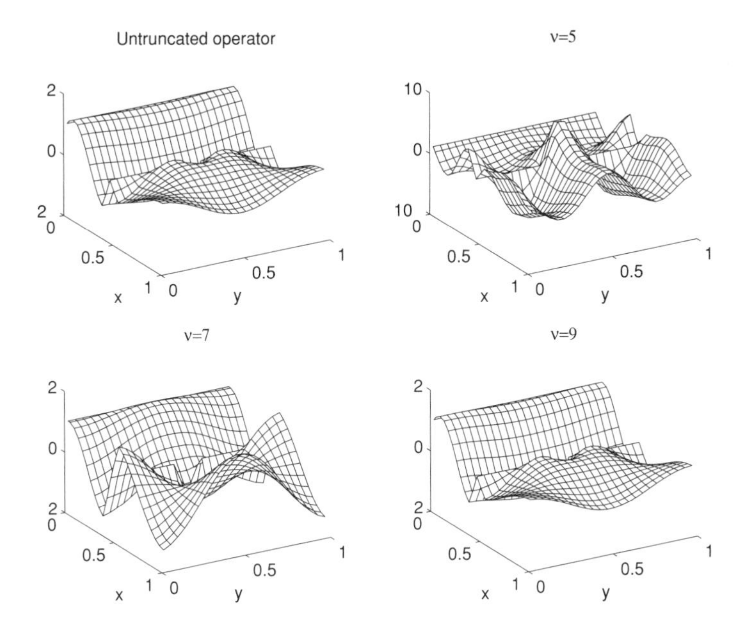

Fig. 8. Results for the 2D Helmholtz example, using the one step homogenized operator, truncated with different ν.

of equations of (4.3) yields

$$\frac{\partial U_{\mathrm{c}}}{\partial t} + \bar{L}_j U_{\mathrm{c}} = \bar{F}_j, \tag{4.5}$$

with $\bar{L}_j$ and $\bar{F}_j$ defined by (3.6).

We found experimentally that using an approximation of the term $-A_j^{-1}\frac{\partial}{\partial t}U_{\mathrm{f}}$ improved the homogenized operator. The correction is derived by taking the time-derivative of (4.4),

$$\frac{\partial U_{\mathrm{f}}}{\partial t} = A_j^{-1}\left(-B_j\frac{\partial U_{\mathrm{c}}}{\partial t} + \frac{\partial F_{\mathrm{f}}}{\partial t} - \frac{\partial^2 U_{\mathrm{f}}}{\partial t^2}\right). \tag{4.6}$$

After a transient mode we have for the model problem (4.2) with $g = \tilde{g}(x/\varepsilon)$ and $\tilde{g}(y)$ 1-periodic,

$$u(x,t) = \bar{u}(x,t) + \varepsilon u_1(x, x/\varepsilon, t) + \mathcal{O}(\varepsilon^2),$$

compare [BLP78]. Hence, $\frac{\partial^2}{\partial t^2}U_{\mathrm{f}}$ is of the same order as $\frac{\partial}{\partial t}U_{\mathrm{f}}$ since the oscillations on the ε-scale are not functions of time and it is reasonable to neglect $-A_j^{-1}\frac{\partial^2}{\partial t^2}U_{\mathrm{f}}$ in (4.6) instead of $-A_j^{-1}\frac{\partial}{\partial t}U_{\mathrm{f}}$ in (4.4). Substituting

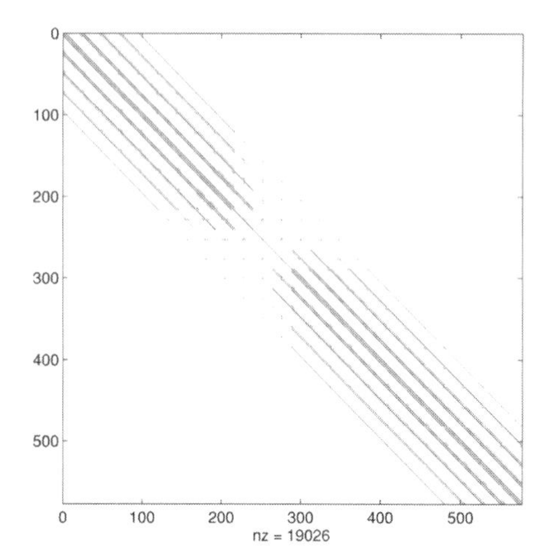

Fig. 9. Structure of the homogenized operator $\bar{L}$, after one homogenization step, for the 2D Helmholtz example. Elements larger than 0.1% of max value shown.

(4.6) after elimination, into (4.3) gives

$$K_j \frac{\partial U_{\mathrm{c}}}{\partial t} + (D_j - C_j A_j^{-1} B_j) U_{\mathrm{c}} = F_{\mathrm{c}} - C_j A_j^{-1} F_{\mathrm{f}} + C_j A_j^{-2} \frac{\partial F_{\mathrm{f}}}{\partial t},$$

with the correction matrix

$$K_j = I + C_j A_j^{-2} B_j.$$

Hence, the corrected homogenized operator $\tilde{L}_j$ and right hand side $\tilde{F}_j$ are

$$\tilde{L}_j = K_j^{-1} \bar{L}_j, \qquad \tilde{F}_j = K_j^{-1} \bar{F}_j + K_j^{-1} C_j A_j^{-2} \frac{\partial F_{\mathrm{f}}}{\partial t}$$

and we will solve (4.5), with $\tilde{L}_j$, $\tilde{F}_j$ substituted for $\bar{L}_j$, $\bar{F}_j$ respectively.

In principle, better approximations could be obtained by reiterating the steps above. After repeated differentiations of (4.6) we could choose to eliminate terms involving successively higher order time derivatives of U_{f}. However, the condition numbers of the correction matrices produced in this manner rapidly deteriorate.

Let $n = 128$, $f \equiv 0$ and

$$b(x) = \begin{cases} \sin^2(4\pi x), & 0 \le x \le 0.25 \\ 0, & 0.25 < x \le 1. \end{cases}$$

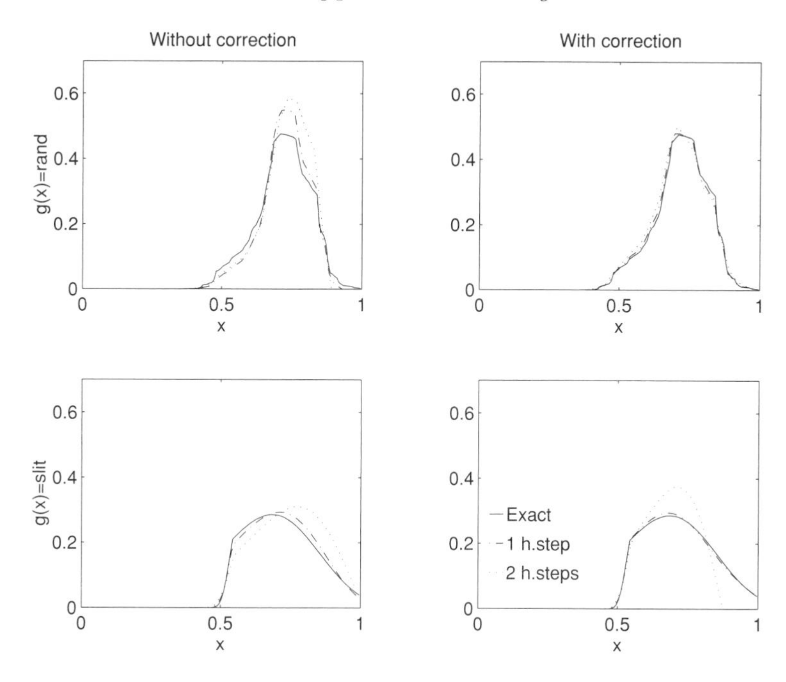

Fig. 10. The solution $u(x)$ at $t = 1$ for the hyperbolic case, using different $g(x)$, with and without first order correction. Solutions shown are the exact solution, computed with the full operator, and solutions computed with the operator homogenized one and two steps.

The result at time $t = 1$, is given in Fig. 10 for the case of $g(x)$ uniformly random distributed in $[0.1, 2]$ and $g(x)$ as in (4.1). The effect of the correction is shown to the right. In these calculations, the time-integration was replaced by a fourth-order Runge-Kutta method.

In view of this we approximate $\bar{L}$ and H, see (3.11), by truncation to lower triangular form. In Fig. 12 results using this truncation is displayed. The same two types of g_ε as in Fig. 10 were used. Like in the elliptic case, truncating the H matrix is more efficient than truncating $\bar{L}$.

It is desirable that the spatial operator corresponds to an upwind discretization of ∂_x, i.e. it should be lower triangular when the coefficient g_ε is positive. If the full operator is lower triangular, the homogenized operator will keep this property in the Haar case. After two homogenization steps the operator, with $g(x)$ a slit, has the structure shown in Fig. 11.

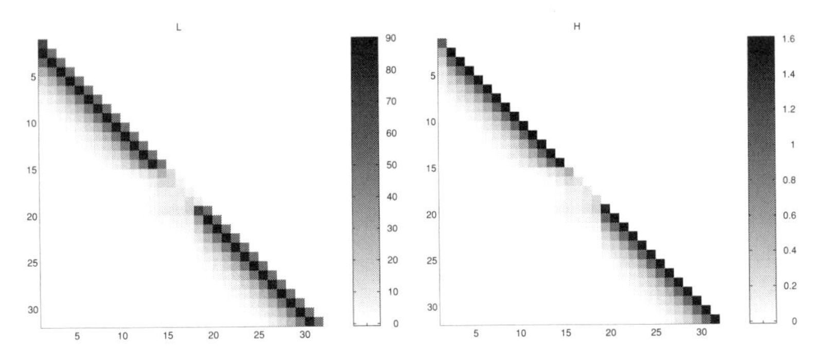

Fig. 11. Structure of the untruncated homogenized operators $\bar{L}$ (left) and H (right) for the hyperbolic problem, $g(x)$ a slit. Gray level indicates absolute value of elements,

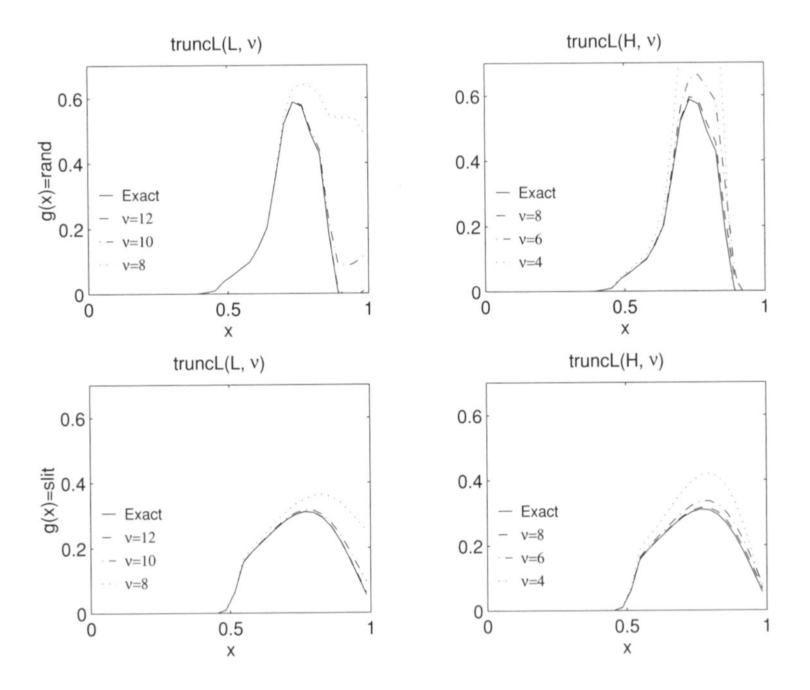

Fig. 12. Solution $u(x)$ for the hyperbolic case, using different $g(x)$ and different approximations of the homogenized operator. The "exact" solution refers to the solution with the full 32×32 homogenized operator,

When the correction method is used it is harder to approximate the resulting homogenized operator, $\tilde{L}$, with a sparse matrix. It will not be as diagonal dominant as in the non-corrected case.

Bibliography

[AB05] G. Allaire & R. Brizzi (2005). A multiscale finite element method for numerical homogenization, *Multiscale Model. Simul.* **4**, 790–812 (electronic).

[And95] J. Anderson (1995). *Computational Fluid Dynamics, The Basics with Applications* (McGraw-Hill, New York)

[AELR99] U. Andersson, B. Engquist, G. Ledfelt & O. Runborg (1999). A contribution to wavelet-based subgrid modeling, *Appl. Comput. Harmon. Anal.* **7**, 151–164.

[Axe94] O. Axelsson (1994). *Iterative Solution Methods* (Cambridge University Press, Cambridge).

[BLP78] A. Bensoussan, J.-L. Lions, & G. Papanicolau (1978). *Asymptotic Analysis for Periodic Structures* (North-Holland, Amsterdam).

[BB94] G. Beylkin & M. Brewster (1995). A multiresolution strategy for numerical homogenization, *Appl. Comput. Harmon. Anal.* **2**, 327–349.

[BBG99] G. Beylkin, M. E. Brewster & A. C. Gilbert (1999). A multiresolution strategy for numerical homogenization of nonlinear ODEs, *Appl. Comput. Harmon. Anal.* **5**, 450–486.

[BC98] G. Beylkin & N. Coult (1998). A multiresolution strategy for reduction of elliptic PDEs and eigenvalue problems, *Appl. Comput. Harmon. Anal.* **5**, 129–155.

[BCR91] G. Beylkin, R. Coifman & V. Rokhlin (1991). Fast wavelet transforms and numerical algorithms I, *Comm. Pure Appl. Math.* **44**, 141–183.

[BDG98] G. Beylkin, J. Dunn & D. L. Gines (1998). LU factorization of non-standard forms and direct multiresolution solvers, *Appl. Comput. Harmon. Anal.* **5**, 156–201.

[CM92] T. Chan & T. Mathew (1992). The interface probing technique in domain decomposition, *SIAM J. Matrix Anal. Appl.* **13**, 212–238.

[CL04] A. Chertock & D. Levy (2004/05). On wavelet-based numerical homogenization. *Multiscale Model. Simul.* **3**, 65–88 (electronic),

[CGM85] C. Concus, G. H. Golub & G. Meurant (1985). Block preconditioning for the conjugate gradient method, *SIAM J. Sci. Stat. Comp.* **6**, 220–252.

[Dau91] I. Daubechies (1991). *Ten Lectures on Wavelets* (SIAM, Philadelphia).

[Dor98] M. Dorobantu & B. Engquist (1998). Wavelet-based numerical homogenization, *SIAM J. Numer. Anal.* **35**, 540–559.

[E92] W. E (1992). Homogenization of linear and nonlinear transport equations, *Comm. Pure Appl. Math.* **45**, 301–326.

[EHW00] Y. R. Efendiev, T. Y. Hou & X. H. Wu (2000). Convergence of a nonconformal multiscale finite element method, *SIAM J. Numer. Anal.* **37**, 888–910.

[EL97] B. Engquist & E. Luo (1997). Convergence of a multigrid method for elliptic equations with highly oscillatory coefficients, *SIAM J. Numer. Anal.* **34**, 2254–2273.

[ER02a] B. Engquist & O. Runborg (2002). Projection generated homogenization. In *Multiscale problems in science and technology (Dubrovnik, 2000)* (N. Antonic, C. J. v. Duijn, W. Jäger & A. Mikelic, eds) 129–150 (Springer, Berlin).

[ER02b] B. Engquist & O. Runborg (2002). Wavelet-based numerical homogenization with applications. In *Multiscale and multiresolution methods* (T.J. Barth, T. Chan & R. Haimes, eds), *Lect. Notes Comput. Sci. Eng.* **20**, 97–148 (Springer, Berlin).

[Gil98] A. C. Gilbert (1998). A comparison of multiresolution and classical

one-dimensional homogenization schemes, *Appl. Comput. Harmon. Anal.* **5**, 1–35.
[HW97] T. Y. Hou & X. H. Wu (1997). A multiscale finite element method for elliptic problems in composite materials and porous media, *J. Comput. Phys.* **134**, 169–189.
[HX92] T. Y. Hou & X. Xin (1992). Homogenization of linear transport equations with oscillatory vector fields, *SIAM J. Appl. Math.* **52**, 34–45.
[Hug95] T. J. R. Hughes (1995). Multiscale phenomena: Green's functions, the Dirichlet-to Neumann formulation, subgrid, scale models, bubbles and the origins of stabilized methods, *Comput. Methods Appl. Mech. Engrg.* **127**, 387–401.
[HFMQ98] T. J. R. Hughes, G. R. Feijóo, L. Mazzei & J.-B. Quicy (1998). The variational multiscale method – a paradigm for computational mechanics, *Comput. Methods Appl. Mech. Engrg.* **166**, 3–24.
[Kel62] J. B. Keller (1962). Geometrical theory of diffraction, *J. Opt. Soc. Amer.* **52**, 113–130.
[Kna99] S. Knapek (1999). Matrix-dependent multigrid-homogenization for diffusion problems, *SIAM J. Sci. Stat. Comp.* **20**, 515–533.
[KRK04] J. Krishnan, O. Runborg & I.G. Kevrekidis (2004). Bifurcation analysis of nonlinear reaction-diffusion problems using wavelet-based reduction techniques, *Comput. Chem. Eng.* **28**, 557–574.
[Leo00] D. D. Leon (2000). *Wavelet Operators Applied to Multigrid Methods*, PhD thesis, Department of Mathematics, UCLA.
[Li94] Z. Li (1994). *The Immersed Interface Method—A Numerical Approach for Partial Differential Equations with Interfaces*, PhD thesis, Department of Applied Mathematics, Univ. of Washington.
[MS00] A.-M. Matache & C. Schwab (2000). Homogenization via p-FEM for problems with microstructure, *Appl. Numer. Math.* **33**, 43–59.
[Neu95] N. Neuss (1995). *Homogenisierung und Mehrgitter*, PhD thesis, Fakultät Mathematik Universität Heidelberg, 1995.
[NJW98] N. Neuss, W. Jäger & G. Wittum (2001). Homogenization and multigrid, *Computing* **66** (1), 1–26.
[OZ07] H. Owhadi & L. Zhang (2007). Metric-based upscaling, *Comm. Pure Appl. Math.* **60**, 675–723.
[PR01] P.-O. Persson & O. Runborg (2001). Simulation of a waveguide filter using wavelet-based numerical homogenization, *J. Comput. Phys.* **166**, 361–382.
[SMM00] B. Z. Steinberg, J. J. McCoy & M. Mirotznik (2000). A multiresolution approach to homogenization and effective modal analysis of complex boundary value problems, *SIAM J. Appl. Math.* **60**, 939–966.
[Taf95] A. Taflove (1995). *Computational Electromagnetics, The Finite-Difference Time-Domain Method*, chapter 10. (Artech House, Boston).
[Wan05] C.-M. Wang (2005). *Wavelet-Based Numerical Homogenization with Application to Flow in Porous Media*, PhD thesis, Department of Mathematics, UCLA.
[Wil93] D. Wilcox (1993). *Turbulence Modeling for CFD* (DCW Industries, Inc.).

6

Plane wave methods for approximating the time harmonic wave equation

Teemu Luostari , Tomi Huttunen
Department of Physics
University of Kuopio
Finland
Email: teemu.luostari@uku.fi *and* tomi.huttunen@uku.fi

Peter Monk
Department of Mathematical Sciences
University of Delaware
United States
Email: monk@math.udel.edu

Abstract

In this paper we shall discuss plane wave methods for approximating the time-harmonic wave equation paying particular attention to the Ultra Weak Variational Formulation (UWVF). This method is essentially a Discontinuous Galerkin (DG) method in which the approximating functions are special traces of solutions of the underlying Helmholtz equation. We summarize the known error analysis for this method, as well as recent attempts to improve the conditioning of the resulting linear system. There are several refinement strategies that can be used to improve the accuracy of the computed solution: h-refinement in which the mesh is refined with a fixed number of basis functions per element, the p-version in which the number of approximating functions per element is increased with a fixed mesh, and a combined hp strategy. We shall provide some numerical results on h and p convergence showing how methods of this type can sometimes provide an efficient solver.

1 Introduction

Traditional methods for discretizing the Helmholtz equation based on using the equation directly suffer from the problem that they become rapidly more expensive as the wave number k (see Eq. (1.1)) increases. For example, finite element, finite difference, finite volume and discontinuous Galerkin methods all suffer from "pollution error" due to the fact that discrete waves have a slightly different wavelength than their exact counterparts (since this error in the wavelength depends on the wave number k, this leads to the "dispersion" of a wave). Pollution error is examined in, for example, [IB95, Ihl98, Coh02] and is proved to exist in general for a certain class of difference method in [BS97]. The practical result of pollution error is that as the wave number k increases, the mesh size in an h-refinement scheme (i.e. mesh refinement without changing the order of convergence on the elements) must decrease faster than $O(1/k)$ to maintain accuracy. Hence the number of unknowns needed for high frequency solutions increases faster than k^d (where d is the dimension of the problem) and such algorithms become prohibitively expensive at high frequency.

One way to mitigate these problems is to use integral equations. For waves propagating in a homogeneous medium this approach can be very effective because the dimension of the domain of the solution is decreased by one compared to, say, a finite element method. In addition pollution error appears to be less severe so that, for standard piecewise polynomial based methods, the dimension of the problem increases roughly as $O(k^{d-1})$ to maintain accuracy. More refined analysis shows how to construct bases that show great promise in controlling the growth of the dimension, and hence the computational effort, still further [GBR05, CL07, ACL07, BG07]. The main restriction of integral equation methods is that they become much more complex in the presence of inhomogeneous media (particularly for Maxwell's equations), and require the use of fast operator evaluation techniques such as the fast multipole method.

An obvious question is therefore if it is possible to improve the pollution error for a partial differential equation based method that can also handle inhomogeneous media. One approach is to use higher order approximation schemes and even p-version methods (in which the grid is fixed, but the order of approximation is increased on each element). The work of Demkowicz [Dem03], Joly and Cohen [CJ96] and Cohen [Coh02] amongst many shows the efficiency of the p-version (and hp-version) approach. In the limit, spectral elements [GO77] can effectively remove pollution error but sampling still requires a minimum number of approximately four collocation points per wavelength so that if h denotes the separation of collocation points and k is the wave number it is still necessary that hk be sufficiently small.

Another approach to this problem is to use exact solutions of the underlying Helmholtz equation on each element to approximate the global solution. This approach, which to date is by no means assured of success, will be the subject of this paper. Note that an important paper on the h-convergence of a general class of such schemes has just appeared in the Isaac Newton Institute report series [GHP07].

We shall use as a model problem the Helmholtz equation on a bounded polyhedral domain $\Omega \subset \mathbb{R}^d, d = 2, 3$. We denote by Γ the boundary of Ω with unit outward normal $\boldsymbol{\nu}$. The acoustic field u then satisfies

$$\nabla \cdot A\nabla u + k^2 n^2 u = 0 \quad \text{in } \Omega \tag{1.1}$$

and is assumed to be subject to the generalized boundary condition

$$\left(\frac{\partial u}{\partial \boldsymbol{\nu}_A} - i\eta u\right) = Q\left(\frac{\partial u}{\partial \boldsymbol{\nu}_A} + i\eta u\right) + g \tag{1.2}$$

where $\partial u/\partial \boldsymbol{\nu}_A = \boldsymbol{\nu} \cdot A\nabla u$. In these equations A and n are bounded piecewise smooth functions of position. The coefficient A is real, strictly positive and in the case of anisotropic media could be a matrix but we will not pursue that idea here (see [HM07]). The function n may be complex and in that case has strictly positive real part and non-negative imaginary part. The boundary function η is real, bounded and strictly positive and Q is real with $|Q| \leq 1$. The data function $g \in H^{1/2}(\Gamma)$ (when $|Q| < 1$ we may allow $g \in L^2(\Gamma)$). When $|Q| < 1$ we may rewrite the boundary condition in the form

$$\frac{\partial u}{\partial \boldsymbol{\nu}_A} - i\eta\frac{(1+Q)}{(1-Q)}u = \frac{g}{1-Q}$$

showing that it is just a standard impedance boundary condition. When $Q = 1$, equation (1.2) is equivalent to a Dirichlet boundary condition and when $Q = -1$ to a Neumann boundary condition.

Next we introduce a mesh $\mathcal{T}_h$ of finite elements. In most of our work $\mathcal{T}_h$ is a standard mesh of regular triangles ($d = 2$) or tetrahedra ($d = 3$) of maximum diameter h. The restriction to simplicial elements is not essential and tests by one of us (Luostari) suggest that rectangular elements may well be more efficient in 2D. In order to use a method based on solutions of the Helmholtz equation on each element we make the simplifying assumption that the functions A and n are constant on each element in the mesh (but can jump from element to element).

We can now consider two different approaches. In the first classical approach, continuous basis functions are constructed and a standard weak formulation of (1.1)-(1.2) is used. This is the approach taken in the Partition of Unity Finite Element Method (PUFEM) in which the basis functions are products of functions in a partition of unity of Ω (typically piecewise linear nodal finite element basis functions) and solutions of the Helmholtz equation (typically plane waves). This method was proposed by Melenk in his important thesis [Mel95, MB96, BM97] and has been tested for example in [PTB02] for homogeneous problems. The method is analyzed in a special case in [HP08].

For problems involving inhomogeneous media (i.e. non constant A or n) however the PUFEM is more tricky to implement. The local solutions of the Helmholtz equation used to construct the basis functions must satisfy the Helmholtz equation on the support of each partition of unity function, and this is difficult to achieve for general piecewise constant media. Alternatively the method can be implemented using a discontinuous basis across discontinuities in the coefficients A or n. Typically a Lagrange multiplier is used to enforce weak continuity in the solution [LBPT05].

The difficulty of constructing solutions of an inhomogeneous problem can be avoided if we assume that A and n are piecewise constant and a discontinuous local basis of solutions of the Helmholtz equation is used element by element (assuming the coefficients are constant on each element). The resulting scheme then needs some way to weakly enforce the appropriate continuity of the solution between elements. At least three ways have been considered in the literature.

• Least squares [Sto98, MW98]: In this method a least squares functional that penalizes jumps in the solution and its normal derivative across element boundaries is minimized. The method gives rise to a Hermitian positive definite matrix problem, but, as usual for all methods based on solutions of the differential equation, the matrix can be ill-conditioned.

• Lagrange multipliers [TF06]: In this method solutions of the differential equation are added to standard finite element functions on each element (rather than multiplied as in the PUFEM). Lagrange multipliers, that are also typically the trace of solutions of the differential equation, are used to weakly enforce continuity between elements. Computational examples (including special preconditioning techniques [LT08]) show impressive performance [TF06].

• Discontinuous Galerkin (DG) methods (see for example [Gab07, AMM06, GHP07, HMM07]): In these methods a suitable bilinear form is derived that enforces weak continuity without a Lagrange multiplier, but typically using stabilizing penalty terms.

In this paper we shall concentrate on a specific discontinuous Galerkin method called the Ultra Weak Variational Formulation (UWVF) due to Cessenat and Després [Ces96, CD98, CD03]. This method can be derived as an upwind discontinuous Galerkin method [HMM07, Gab08, AMM06, GHP07]. For the more general case of an interior penalty discontinuous Galerkin scheme see [GHP07, Gab07]. In [AMM06] we considered a large class of DG methods and analyzed their dispersion error. We noted in particular that the UWVF is a DG method with a special choice of coefficients. Recently Gittelson, Hiptmair and Perugia [GHP07] have provided a very insightful analysis of the h-convergence of such methods when plane waves are used to discretize the problem element by element. They show that, at least for low wave number, the general DG framework can be used to construct methods that give improved h-convergence compared to the standard UWVF. Nevertheless we shall only consider the standard UWVF here because we shall show that it has good performance when the number of basis functions per element is also carefully controlled.

The UWVF was originally developed for Maxwell's equations and the Helmholtz equation [Ces96, CD98], but has been extended to linear plane elasticity [HMCK04] and to the fluid-structure interaction problem [HKM08].

The plan of this paper is as follows. In the next section we shall derive the UWVF following Cessenat and Després. We shall then summarize the choice of basis functions and the known error estimates for the UWVF. In Section 4 we shall present several numerical experiments. In particular we shall show that in some special cases the method can approximate solutions of the Helmholtz equation with a density of unknowns that increases much less rapidly then standard finite element methods. Finally we shall draw some conclusions.

2 Derivation of the UWVF

The UWVF can be derived from an appropriate symmetric hyperbolic system equivalent to the wave-equation after taking the Fourier transform in time [HMM07, Gab08]. It can also be derived as a special discontinuous Galerkin method [AMM06, GHP07, BM07] or as a least squares method (as we shall discuss later in this paper). However a quicker approach is provided in the original thesis of Cessenat (see [Ces96, CD98]) and we use it here. We first derive an equality termed by Cessenat and Després the "Isometry Lemma".

Lemma 1 *Suppose ξ is a smooth solution of the conjugated Helmholtz equation*

$$\nabla A \nabla \xi + k^2 \overline{n}^2 \xi = 0 \tag{2.1}$$

in an element $K \in \mathcal{T}_h$ where $\bar{n}$ is the complex conjugate of n. Then if u satisfies (1.1) and η is any strictly positive real function in $L^\infty(\partial K)$ the following equality holds:

$$\begin{aligned} & \int_{\partial K} \frac{1}{\eta}\left(\frac{\partial u}{\partial \boldsymbol{\nu}_A} + i\eta u\right)\overline{\left(\frac{\partial \xi}{\partial \boldsymbol{\nu}_A} + i\eta\xi\right)}\, d\sigma \\ = & \int_{\partial K} \frac{1}{\eta}\left(-\frac{\partial u}{\partial \boldsymbol{\nu}_A} + i\eta u\right)\overline{\left(-\frac{\partial \xi}{\partial \boldsymbol{\nu}_A} + i\eta\xi\right)}\, d\sigma. \end{aligned} \tag{2.2}$$

Proof The proof uses Green's theorem. Direct calculation shows that if $\boldsymbol{\nu}$ is the unit outward normal to K,

$$\begin{aligned} & \int_{\partial K} \frac{1}{\eta}\left(\frac{\partial u}{\partial \boldsymbol{\nu}_A} + i\eta u\right)\overline{\left(\frac{\partial \xi}{\partial \boldsymbol{\nu}_A} + i\eta\xi\right)}\, d\sigma \\ = & \int_{\partial K} \frac{1}{\eta}\left(-\frac{\partial u}{\partial \boldsymbol{\nu}_A} + i\eta u\right)\overline{\left(-\frac{\partial \xi}{\partial \boldsymbol{\nu}_A} + i\eta\xi\right)}\, d\sigma \\ - & \ 2\int_{\partial K} \frac{\partial u}{\partial \boldsymbol{\nu}_A}\bar{\xi} - u\frac{\partial \bar{\xi}}{\partial \boldsymbol{\nu}_A} d\sigma. \end{aligned}$$

But using Green's theorem

$$\int_{\partial K} \frac{\partial u}{\partial \boldsymbol{\nu}_A}\bar{\xi} - u\frac{\partial \bar{\xi}}{\partial \boldsymbol{\nu}_A} d\sigma = \int_K \nabla \cdot A \nabla u \bar{\xi} - u \nabla \cdot A \nabla \bar{\xi} dx$$

and using (1.1) and (2.1) shows that this term vanishes and completes the proof. □

Using this lemma we can now formulate the UWVF. Let η now represent an L^∞ function on the skeleton (union of the faces) of the mesh $\mathcal{T}_h$. On Γ, the function η is given by the data in (1.2) but elsewhere η needs only to be chosen to be real, strictly positive and bounded. Usually, when $A = 1$ and $n = 1$ we choose, like Cessenat and Després [CD98], $\eta = k$. However it may be that when k is small we should choose $\eta > k$ or even choose it to be a function of h as suggested in [GHP07]. We shall return to this question in Section 5.2.

Enumerating the elements in $\mathcal{T}_h$ by $\{K_j\}_{j=1}^{N_h}$ we can define the solution

$$\mathcal{X}_j = \left(\frac{\partial u}{\partial \boldsymbol{\nu}_A^{K_j}} + i\eta u\right)\Bigg|_{\partial K_j} \in L^2(\partial K_j)$$

where $\partial u/\partial \boldsymbol{\nu}_A^K = \boldsymbol{\nu}^K \cdot A\nabla u$ and $\boldsymbol{\nu}^K$ is the unit outward normal to K. Let $\Sigma_{j\ell} = \bar{K}_j \cap \bar{K}_\ell$ with normal $\boldsymbol{\nu}^{K_j}$. We also define $\Gamma_j = \partial K_j \cap \Gamma$. Then for a given element K_j in the mesh and for any smooth solution ξ^{K_j} of (2.1) on K_j, we define

$$\mathcal{Y}_j = \left(\frac{\partial \xi^{K_j}}{\partial \boldsymbol{\nu}_A^{K_j}} + i\eta \xi^{K_j} \right)\Bigg|_{\partial K_j} \in L^2(\partial K_j)$$

and define

$$F_j(\mathcal{Y}_j) = \left(-\frac{\partial \xi^{K_j}}{\partial \boldsymbol{\nu}_A^{K_j}} + i\eta \xi^{K_j} \right)\Bigg|_{\partial K_j}$$

so $F_j : L^2(\partial K_j) \to L^2(\partial K_j)$. Then the result of Lemma 1 shows that

$$\begin{aligned} \int_{\partial K_j} \frac{1}{\eta} \mathcal{X}_j \bar{\mathcal{Y}}_j \, d\sigma &= \sum_{\ell=1,\, \ell\neq j}^{N_h} \int_{\Sigma_{j\ell}} \frac{1}{\eta} \left(-\frac{\partial u}{\partial \boldsymbol{\nu}_A^{K_j}} + i\eta u \right) \overline{F_j(\mathcal{Y}_j)} d\sigma \\ &+ \int_{\Gamma_j} \frac{1}{\eta} \left(-\frac{\partial u}{\partial \boldsymbol{\nu}_A^{K_j}} + i\eta u \right) \overline{F_j(\mathcal{Y}_j)} d\sigma. \end{aligned}$$

But by the continuity of the solution and its normal derivative across $\Sigma_{j\ell}$,

$$\left(-\frac{\partial u}{\partial \boldsymbol{\nu}_A^{K_j}} + i\eta u \right)\Bigg|_{\Sigma_{j\ell}} = \left(\frac{\partial u}{\partial \boldsymbol{\nu}_A^{K_\ell}} + i\eta u \right)\Bigg|_{\Sigma_{\ell j}} = \mathcal{X}_\ell.$$

On the boundary, using (1.2),

$$\left(-\frac{\partial u}{\partial \boldsymbol{\nu}} + i\eta u \right)\Bigg|_{\Gamma_j} = -Q\mathcal{X}_j - g$$

so (2.2) may be written

$$\begin{aligned} \int_{\partial K_j} \frac{1}{\eta} \mathcal{X}_j \mathcal{Y}_j \, d\sigma &= \sum_{\ell=1,\, \ell\neq j}^{N_h} \int_{\Sigma_{\ell j}} \frac{1}{\eta} \mathcal{X}_\ell \overline{F_j(\mathcal{Y}_j)} d\sigma \\ &\quad - \int_{\Gamma_j} \frac{1}{\eta} (Q\mathcal{X}_j + g) \overline{F_j(\mathcal{Y}_j)} d\sigma. \end{aligned}$$

Adding this result over all elements we are led to a variational problem posed on the space

$$X = \prod_{j=1}^{N_h} L^2(\partial K_j).$$

If $\mathcal{X} = (\mathcal{X}_1, \mathcal{X}_2, \ldots, \mathcal{X}_{N_h}) \in X$, the norm of $\mathcal{X} \in X$ is defined by

$$\|\mathcal{X}\|_X^2 = \sum_{j=1}^{N_h} \int_{\partial K_j} \frac{1}{\eta} |\mathcal{X}_j|^2 \, d\sigma$$

and the inner product is defined, for all $\mathcal{X}, \mathcal{Y} \in X$, by

$$\langle \mathcal{X}, \mathcal{Y} \rangle = \sum_{j=1}^{N_h} \int_{\partial K_j} \frac{1}{\eta} \mathcal{X} \overline{\mathcal{Y}} \, d\sigma.$$

Using this notation, the UWVF of the Helmholtz equation before discretization is to seek $\mathcal{X} \in X$ such that

$$a(\mathcal{X}, \mathcal{Y}) = G(\mathcal{Y}) \quad \text{for all} \quad \mathcal{Y} \in X, \tag{2.3}$$

where

$$\begin{aligned} a(\mathcal{X}, \mathcal{Y}) &= \langle \mathcal{X}, \mathcal{Y} \rangle - \sum_{j=1}^{N_h} \sum_{\ell=1, \ell \neq j}^{N_h} \int_{\Sigma_{\ell j}} \frac{1}{\eta} \mathcal{X}_\ell \overline{F_j(\mathcal{Y}_j)} \, d\sigma \\ &\quad + \sum_{j=1}^{N_h} \int_{\Gamma_j} \frac{1}{\eta} Q \mathcal{X}_j \overline{F_j(\mathcal{Y}_j)} \, d\sigma, \end{aligned}$$

and

$$G(\mathcal{Y}) = -\sum_{j=1}^{N_h} \int_{\Gamma_j} \frac{1}{\eta} g \overline{F_j(\mathcal{Y}_j)} \, d\sigma.$$

The well-posedness of this problem is established in [CD98]. Furthermore it is shown that the solution $\mathcal{X}_j$, $j = 1, \ldots, N_h$ of (2.3) provides the impedance boundary data for u on the faces of the elements, so that solving (2.3) effectively provides a solution to the original scattering problem.

We can now follow Cessenat and Després by introducing the operators $\pi : X \to X$ and $F : X \to X$ where

$$(\pi \mathcal{X})_j|_{\Sigma_{j,\ell}} = \mathcal{X}_\ell|_{\Sigma_{j,\ell}} \text{ if } \Sigma_{j,\ell} \neq \emptyset$$

and

$$(\pi \mathcal{X})_j|_{\Gamma_j} = -Q \mathcal{X}_j|_{\Gamma_j} \text{ if } \Gamma_j \neq \emptyset.$$

The operator F is defined componentwise by

$$F(\mathcal{X}) = (F_1(\mathcal{X}_1), F_2(\mathcal{X}_2), \ldots, F_{N_h}(\mathcal{X}_{N_h})).$$

With this notation

$$a(\mathcal{X},\mathcal{Y}) = \langle \mathcal{X},\mathcal{Y}\rangle - \langle \pi\mathcal{X}, F(\mathcal{Y})\rangle.$$

If $\tilde{g} \in X$ is such that $\tilde{g} = g$ on Γ and $\tilde{g} = 0$ on faces away from Γ we may write (2.3) in the operator form

$$(I - F^*\pi)\mathcal{X} = F^*\tilde{g}. \tag{2.4}$$

where F^* is the X adjoint of F.

In the special case when A and n are real, F is an isometry [CD98] (this follows from Lemma 1) and so in that case (2.4) becomes

$$(F - \pi)\mathcal{X} = \tilde{g}$$

and hence

$$(F - \pi)^*(F - \pi)\mathcal{X} = (F - \pi)^*\tilde{g}.$$

But this is just the normal equation for the problem of finding $\mathcal{X} \in X$ that minimizes

$$\|F\mathcal{X} - \pi\mathcal{X} - \tilde{g}\|_X^2$$

which is a standard least squares method for minimizing the difference in impedance fluxes across the faces between elements. Thus we see that, before discretization, the UWVF is equivalent to a factorization of the normal equations for the least squares method and after discretization should therefore give a better conditioned linear system. This seems to be a crucial advantage over the least squares method mentioned earlier.

3 Discretization of the UWVF

In principle, the discretization of the UWVF simply requires a suitable family of finite dimensional subspaces X_h of X for each $h > 0$. Once this approximating subspace is chosen we can pose the problem of computing a discrete solution $\mathcal{X}_h \in X_h$ that solves

$$\langle \mathcal{X}_h,\mathcal{Y}\rangle - \langle \pi\mathcal{X}_h, F(\mathcal{Y})\rangle = \langle \tilde{g}, F(\mathcal{Y})\rangle \quad \text{for all} \quad \mathcal{Y} \in X_h \tag{3.1}$$

This problem is shown to have a unique solution in [CD98] when A and n are real and we shall also sketch a proof of this result later in the paper. When n is complex, existence and uniqueness can also be verified [BM07].

Unfortunately this general choice of X_h is not practical since, for any $\mathcal{Y} \in X_h$, we need to evaluate $F(\mathcal{Y})$. In general this problem is almost as difficult to solve as the original scattering problem. However it is easy to evaluate F if X_h is constructed from impedance traces of solutions of (2.1) as we describe next [CD98].

For each element K let M_P^K denote a finite dimensional space of P smooth solutions of (2.1) (here $P = P(K)$ may vary from element to element). Then $X_h = \Pi_{j=1}^{N_h} X_{h,j}$ where

$$X_{h,j} = \left\{ \mathcal{Y}_j \;\middle|\; \mathcal{Y}_j = \left(\frac{\partial \xi}{\partial \boldsymbol{\nu}_A^{K_j}} + i\eta\xi \right)\bigg|_{\partial K_j}, \; \xi \in M_{P_j}^{K_j} \right\}.$$

Obviously it is then easy to compute $F(\mathcal{Y})$ for any $\mathcal{Y} \in X_h$ just by reversing the sign of the normal derivative term in the definition of $X_{h,j}$ above.

Two families of solutions of the Helmholtz equation are obvious candidates for constructing M_P^K. To simplify our discussion we shall assume that the centroid of K is at the origin (if not a suitable translation needs to be added to the definition of M_P^K) and we shall assume that we are solving the problem in $\mathbb{R}^2$ and provide a few comments about $\mathbb{R}^3$ later. Let the local wave number on K be defined by

$$\kappa = k\bar{n}^K / \sqrt{A^K}$$

where n^K and A^K are the constant values of n and A respectively on K. We choose P linearly independent directions $\boldsymbol{d}_j, |\boldsymbol{d}_j| = 1, \; j = 1, \ldots, P$ and in practice the directions are uniformly distributed (one interesting question is how to choose a better set of directions for a given problem adaptively) so that

$$\boldsymbol{d}_j = (\cos(\theta_j), \sin(\theta_j))^T, \quad \theta_j = 2j\pi/P, \quad 1 \le j \le P. \tag{3.2}$$

In this case

$$M_P^K = \text{span} \left\{ \xi_j^{PW} := \exp(i\kappa \boldsymbol{x} \cdot \boldsymbol{d}_j), \; 1 \le j \le P \right\}.$$

This choice of local basis is very convenient because various integrals needed in the variational problem can be computed in closed form (on straight edges) and the basis functions are easy to evaluate.

An alternative basis is given by the Bessel functions. Let $r = |\boldsymbol{x}|$ and θ denote polar coordinates so $\boldsymbol{x} = (r\cos\theta, r\sin\theta)$. Then, in this case,

$$M_P^K = \text{span} \left\{ \xi_j^B := J_{j-p-1}(\kappa r) \exp(i(j-p-1)\theta), \quad 1 \le j \le 2p+1 \right\}$$

so that $P = 2p + 1$. This basis is less convenient than the plane wave basis since Bessel functions are more expensive to evaluate than plane wave functions and inner products of basis functions on edges must be computed by quadrature. This markedly increases the time needed to evaluate the discrete UWVF matrix system.

Limited experiments in [MW98] showed that for higher wave number k there is little to be gained from the Bessel function basis in terms of accuracy (for the same P). However for a smaller wave number k the plane wave basis gives rise to very ill-conditioned matrices since

$$\exp(i\kappa \boldsymbol{x} \cdot \boldsymbol{d}) = 1 + O(\kappa \boldsymbol{x} \cdot d) \quad \text{as } \kappa \to 0.$$

Note that the same loss of conditioning occurs for fixed κ as h decreases. If we write $\boldsymbol{x} = h\hat{\boldsymbol{x}}$ where $\|\hat{\boldsymbol{x}}\| = 1$ we obtain a similar asymptotic expansion for the plane wave basis function as $h \to 0$.

Gittelson, Hiptmair and Perugia [GHP07] suggest that an appropriate change of basis can help conditioning (although the best way to compute with this new basis is not obvious). As they point out, one way to understand their idea is to recall that Bessel functions do not have the same difficulty for low κ (or equivalently low k). In particular

$$J_j(\kappa r) = \frac{1}{j!}\frac{\kappa^j r^j}{2^j} + O(\kappa r)^{j+2} \text{ as } \kappa \to 0.$$

So

$$r^j \exp(ij\theta) = \frac{2^j j!}{\kappa^j} J_j(\kappa r)\exp(ij\theta) + O(\kappa r)^{j+2}.$$

Thus as $\kappa \to 0$, suitably weighted Bessel functions give rise to harmonic polynomials that are appropriate basis functions for approximating Laplace's equation. Thus it appears that a scaled Bessel function basis may be a good choice at low κ (equivalently, low k).

The plane wave and Bessel function bases are related by the Jacobi-Anger expansion [CR98] and evaluating the Fourier coefficients of both sides of this expansion shows the 2D analogue of the Funk–Hecke formula:

$$2\pi i^j J_j(\kappa r)\exp(ij\theta) = \int_0^{2\pi} \exp(i\kappa \boldsymbol{x} \cdot \boldsymbol{d}(\phi))\exp(ij\phi)\, d\phi \qquad (3.3)$$

where $\boldsymbol{d}(\phi)$ is defined as in (3.2) with ϕ replacing θ. Now applying the trapezoidal rule using P points to (3.3) gives

$$i^j J_j(\kappa r)\exp(ij\theta) = \frac{1}{P}\sum_{\ell=1}^{P} \exp(i\kappa \boldsymbol{x} \cdot \boldsymbol{d}_\ell)\exp(ij\phi_\ell) + O(\kappa^2/P^2).$$

Thus if $\vec{\xi}^{PW} = (\xi_1^{PW}, \ldots, \xi_p^{PW})^T$ and $\vec{\xi}^{B} = (\xi_1^{B}, \ldots, \xi_p^{B})^T$ we have the relation

$$\vec{\xi}^B = M\vec{\xi}^{PW} + O(\kappa^2/P^2) \tag{3.4}$$

where the $P \times P$ matrix M has entries $M_{j,\ell} = i^{-j}\exp(ij\phi_\ell)/P$ provided P is odd. Using this change of basis element by element should improve conditioning as κ decreases, but the application of M involves cancellation (essentially numerical differentiation) and so for low κ the Bessel function basis should, perhaps, be used directly.

This change of basis can be extended to three dimensions using the Funk–Hecke formula and choosing directions on the unit sphere that provide an accurate evaluation of spherical harmonics (spherical designs for example [SW01]).

A different mechanism for ill-conditioning occurs if we increases P on a fixed mesh with a fixed wave number κ. If P is too large compared to $h\kappa$ the corresponding diagonal block of D becomes ill-conditoned. To see this, recall the large argument asymptotics of the Bessel functions:

$$J_j(\kappa r) = \frac{(\kappa r)^j}{2^j j!}\left(1 + O\left(\frac{1}{j}\right)\right) \quad \text{as } j \to \infty$$

From (3.3) we see that

$$\int_0^{2\pi} \exp(i\kappa \boldsymbol{x} \cdot \boldsymbol{d}(\phi)) \exp(ij\phi)\, d\phi = 2\pi i^j \frac{(\kappa r)^j}{2^j j!}\left(1 + O\left(\frac{1}{j}\right)\right) \to 0$$

as $j \to \infty$. Applying P point trapezoidal rule quadrature to this integral, the above estimate implies that for P large enough

$$\frac{1}{P}\sum_{\ell=1}^{P} \exp(i\kappa \boldsymbol{x} \cdot \boldsymbol{d}_\ell) \exp(ij\phi_\ell) \approx 0$$

so that the plane waves are almost linearly dependent when P is large. The change of variables (3.4) is unlikely to help with this sort of ill-conditioning since it is based on the large argument asymptotics of the Bessel functions. We shall return to the problem of choosing P element by element in Section 5.

4 Analysis of the UWVF

The analysis of the UWVF is by no means complete particularly when n is complex. In that case, virtually nothing is known [BM07]. When n

and A are real, the situation is slightly better in the case of h convergence, while P-convergence is again not yet analyzed in 3D (the analysis of Gittelson et al. [GHP07] does not cover the standard UWVF).

In [BM07] the following estimate is proved by reinterpreting UWVF as a special discontinuous Galerkin method. Of course this reinterpretation is only a guide and the estimates can be derived directly.

We assume that A and n are real and $Q = 0$ (the case of general $|Q| < 1$ is easy to handle also). Let u_j denote the solution of

$$\nabla \cdot A \nabla u_j^h + k^2 n^2 u_j^h = 0 \quad \text{on} \quad K_j \tag{4.1}$$

$$\frac{\partial u_j^h}{\partial \nu_A^{K_j}} + i\eta u_j^h = \mathcal{X}_j^h \quad \text{on} \quad \partial K_j \tag{4.2}$$

and let $\boldsymbol{v}_j^h = -(A/ik)\nabla u_j^h$. Let P_h denote the projection operator from X to X_h in the X inner product. Let E_I denote the set of interior edges of the mesh.

Suppose K_j and K_ℓ meet at the edge $\Sigma_{j\ell}$ then we define

$$[u] = |u_j - u_\ell| \text{ on } \Sigma_{j\ell},$$

and

$$[\boldsymbol{v}] = |\boldsymbol{v}_\ell \cdot \boldsymbol{\nu}_j - \boldsymbol{v}_j \cdot \boldsymbol{\nu}_j| \text{ on } \Sigma_{j\ell}.$$

The following lemma holds:

Lemma 2 *[BM07] Let u satisfy (1.1) - (1.2) and $u^h = (u_1^h, u_1^h, \ldots, u_{N_h}^h)$ denote the piecewise defined solution of (3.2) - (3.3) (similarly for $\boldsymbol{v}^h$). Then the following estimate holds if $Q = 0$,*

$$\sum_{e \in E_I} \int_e \frac{\eta}{2} |[u - u^h]|^2 + \frac{1}{2\eta} |[\boldsymbol{v} - \boldsymbol{v}^h]|^2 \, d\sigma \tag{4.3}$$

$$+ \sum_j \int_{\Gamma_j} \frac{1}{2\eta} |F_j(\chi_j - \chi_j^h)|^2 \, d\sigma \leq 2\|(I - P_h)\mathcal{X}\|_X^2.$$

In addition [Ces96, CD98],

$$\sum_j \int_{\Gamma_j} \frac{1}{2\eta} |\chi_j - \chi_j^h|^2 ds \leq 2\|(I - P_h)\chi\|_X^2. \tag{4.4}$$

This result can be used to prove uniqueness of the discrete solution. To see this note that if $\mathcal{X} = 0$ then u^h is a continuous solution of the Helmholtz equation since $[u - u^h] = -[u^h] = 0$ on each interior edge.

The vanishing boundary data on Γ (from (4.4)) shows that $u^h = 0$ and hence that $\mathcal{X}_h = 0$.

Lemma 2 does not provide a direct estimate of the error under h-refinement since the left hand side is a mesh dependent norm. Using a result from [MW98] we are then able to prove the following theorem regarding global convergence.

Theorem 3 *If A and n are real, $Q = 0$ and $\mathcal{T}_h$ is quasi-uniform:*

$$\|u - u^h\|_{L^2(\Omega)} \leq Ch^{-1/2}\|(I - P_h)\mathcal{X}\|_X \tag{4.5}$$

where C is independent of h and u.

An estimate for $(I - P_h)\mathcal{X}$ is proved in [CD98] (see also [Mel95]) in the case of h refinement (or mesh refinement). In particular they show that in 2D, if $P = 2p + 1$, then

$$\|(I - P_h)\chi\|_X \leq Ch^{p-1/2}\|u\|_{C^{p+1}(\Omega)}$$

hence using (4.5) we can prove that

$$\|u - u^h\|_{L^2(\Omega)} \leq Ch^{p-1}\|u\|_{C^{p+1}(\Omega)}.$$

Numerical experiments in [CD98] and [BM07] suggest that, at least for waves without a significant evanescent component, the estimate is not optimal and that $O(h^p)$ convergence is seen. A more refined analysis is clearly desirable (or an example exhibiting the predicted convergence rate!).

When n is complex (recall that $\Im(n) > 0$) we find that F is no longer an isometry. In fact for any $\mathcal{Y} \in X$,

$$\|F_j(\mathcal{Y})\|^2_{L^2_\eta(\partial K_j)} = \|\mathcal{Y}_j\|^2_{L^2_\eta(\partial K_j)} + 2ik^2 \int_{K_j} |\xi_j|^2 (n - \bar{n})\, dA$$

where ξ_j is the solution of

$$\begin{aligned} \nabla \cdot A\nabla\xi_j + k^2\bar{n}\xi_j = 0 \quad & \text{in} \quad K_j, \\ \frac{\partial \xi_j}{\partial n_A} + i\eta\xi_j = \mathcal{Y}_j \quad & \text{on} \quad \partial K_j. \end{aligned}$$

The fact that F_j is now a contraction when $\Im(n) > 0$ can be used to derive an error estimate (in an unusual norm), see [BM07] for details.

For an analysis of error in an interior penalty discontinuous Galerkin formulation of the Helmholtz equation related to the UWVF but with a different choice of penalty terms to enforce inter element continuity see [GHP07].

5 Numerical implementation and results for the UWVF

The discrete problem (3.1) can be implemented in the usual finite element fashion. Enumerating the coefficients of the basis functions for $\mathcal{X}_h$ element by element we must compute a vector $\vec{x}$ of degrees of freedom of dimension $\sum_{j=1}^{N_h} P_{K_j}$ where P_{K_j} is the number of basis functions on K_j. The resulting matrix problem has the form

$$(D - C)\vec{x} = \vec{g}$$

where D is the inner product matrix corresponding to $\langle \mathcal{X}, \mathcal{Y} \rangle$ and C is the matrix resulting from $\langle \pi\mathcal{X}, F(\mathcal{Y}) \rangle$. The right hand side $\vec{g}$ is computed from $\langle \tilde{g}, F(\mathcal{Y}) \rangle$. Clearly D is block diagonal, Hermitian and positive definite. The matrix C is more complicated. On straight edges (or flat faces in 3D) the integrals needed to compute C and D can be performed analytically [Ces96], as can those for $\vec{g}$ if the boundary data is due to a plane wave (for example, in plane wave scattering problems). Otherwise quadrature is needed which greatly slows down matrix assembly.

In practice we solve

$$(I - D^{-1}C)\vec{x} = D^{-1}\vec{g}. \tag{5.1}$$

Cessenat and Després show that the eigenvalues of $D^{-1}C$ (when A and n are real) lie in the closure of the unit disk with 1 removed and propose to use a damped Richardson scheme to solve (5.1). We have found that BiCGStab is usually faster and use it exclusively in 3D (in 2D we generally use a direct method).

The convergence of the iterative scheme for (5.1) depends on being able to invert D and on the conditioning of $(I - D^{-1}C)$. We have found that the conditioning of D (hence of the diagonal blocks of D) gives a good indication of the conditioning of the global matrix $(I - D^{-1}C)$ [HMK02]. Cessenat and Després [CD98] show that each diagonal block of D has a condition number proportional to h^{-2p} as $h \to 0$ for fixed p which is a very rapid growth rate for practical numbers of directions (say $p = 5$). It is thus critical not to pick too many directions on any given element. In [HMK02] we pick the number of directions per element precisely to control the condition number of the corresponding diagonal block of D. A threshold is set by the user and the number of directions on a given element is chosen to give the maximum condition number of the diagonal blocks of D less than this threshold. Thus P varies from element to element.

In the threshold scheme, accuracy is controlled indirectly by the choice of the previously mentioned threshold parameter. Analysis of the Jacobi-Anger exapansion (Remark 3.3 of [CC04]) suggests that high accuracy can be achieved by choosing, element by element,

$$P = Ch\kappa \tag{5.2}$$

when κ is large, where h is the diameter of the element and C is a sufficiently large constant as $\kappa \to \infty$ (the precise choice for C depends on how h is defined element by element). Similar results are shown in [Per06]. We use (5.2) as a heuristic choice relating P and $h\kappa$ for more general fields and test this choice later, but note that this choice will only be useful away from singularities of the solution and at higher wave numbers.

Besides the issue of conditioning which is particularly difficult at low wave number where (5.2) is not likely to be applicable, the plane wave UWVF does not approximate the solution u of (1.1)-(1.2) well near singularities (for example, near reentrant corners). This has been investigated both for the UWVF and PUFEM in [HGA08]. Near singularities it may well be that finite elements should be used in a combined plane wave and finite element method. Perry-Debain [Per06] has investigated the approximation of solutions of the Helmholtz equation by plane waves on the unit disk and has shown that plane waves do converge for evanescent components, but that the expansion coefficients grow very rapidly and thus limit overall accuracy because of cancellation error in the floating point system. It appears that very small elements are needed near the singularity.

The simple absorbing boundary condition (1.2) is not efficient for scattering problems in practice. We have shown that the perfectly matched layer (PML) can be used with the UWVF [HKM04] or that the UWVF can be coupled to the fast multipole method [DM07] to provide greater efficiency for scattering problems.

We shall now present a few numerical results that illuminate the discussion in this paper. For other numerical results including tests of the order of convergence for the UWVF see for example [Ces96, CD98, HMK02, HKM04, GHP07].

5.1 Convergence under mesh refinement

An obvious question is whether the UWVF suffers from dispersion or pollution error. In general our experience and the results reported in [GHP07] show that it does indeed exhibit pollution error under h-refinement. However in some interesting cases this source of error appears to be absent or slowly growing. We now investigate a special example drawn from scattering theory. Consider an annulus $0 < a < r < b$ where $r = |\boldsymbol{x}|$. On the boundary $r = a$ we enforce

$$u(\boldsymbol{x}) = -\exp(ik\boldsymbol{x} \cdot \boldsymbol{d}), \quad |\boldsymbol{x}| = a, \tag{5.3}$$

via an appropriate choice of g in (1.2) with $\eta = k$ and $Q = 1$. The vector $\boldsymbol{d} = (1, 0)^T$. On the boundary $r = b$ we impose the approximate absorbing boundary condition

$$\frac{\partial u}{\partial r} - iku = 0, \quad r := |\boldsymbol{x}| = b, \tag{5.4}$$

by choosing $\eta = k$, $Q = 0$ and $g = 0$ in (1.2) there. In addition $A = n = 1$. The exact solution of (1.1), (5.3), (5.4) can be derived in terms of Bessel functions (it is not exactly the solution of the scattering problem due to the inexact absorbing boundary condition (5.4)). We can compute u via the Bessel function series expansion or approximate it via the UWVF for different choices of the wave number k.

A typical grid is shown in Fig. 1, left panel. Note that we actually use curvilinear triangles near the inner and outer boundary to produce an exact fit to the boundary. This is important when considering convergence with respect to the number of directions p and is necessary to avoid boundary approximation error.

If there is no dispersion error we expect the global error to depend only on hk. Thus a plot of error against $1/hk$ for a variety of meshes and wave numbers can be used to diagnose pollution error. In Fig. 1, right panel, we show results of such a study when $a = 1$, $b = 2$ and we fix the number of basis functions at $p = 9$ on every element in the grid. The mesh size h is varied to obtain convergence.

The onset of convergence occurs at the same point ($1/hk \approx 0.08$) suggestive of little pollution error, on the other hand convergence curves for low k are shifted to the right. This somewhat equivocal result does not suggest a strong effect of pollution, particularly for high k.

However care should be taken in extrapolating far from this result. The circle in 2D is a very special (convex) smooth scatterer. More important, we always choose the incident field $\exp(ikx_1)$ to be one of the plane wave basis functions. A tentative conclusion may be that it is possible in some cases to avoid dispersion error by a careful choice of directions in the basis.

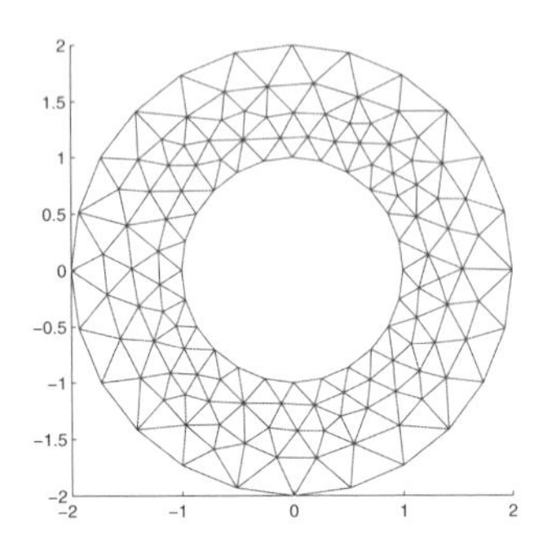

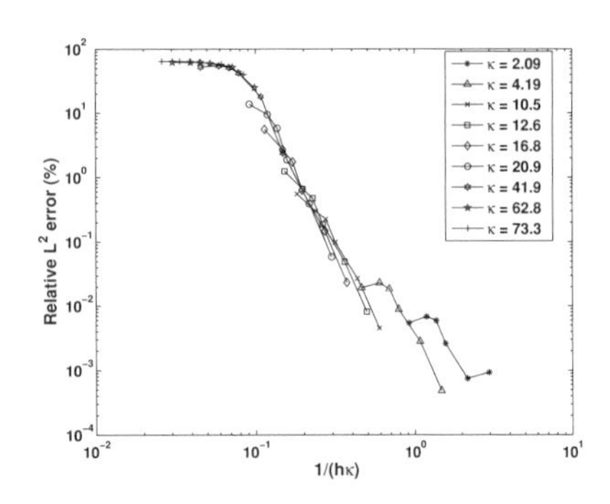

Fig. 1. Left: An example of the triangular mesh used in this study. Right: The relative $L^2(\Omega)$ error of the solution agaist $1/hk$ where h is the diameter of the largest element in the mesh and k is the wave number. The onset of convergence (i.e. the point where the graph turns down) is independent of hk suggesting that the method does suffer from pollution error for this problem.

The results in Fig. 1 are consistent with an order of approximation of $O(h^5)$ whereas our theory, with $p = 9$ predicts $O(h^4)$ convergence. This suggests a loss of optimality in the error estimates, and that a new method of proof is needed.

5.2 Choice of the coupling parameter η

The choice of the coupling parameter η on interior edges of the mesh is not a priori fixed. Cessenat and Després [CD98] suggest the choice $\eta = \kappa$ (or $\eta = k$ since $A = n = 1$ in our examples) since this gives rise to typical coupling across inter-element boundaries that is the frequency domain analogue of upwind (or characteristic) coupling in the time domain. Gittelson et al [GHP07] suggest that coupling parameters involving the mesh size and involving extra stabilization terms give better solutions than the classical choice in the UWVF. In this section we investigate the optimal choice of the coupling parameter. We again use the annular problem of the previous section that approximates scattering by a disk. We fix a mesh as shown in Fig. 1 (top left) and fix P. Then we vary η on internal edges and report the behavior of the global relative $L^2(\Omega)$ error for two different choices of k: $k = 0.0419$ with $P = 5$ or $k = 73.3038$ and $P = 27$. To remove the influence of changes in P, we use a constant number of directions on all triangles, and so use a more uniform mesh as shown in Fig. 2 and we also show the results of this study in the same figure.

The results suggest that the choice of η is very important and that even within the standard UWVF framework a more general choice than that used by Cessenat and Després is useful. At the low wave number of $k = 0.0419$ the choice $\eta = k$ gives a error of 1.8%. Whereas a substantially larger choice of $\eta = 50k \approx 2.09...$ gives an error of 0.1711%. However at a higher wave number (i.e. when the wavelength is short compared to the triangle diameter) the choice $\eta = k$ gives an error of 0.87% and the optimal choice $\eta = 1.9k$ gives an error of 0.84%. In the two examples here, for each fixed k, the condition number of the problem increases with η and so the optimal choice of η is not due to minimizing ill-conditioning.

As we expect, for a fixed mesh and higher wave number the choice of $\eta = k$ works well. For lower wave number or under h-convergence it is worthwhile to consider a different choice of η. This needs further practical and theoretical study.

5.3 Convergence under increasing number of directions

The density of X_h in X as $P \to \infty$ for fixed h (proved in one case for Maxwell's equations in 3D in [CN03] and following from the estimates of Melenk in 2D [Mel95]) show that for a fixed grid $u_h \to u$ in $L_2(\Omega)$ as $P \to \infty$ (provided the grid does not introduce any geometric approximation).

This is particularly useful if results are required for several different wave numbers k. We can use a single mesh and vary P on each element to provide a solution for each k. It is therefore of interest to consider convergence with respect to P refinement.

In our first investigation of this case we want to minimize the effect of the grid on the solution and so choose a uniform triangulation of the square (see Fig. 3, left panel) to reduce any effects of geometry. We choose $\Omega = [-1/2, 1/2]^2$ and the exact solution is $u(\boldsymbol{x}) = i/4\, H_0^{(1)}(|\boldsymbol{x} - \boldsymbol{y}|)$ where $\boldsymbol{y} = (-3/4, 0)^T$ and $H_0^{(1)}$ is the Hankel function of first kind and order zero. We are interested to test convergence as a function of P/hk where we now set h to be the length of the longest side of the triangles. In Fig. 3 (right panel) we plot the relative $L_2(\Omega)$ error as a function of $P/(hk)$ for various k, P and h using uniform submeshes of the mesh shown in the left panel of Fig. 3. It is clear that for this solution the choice $P = Chk$ gives an increasingly accurate solution as k increases. For example, in our experiment, the choice $P = 1.4hk$ gives a global L^2 relative error of approximately 10% when $k = 20$, approximately 1%

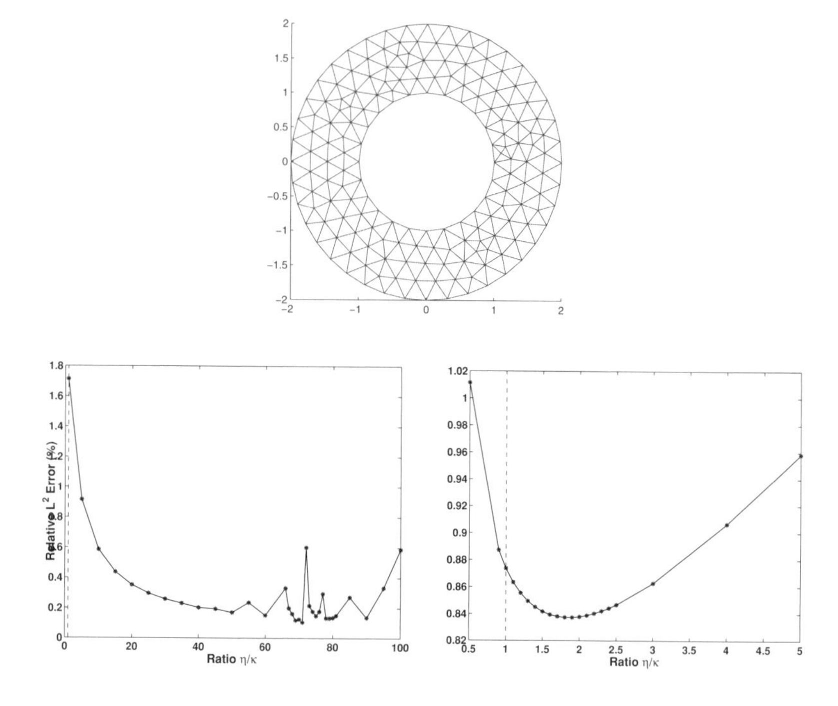

Fig. 2. Error versus the ratio of impedance parameter to wave number η/κ for $k = 0.0419$ and $P = 5$ (bottom left), and $k = 73.3038$ and $P = 27$ (bottom right). The dotted line in each figure marks the choice $\eta = k$. We also show the mesh used in this study in the top frame. The results show that the choice $\eta = k$ is close to optimal for higher frequencies, but that for lower frequencies it can be desirable to use η much larger than k.

when $k = 40$ and better than 0.01% for $k \geq 80$. The choice $P = 1.1hk$ provides less than 0.01% error when $k = 160$.

Thus, at least in this case, it seems that $P = Chk$ is a good choice for large k (the choice of C depends on the desired error). We now want to compare the density of unknowns in this case with a more standard difference method. Usually the number of grid points per wave length is an important parameter to assess a finite element or finite difference method. The UWVF concentrates many unknowns on an element so this has to be taken into account in comparing the density of the number of degrees of freedom (DOF). In the numerical experiment of this section, having a uniform mesh on a square, the total number of DOF is $2P/h^2$. Using our choice of $P = Chk$, the total number of DOF is $2Ck/h$. This

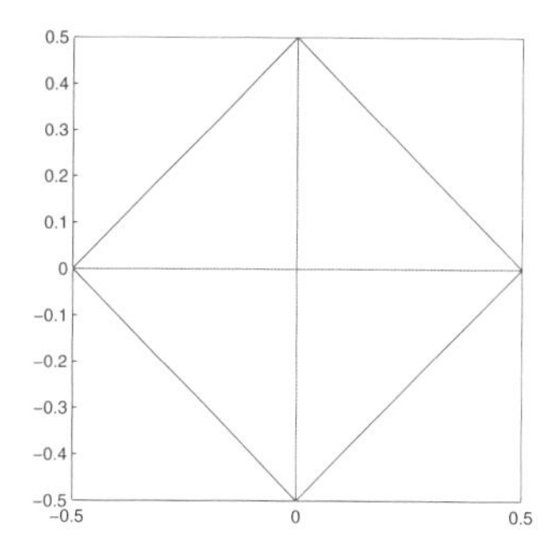

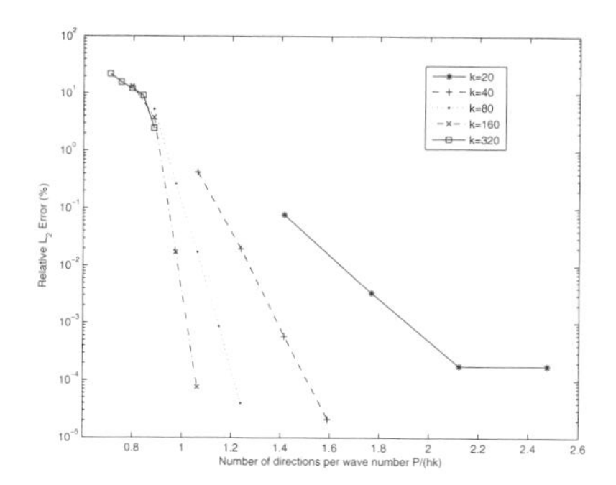

Fig. 3. Here we show the relative global L_2 error as a function of $P/(hk)$ for various choices of k on a a family of fixed meshes obtained by uniform subdivision of the mesh in the left panel. In the right we show the results for various k and P. Clearly, as k increases the choice $P = Chk$ with $C \approx 1$ gives better and better approximation of the solution of our test problem since the convergence curves become increasingly steep as k increases.

corresponds to approximately $\sqrt{2Ck}/\sqrt{h}$ DOF along one dimension to discretize $k/2\pi$ wave lengths (the wavelength of the solutions is $2\pi/k$) and hence to $2\pi\sqrt{2C}/\sqrt{hk}$ DOF per wavelength. Clearly as k increases it is possible, in principle, to obtain high accuracy at less than one DOF per wavelength.

The experiment discussed so far in this section shows that, at least for some fields of the type encountered in scattering calculations, the above estimation of the number of degrees of freedom per wavelength is valid, so that the UWVF can achieve very low grid densities in some cases.

We have also performed a numerical experiment on the more realistic annular "scattering" problem used in the previous sections. The results are shown in Fig. 4. Results are broadly similar to those for the mesh of right triangles. For higher values of k, the choice $P/(hk) = C$ will give high accuracy for moderate values of C. We perform an initial test of this idea using the meshes from Section 5.1 and varying P choosing $P = round(Chk)$ element by element with $C = 1.5$ where h is take to be the longest edge of the local triangle. Results are shown in Table 1. Clearly this choice of C over estimates P for higher wave numbers (the error decreases with k and the condition number rises) and under estimates for lower wave numbers (this is not surprising since the estimate $P = Chk$ is only expected to be a good heuristic for large k). In the next section we improve the choice of the rule for P.

Table 1. *Results for the annular 2D scattering problem when the mesh size is fixed ($h = 0.5227$) and $p = Chk$ with $C = 1.5$.*

k	error (%)	P	#DOF	cond(D)
4.1888	26.7110	3	708	6.2688
6.2832	6.4191	5	1180	108.6251
8.3776	1.5420	7	1652	$2.9538{\cdot}10^3$
10.4720	1.8439	8	1888	$1.1021{\cdot}10^4$
12.5664	0.6333	10	2360	$2.1493{\cdot}10^5$
14.6608	0.5179	11	2596	$1.4051{\cdot}10^5$
16.7552	0.2262	13	3068	$2.9585{\cdot}10^6$
18.8496	0.1032	15	3540	$3.4435{\cdot}10^7$
20.9440	0.1205	16	3776	$2.2501{\cdot}10^8$
23.0383	0.0703	18	4248	$3.1356{\cdot}10^9$
25.1327	0.0232	20	4720	$7.5466{\cdot}10^{10}$
27.2271	0.0227	21	4956	$4.5693{\cdot}10^{10}$
29.3215	0.0118	23	5428	$7.8641{\cdot}10^{11}$
31.4159	0.0072	25	5900	$1.5650{\cdot}10^{13}$

5.4 Error control at higher wave numbers

We have seen in the previous section that for the annular "scattering" problem, the choice $P = round(Chk)$ with $C = 1.5$ does control the error at higher wave numbers but that the condition number rises, and the error is not uniform as k varies (this choice of C over estimates the number of directions needed for higher k and underestimates P for lower k), as might be expected from Fig. 3 right panel. We can improve the prediction of P by using a slightly more general formula

$$P = round(C_1 h\kappa + C_2)$$

element by element (recalling again that $\kappa = k$ in our examples). Using the results from several calculations using control of the condition number to choose the number of directions per element, and choosing a desired accuracy level, we can then compute an "optimized" pair of coefficients C_1 and C_2 that should give the desired error (in our case we

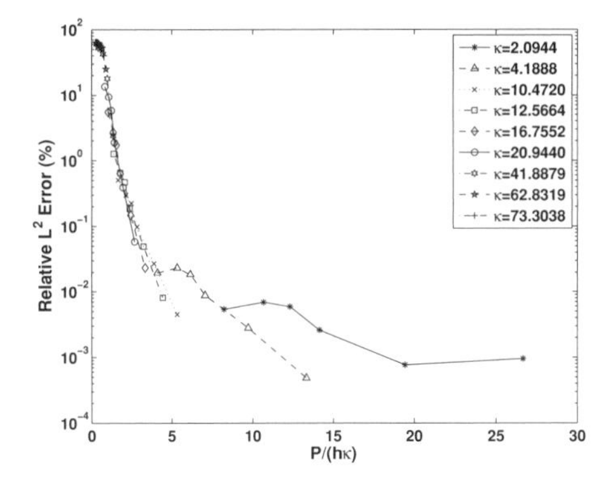

Fig. 4. Here we show the relative global L_2 error as a function of $P/(hk)$ for various choices of k on different meshes so that h, P and k are varied for the annular scattering problems. Although less clear than the results in Fig. 3 (perhaps because the mesh is no longer uniform), this figure again supports the conclusion that as k increases the choice $P = Chk$ for moderate values of C gives better and better approximation of the solution of our test problem since the convergence curves become increasingly steep as k increases.

choose 1% relative error). The linear fit gives an element by element formula

$$P = round\,(1.2485h\kappa + 3.7917)\,, \tag{5.5}$$

where h is maximum length of the triangle edges. Results are shown in tables 2 and 3. It is clear that the optimized formula produces a more uniform error for a larger range of k than in Table 1, and that the condition number is better controlled. Nevertheless the error is still smaller for higher k.

6 Conclusion

Plane waves can be used as the basis for several methods in computational acoustics. The work of Gittelson et al. [GHP07] shows that these methods generally do suffer from pollution error when a mesh or h-refinement strategy is used. Our results suggest that in some cases this can be avoided or decreased by a special choice of directions of the plane waves. A more interesting general refinement strategy is to increase the number of plane waves per element as the wave number increases in order to control accuracy. Our results suggest that, in our 2D "scattering" example, it suffices to take the number of directions to be an affine

Table 2. *Results for the annular 2D scattering problem for the coarsest mesh in our study when the number of elements is 236 using the optimized relationship between P, k and h element by element.*

k	error (%)	Range of P	# DOF	cond(D)
2.0944	1.3442	4..5	1124	$1.2261 \cdot 10^3$
4.1888	1.8478	5..7	1305	$4.5798 \cdot 10^3$
10.4720	1.0893	6..11	1946	$7.4586 \cdot 10^3$
12.5664	1.1766	7..12	2150	$8.0730 \cdot 10^3$
16.7552	0.7322	8..15	2566	$2.5783 \cdot 10^4$
20.9440	0.5234	9..17	2990	$5.2705 \cdot 10^4$
41.8879	0.2928	14..31	8076	$1.0131 \cdot 10^7$

Table 3. *Results for the annular 2D scattering problem for the finest mesh used in this study when the number of elements is 946 using the optimized relationship between P, k and h element by element.*

k	error (%)	Range of P	# DOF	cond(D)
2.0944	0.8846	4	3784	$2.2345 \cdot 10^3$
4.1888	0.7231	4..5	4668	$1.2648 \cdot 10^3$
10.4720	1.7098	5..7	5667	$5.0091 \cdot 10^3$
12.5664	1.3961	5..7	6098	$2.8137 \cdot 10^3$
16.7552	1.0064	6..8	6951	$7.2533 \cdot 10^3$
20.9440	0.9719	7..10	7781	$6.6065 \cdot 10^3$
41.8879	0.3810	9..15	11959	$4.1638 \cdot 10^4$

function of hk when k is large. This heuristic approach results in a decreasing number of degrees of freedom per wavelength as k increases and shows that such methods become more efficient at higher wave number (notwithstanding the total number of degrees of freedom needed to gain a fixed accuracy does increase with k so the UWVF or similar methods are not a panacea). Clearly a better theoretical understanding of this heuristic is needed. In particular, p-version approximation estimates for plane waves in 3D are needed.

Our results are for one simple convex and smooth scatterer. In the presence of corners the solution will loose regularity and small elements are needed to gain accuracy [HGA08]. In this case plane wave methods do not offer an advantage over standard finite element methods near the corner. This motivates the need for a scheme that combines both plane wave basis functions on some (larger) elements away from singularities with a regular finite element method on (smaller) elements near the singularity. Such methods are easy to define using discontinuous Galerkin techniques, and one that combines the Raviart–Thomas elements and plane waves within the UWVF framework is currently under development.

Acknowledgments

The research of PM is partially supported by a grant from the US AFOSR under grant number F49620-02-1-0071. We are indebted to Bruno Després for pointing out the connection between the UWVF and the least squares method.

Bibliography

[AMM06] M. Ainsworth, P. Monk & W. Muniz (2006). Dispersive and dissipative properties of discontinuous Galerkin methods for the wave equation, *J. Sci. Comput.* **27**, 5–40.

[ACL07] S. Arden, S. Chandler-Wilde & S. Langdon (2007). A collocation method for high-frequency scattering by convex polygons, *J. Comput. Appl. Math.* **204**, 334–343.

[BM97] I. Babuška & J. Melenk (1997). The partition of unity method. *Int. J. Numer. Meth. Eng.* **40**, 727–758.

[BS97] I. Babuška & S. Sauter (1997). Is the pollution effect of the FEM avoidable for the Helmholtz equation considering high wave numbers?, *SIAM J. Numer. Anal.* **34**, 2392–423.

[BG07] O. Bruno & C. Geuzaine (2007). An $o(1)$ integration scheme for three-dimensional surface scattering problems, *J. Comput. Appl. Math.* **204**, 463–476.

[BM07] A. Buffa & P. Monk (2007). Error estimates for the Ultra Weak Variational Formulation of the Helmholtz equation, submitted.

[CC04] Q. Caryol & F. Collino (2004). Error estimates in the fast multipole method for scattering problems part 1: Truncation of the Jacobi–Anger series, *ESAIM: Mathematical Modeling and Numerical Analysis* **38**, 371–394.

[Ces96] O. Cessenat (1996). *Application d'une nouvelle formulation variation-*

nelle aux équations d'ondes harmoniques. Problèmes de Helmholtz 2D et de Maxwell 3D., PhD thesis (Université Paris IX Dauphine).

[CD98] O. Cessenat & B. Després (1998). Application of the ultra-weak variational formulation of elliptic PDEs to the 2-dimensional Helmholtz problem, *SIAM J. Numer. Anal.* **35**, 255–299.

[CD03] O. Cessenat & B. Després (2003). Using plane waves as base functions for solving time harmonic equations with the Ultra Weak Variational Formulation, *J. Comput. Acoustics* **11**, 227–238.

[CL07] S. Chandler-Wilde & S. Langdon (2007). A Galerkin boundary element method for high frequency scattering by convex polygons, *SIAM J. Numer. Anal.* **45**, 610–640.

[Coh02] G. Cohen (2002). *Higher-order numerical methods for transient wave equations* (Springer, Berlin).

[CJ96] G. Cohen & P. Joly (1996). Construction and analysis of fourth-order finite difference schemes for the acoustic wave equation in nonhomogeneous media, *SIAM J. Numer. Anal.* **33**, 1266–1302.

[CR98] D. Colton & R. Kress (1998). *Inverse Acoustic and Electromagnetic Scattering Theory*, 2nd edition (Springer-Verlag, New York).

[CN03] D. Colton & P. Monk (2003). Herglotz wave functions in inverse electromagnetic scattering theory, in *Topics in Computational Wave Propagation: Direct and Inverse Problems* (M. Ainsworth, P. Davies, D. Duncan, P. Martin & B. Rynne, eds), Lecture Notes in Computational Science and Engineering **31**, 267–294 (Springer, Berlin).

[DM07] E. Darrigrand & P. Monk (2007). Coupling of the Ultra-Weak Variational Formulation and an integral representation using a Fast Multipole Method in electromagnetism, *J. Comput. Appl. Maths* **204**, 400–407.

[Dem03] L. Demkowicz (2003). *hp*-adaptive finite elements for time-harmonic Maxwell equations, in *Topics in Computational Wave Propagation: Direct and Inverse Problems* (M. Ainsworth, P. Davies, D. Duncan, P. Martin & B. Rynne, eds), Lecture Notes in Computational Science and Engineering **31** (Springer, Berlin).

[Gab07] G. Gabard (2006). Discontinuous Galerkin methods with plane waves for the displacement-based acoustic equation, *Int. J. Numer. Meth. Eng.* **66**, 549–569.

[Gab08] G. Gabard (2008). Discontinuous Galerkin methods with plane waves for time-harmonic problems, to appear.

[GBR05] C. Geuzaine, O. Bruno & F. Reitich (2005). On the $o(1)$ solution of multiple-scattering problems, *IEEE Trans. Mag.* **41**, 1488–1491.

[GHP07] C. Gittelson, R. Hiptmair & I. Perugia (2007). Plane wave discontinuous Galerkin methods, Preprint NI07088-HOP, Isaac Newton Institute Cambridge, (Cambridge, UK), `http://www.newton.cam.ac.uk/preprints/NI07088.pdf`.

[GO77] D. Gottlieb & S. Orszag (1977). *Numerical Analysis of Spectral Methods: Theory and Applications*,Regional Conference Series in Applied Mathematics **26** (SIAM, Philadelphia).

[HP08] B. Heubeck & C. Pflaum (2008). Convergence analysis of non-conforming trigonometric finite wave elements, to appear in *J. Comput. & Appld Maths*.

[HGA08] T. Huttunen, P. Gamallo & R. Astley (2008). Comparison of two wave element methods for the Helmholtz problem, to appear in *Comm. Numer. Methods Engng*.

[HKM04] T. Huttunen, J. Kaipio & P. Monk (2004). The perfectly matched layer for the ultra weak variational formulation of the 3D Helmholtz equation, *Int. J. Numer. Meth. Eng.* **61**, 1072–1092.

[HKM08] T. Huttunen, J. Kaipio & P. Monk (2008). An ultra-weak method for acoustic fluid-solid interaction, *J. Comput. Appl. Maths* **213**, 166–185.

[HMM07] T. Huttunen, M. Malinen & P. Monk (2007). Solving Maxwell's equations using the Ultra Weak Variational Formulation, *J. Comput. Phys.* **223**, 731–758.

[HM07] T. Huttunen & P. Monk (2007). The use of plane waves to approximate wave propagation in anisotropic media, *J. Comput. Maths* **25**, 350–367.

[HMCK04] T. Huttunen, P. Monk, F. Collino & J. Kaipio (2004). The Ultra Weak Variational Formulation for elastic wave problems, *SIAM J. Sci. Comput.* **25**, 1717–1742.

[HMK02] T. Huttunen, P. Monk & J. Kaipio (2002). Computational aspects of the Ultra Weak Variational Formulation, *J. Comput. Phys.* **182**, 27–46.

[Ihl98] F. Ihlenburg (1998). *Finite Element Analysis of Acoustic Scattering*, Applied Mathematical Sciences **132** (Springer, Berlin).

[IB95] F. Ihlenburg & I. Babuv̌ka (1995). Finite element solution of the Helmholtz equation with high wavenumber Part I: The h-version of the FEM, *Computers Math. Applic.* **30**, 9–37.

[LBPT05] O. Laghrouche, P. Bettess, E. Perrey-Debain, & J. Trevelyan (2005). Wave interpolation finite elements for Helmholtz problems with jumps in the wave speed, *Comput. Meth. Appl. Mech. Eng.* **194**, 367–381.

[LT08] J. Li & X. Tu (2008). Convergence analysis of a balancng domain decomposition methods for solving interior Helmholtz equations, preprint.

[Mel95] J. Melenk (1995). On generalized finite element methods, PhD thesis, (University of Maryland, College Park, MD).

[MB96] J. Melenk & I. Babuv̌ka (1996). The partition of unity finite element method: Basic theory and applications, *Comput. Meth. Appl. Mech. Eng.* **139**, 289–314.

[MW98] P. Monk & D. Wang (1999). A least squares method for the Helmholtz equation, *Comput. Meth. Appl. Mech. Eng.* **175**, 121–136.

[Per06] E. Perrey-Debain (2006). Plane wave decomposition in the unit disc: Convergence estimates and computational aspects, *J. Comput. Appld Maths* **193**, 140–156.

[PTB02] E. Perrey-Debain, J. Trevelyan & P. Bettess (2002). Use of wave boundary elements to extend discrete methods in acoustics computations to higher frequencies, Proceedings of WCCM-V, preprint.

[SW01] I. Sloan & R. Womersley (2001). Extremal systems of points and numerical integration on the unit sphere, Appld Maths Reports AMR 15-01 (University of New South Wales, Sydney, Australia), `http://web.maths.unsw.edu.au/~rsw/Sphere/Extremal`.

[Sto98] M. Stojek (1998). Least-squares Trefftz-type elements for the Helmholtz equation, *Int. J. Numer. Meth. Eng.* **41**, 831–849.

[TF06] R. Tezaur & C. Farhat (2006). Three-dimensional discontinuous Galerkin elements with plane waves and Lagrange multipliers for the solution of mid-frequency Helmholtz problems, *Int. J. Numer. Meth. Eng.* **66**, 796–815.

7

Boundary integral methods in high frequency scattering

Simon N. Chandler-Wilde

Department of Mathematics
University of Reading
Whiteknights, P.O. Box 220, Berkshire RG6 6AX
United Kingdom
Email: S.N.Chandler-Wilde@reading.ac.uk

Ivan G. Graham

Department of Mathematical Sciences
University of Bath
Claverton Down, Bath BA2 7AY
United Kingdom
Email: I.G.Graham@bath.ac.uk

Abstract

In this article we review recent progress on the design, analysis and implementation of numerical-asymptotic boundary integral methods for the computation of frequency-domain acoustic scattering in a homogeneous unbounded medium by a bounded obstacle. The main aim of the methods is to allow computation of scattering at arbitrarily high frequency with finite computational resources.

1 Introduction

There is huge mathematical and engineering interest in acoustic and electromagnetic wave scattering problems, driven by many applications such as modelling radar, sonar, acoustic noise barriers, atmospheric particle scattering, ultrasound and VLSI. For time harmonic problems in infinite domains and media which are predominantly homogeneous, the boundary element method is a very popular solver, used in a number of large commercial codes, see e.g. [CSCVHH04]. In many practical applications the characteristic length scale L of the domain is large compared to the wavelength λ. Then the small dimensionless wavelength λ/L induces oscillatory solutions, and the application of conventional (piecewise polynomial) boundary elements for this multiscale problem yields full matrices of dimension at least $N = (L/\lambda)^{d-1}$ (in $\mathbb{R}^d$). (Domain finite elements lead to sparse matrices but require even larger N.) Since this "loss of robustness" as $L/\lambda \to \infty$ puts high frequency problems outside the reach of many standard algorithms, much recent research has been devoted to finding more robust methods.

One approach is to seek faster implementations of standard methods. Fast multipole methods have allowed conventional BEM solutions for much larger N (e.g. [Dar02, DH04]), but it remains impossible to compute with L/λ much beyond a few hundred in 3D. To allow larger L/λ, a highly promising new direction is the development of "hybrid" algorithms, which incorporate asymptotic information about the oscillation of the solution into the approximation space [ANZ94, ANZ95, BGMR04, GK04, LC06, CLR04, PLBT04, ACL07, CL07, DGS07, BG07, ER06, HTH06]. Initial experiments using geometric-optics type approximations on simple model problems indicate the possibility of delivering almost uniform accuracy for N fixed as $L/\lambda \to \infty$. This review will explain the key ideas behind these methods and the mathematical tools which have been so-far developed for their analysis. We also highlight some important open problems which are the focus of current research in this very active area. Another approach to high frequency problems, involving the solution of appropriate limiting problems, is dealt with elsewhere in this volume [MR08].

Throughout this review we will focus on the specific physical situation of time harmonic acoustic scattering ($\mathrm{e}^{-\mathrm{i}\omega t}$ time dependence for some $\omega > 0$); indeed for most of the paper on the case of a sound soft obstacle. This focus is made partly for brevity and simplicity (the algorithms we discuss should generalise to other boundary conditions and to, e.g., elastic and electromagnetic waves), but also because most development of algorithms and most analysis of those algorithms has focused so far on this simplest case. Thus, we suppose an incident plane wave $u^I(x) = \exp(\mathrm{i}kx \cdot \hat{a})$, $x \in \mathbb{R}^d$, with direction given by the unit vector $\hat{a}$ and k denoting wavenumber ($k = 2\pi/\lambda = \omega/c$, where c is the wave speed), is scattered by a bounded object $\Omega \subset \mathbb{R}^d$ to produce a radiating scattered wave u^S. The total wave $u = u^I + u^S$ satisfies the Helmholtz equation:

$$\Delta u + k^2 u = 0 \quad \text{in} \quad D := \mathbb{R}^d \setminus \Omega \quad (d = 2 \text{ or } 3). \tag{1.1}$$

Let $\Phi(x, y)$ denote the standard free-space fundamental solution of the

Helmholtz equation, given, in the 2D and 3D cases, by

$$\Phi(x,y) := \begin{cases} \frac{\mathrm{i}}{4} H_0^{(1)}(k|x-y|), & d=2, \\ \dfrac{\exp(\mathrm{i}k|x-y|)}{4\pi|x-y|}, & d=3, \end{cases} \tag{1.2}$$

for $x, y \in \mathbb{R}^d$, $x \neq y$, where $H_\nu^{(1)}$ denotes the Hankel function of the first kind of order zero. Then in the simplest Dirichlet case ($u = 0$ on the boundary Γ), starting from Green's representation theorem (see e.g. [CL07] for details in the general Lipschitz case) we obtain

$$u(x) = u^I(x) - \int_\Gamma \Phi(x,y) \frac{\partial u}{\partial n}(y)\, \mathrm{d}s(y), \quad x \in D, \tag{1.3}$$

and the scattering problem can be reformulated as the boundary integral equation (see e.g. [CK83])

$$\frac{\partial u}{\partial n}(x) + 2\int_\Gamma \left(\frac{\partial \Phi(x,y)}{\partial n(x)} - \mathrm{i}\eta \Phi(x,y) \right) \frac{\partial u}{\partial n}(y)\, \mathrm{d}s(y) = f(x), \quad x \in \Gamma. \tag{1.4}$$

Here $\partial/\partial n$ denotes the normal derivative (outward from Ω), $\eta > 0$ is a *coupling parameter* (which ensures that (1.4) is well-posed),

$$f(x) := 2\frac{\partial u^I}{\partial n}(x) - 2\mathrm{i}\eta u^I(x), \quad x \in \Gamma,$$

and $\partial u/\partial n$ is to be determined. Standard boundary element methods approximate the whole (oscillatory) $\partial u/\partial n$ by (piecewise) polynomials. By contrast the *hybrid methods* which we shall discuss in the following section employ asymptotic analysis to obtain analytic information about the oscillations in $\partial u/\partial n$. This information is then exploited directly in the numerical method: only slowly-varying components are approximated and this yields a method which is more "robust" as the frequency increases.

Throughout the review we shall make use of the single-layer, double-layer, adjoint double-layer and hypersingular operators S, D, D' and H, defined respectively by:

$$S\psi = 2\int_\Gamma \Phi(x,y)\psi(y)ds(y), \qquad D\psi = 2\int_\Gamma \frac{\partial \Phi(x,y)}{\partial n(y)}\psi(y)ds(y)$$

$$D'\psi = 2\int_\Gamma \frac{\partial \Phi(x,y)}{\partial n(x)}\psi(y)ds(y), \qquad H\psi = 2\int_\Gamma \frac{\partial^2 \Phi(x,y)}{\partial n(x)\partial n(y)}\psi(y)ds(y).$$

The particular equation (1.4) can then be written as

$$A'v := (I + D' - \mathrm{i}\eta S)v = f, \quad \text{where} \quad v = \partial u/\partial n. \tag{1.5}$$

This integral equation formulation is well known and is attributed to Burton and Miller in [CK83]. There one can find a proof that (1.4) is uniquely solvable in $C(\Gamma)$ in the case when Γ is sufficiently smooth (a proof of well-posedness in $L^2(\Gamma)$, indeed in the Sobolev space $H^s(\Gamma)$ for $-1 \leq s \leq 0$, for the case of general Lipschitz Γ is given recently in [CL07]). It is a *direct* integral equation formulation, meaning that it arises directly from applying Green's theorems to the solution of the scattering problem, so that the unknown in the integral equation is the unknown part of the Cauchy data of problem (1.1) on the boundary. A closely-related *indirect* formulation, due to Brakhage & Werner [BW65], Leis [Lei65], and Panič [Pan65], obtained by seeking the solution as a linear combination of single- and double-layer potentials with some unknown density ϕ, can be written in operator form as

$$A\phi = (I + D - \mathrm{i}\eta S)\phi = -2u^I|_\Gamma. \tag{1.6}$$

Note that equations (1.6) and (1.5) are intimately related; indeed A' is the formal adjoint of A, as a consequence of which, as operators on $L^2(\Gamma)$, A and A' have the same spectrum, norm and condition number (see [CGLL07]). We shall focus more on (1.5) in this review. As noted by Bruno et al. [BGMR04], this equation seems better behaved in the high frequency regime, since its solution is the normal derivative on Γ of the solution of the original scattering problem, while it can be shown that the solution ϕ of (1.6) is the difference between solutions to interior and exterior boundary value problems. For this reason the solution of (1.5) is less oscillatory and its high frequency behaviour is better understood, especially for convex scatterers.

An important issue for (1.6) and (1.5), which we will address in section 3, is how to choose the coupling parameter $\eta > 0$ optimally, e.g. to minimise the condition number of A'. Discussion of this issue goes back to Kress and Spassow [KS83] (and see [Kre85, Ami90, Ami93, Gie97, BS06, DGS07, BS07, CM08, CGLL07]). We will see in section 3 that a correct k-dependent choice is essential in the high frequency limit.

Equation (1.5) is a second-kind integral equation which determines the unknown solution $v := \frac{\partial u}{\partial n}$, and there is a huge literature on equations of this form. When the boundary Γ is sufficiently smooth (C^1 is sufficient [FJR78]) the integral operators D' and S in (1.5) are compact on standard function spaces, so that A' is a compact perturbation of the identity operator. Using classical arguments based on this property, one can show that standard numerical techniques like Galerkin and collocation methods using piecewise polynomial basis functions lead to uniquely determined numerical solutions v_N satisfying quasi-optimal error estimates of the form

$$\|v - v_N\| \leq C \inf_{\phi_N \in \mathcal{S}_N} \|v - \phi_N\|, \tag{1.7}$$

where $\mathcal{S}_N$ denotes the finite-dimensional approximation space being used (and N is the discretisation parameter, e.g. the dimension of the space $\mathcal{S}_N$). More precisely, for properly-designed Galerkin method and collocation methods, these classical arguments (e.g. Atkinson [Atk97]) tell us that there exists a $C > 0$ and $N_0 > 0$ such that (1.7) holds for all $N \geq N_0$ (see subsection 3.1 for a little more detail).

Based on (1.7) one can think of the numerical analysis of robust methods for scattering problems as requiring research on three related questions:

Q1 The design of good, k-dependent, finite-dimensional approximation spaces $\mathcal{S}_N$, so that the best approximation error $\inf_{\phi_N \in \mathcal{S}_N} \|v - \phi_N\|$ is growing as slowly as possible as $k \to \infty$. These spaces will normally depend on k and so we denote them $\mathcal{S}_{N,k}$.

Q2 The proof of sharp estimates for the dependence of the "stability constant" C in (1.7) on k, hopefully showing that these again indicate boundedness or mild growth as $k \to \infty$.

Q3 The design of good methods of implementing the numerical methods using the optimal approximation spaces in item 1; ideally show that these are realisable in a computation time which remains bounded as $k \to \infty$.

For **Q1**, an "ideal" aim might be that when v is the solution of (1.4), the best approximation error should remain constant for each fixed N as $k \to \infty$. Recent results on the analysis of this problem are given in section 2.

For **Q2**, the classical error analysis results for second-kind integral equations tell us that (1.13) holds for all sufficiently large N ($N \geq N_0$). However, because the wavenumber k appears non-linearly inside the kernel of the operator A' in (1.5), they give us no clear quantitative information on either: (i) how, for fixed N, the constant C depends on the parameter k; or (ii) how, for fixed C, the threshold N_0 depends on k. An alternative method of analysis starts from the following variational formulation of (1.5):

$$\text{Seek } v \in L^2(\Gamma) \quad \text{such that} \quad a(v,w) = (f,w)_{L^2(\Gamma)} \tag{1.8}$$

for all $w \in L^2(\Gamma)$, where $a(v,w) = (A'v,w)_{L^2(\Gamma)}$. Then the standard abstract theory of variational methods shows, for example, that, provided a satisfies for all $v,w \in L^2(\Gamma)$ the two conditions

$$\left.\begin{array}{rll} |a(v,w)| & \leq B\|v\|_{L^2(\Gamma)}\ \|w\|_{L^2(\Gamma)} & \text{(continuity)} \\ |a(v,v)| & \geq \alpha\|v\|^2_{L^2(\Gamma)} & \text{(coercivity)} \end{array}\right\} \tag{1.9}$$

for some positive constants B and α, then the equation (1.8) is uniquely solvable. Moreover if the Galerkin (variational) method of approximation is applied to (1.8) in *any* finite dimensional subspace $\mathcal{S}_{N,k} \subset L^2(\Gamma)$, i.e. seek $v_N \in \mathcal{S}_{N,k}$ such that

$$a(v_N, w_N) = (f, w_N)_{L^2(\Gamma)}, \quad \text{for all} \quad w_N \in \mathcal{S}_{N,k}, \tag{1.10}$$

then we have the error estimate (1.7) with $C = B/\alpha$. Therefore one potential way to answer **Q2** is to show that a is coercive and to estimate the dependence of B and α on k. Results on this and related problems are discussed in subsection 3.

Finally, with regard to **Q3**, in subsection 4 we discuss recent work on the key implementation issue of computation of the oscillatory integrals which arise in the assembly of stiffness matrices arising from hybrid methods. We also discuss briefly linear algebra issues relating to the fast solution of the dense linear systems arising from hybrid methods.

Before continuing, we would like to explore, a little more carefully, reasonable ways of measuring the accuracy of v_N. Since v itself depends on k, rather than controlling the absolute error in some norm (as in (1.7)), it would seem more sensible to control relative error measures such as

$$\frac{\|v - v_N\|_{L^2(\Gamma)}}{\|v\|_{L^2(\Gamma)}} \quad \text{or} \quad \frac{\|v - v_N\|_{L^2(\Gamma)}}{\|v^I\|_{L^2(\Gamma)}},$$

where $v^I = \partial u^I/\partial n$. The attraction of the second of these is that the behaviour of $\|v^I\|_{L^2(\Gamma)}$ is clear, in particular it grows proportional to k as $k \to \infty$ for an arbitrary obstacle. For smooth convex obstacles, for which we know (via the Kirchhoff approximation) that $v \approx 0$ on the shadow side, $v \approx 2v^I$ on the lit side, it is clear that $\|v\|_{L^2(\Gamma)}$ grows in proportion to $\|v^I\|_{L^2(\Gamma)}$, and so in proportion to k. Thus, for the second measure of error, and also for the first in the convex case, controlling the above measures of error, for a fixed obstacle, amounts to controlling

$$k^{-1}\|v - v_N\|_{L^2(\Gamma)}. \tag{1.11}$$

A reasonable alternative is to take the view that the computation of $v = \partial u/\partial n$ is an intermediate step, and that the real goal is to compute u accurately in the domain D, by substituting the approximation v_N to $\partial u/\partial n$ into equation (1.3). Denoting by u_N the resulting approximation to u, we see that

$$u(x) - u_N(x) = \int_\Gamma \Phi(x,y)(v_N(y) - v(y))\,\mathrm{d}s(y), \quad x \in D. \tag{1.12}$$

In this context we may seek to control

$$\frac{\|u - u_N\|_{L^p(G)}}{\|u\|_{L^p(G)}} \quad \text{or} \quad \frac{\|u - u_N\|_{L^p(G)}}{\|u^I\|_{L^p(G)}}, \tag{1.13}$$

where $\|\cdot\|_{L^p(G)}$, for $1 \le p \le \infty$, is the standard L^p norm on some region $G \subset D$ (e.g. one might choose $p = 2$ or ∞, and, in the latter case, choose $G = D$ (as in (2.28) below)). Applying the Cauchy-Schwarz inequality to (1.12), we obtain the upper bound:

$$|u(x) - u_N(x)| \le c(x)\|v - v_N\|_{L^2(\Gamma)}, \quad c(x) := \left\{\int_\Gamma |\Phi(x,y)|^2\,\mathrm{d}s(y)\right\}^{1/2}. \tag{1.14}$$

Thus small relative error in u can be achieved by controlling $\|v - v_N\|_{L^2(\Gamma)}$. However, the value of $\|u^I\|_{L^p(G)}$ is independent of k, and, in the 3D case, $(c(x))^2 = (4\pi)^{-2}\int_\Gamma |x-y|^{-2}\,\mathrm{d}s(y)$ has a value independent of k, while, in 2D, $(c(x))^2 \sim (\pi/(8k))\int_\Gamma |x-y|^{-2}\,\mathrm{d}s(y)$ as $k \to \infty$. Thus to achieve small values for the measures of relative error (1.13) by controlling $\|v - v_N\|_{L^2(\Gamma)}$ one needs to ensure that

$$k^{-(3-d)/2}\|v - v_N\|_{L^2(\Gamma)} \tag{1.15}$$

is small. Of course, in the high frequency limit, especially in 3D ($d = 3$), this is a significantly stronger requirement than (1.11). We remark that the scaling by $k^{-(3-d)/2}$ in (1.15) is rather natural in that it makes the expression (1.15) dimensionless.

2 Hybrid approximation spaces

Instead of approximating $v := \partial u/\partial n$ in (1.4) directly by piecewise polynomials, the hybrid numerical-asymptotic methods which we are interested in here use approximations with the general form (where we highlight the dependence on k in the notation):

$$v(x,k) \approx \sum_{m=1}^{M} k \exp(\mathrm{i}k\gamma_m(x)) V_m(x,k), \quad x \in \Gamma, \tag{2.1}$$

with the phase functions $\gamma_m(x)$ chosen *a priori* and only the unknowns $V_m(x,k)$ approximated by piecewise polynomials. The key point is that asymptotic analysis can be used to determine the γ_m in such a way that the V_m are very much less oscillatory than the original $\partial u/\partial n$.

Some of the pioneering work in the development of hybrid boundary element methods for scattering problems was carried out by Abboud et. al. [ANZ94, ANZ95], who considered the problem (1.1), subject to the impedance boundary condition : $\frac{\partial u}{\partial n} + ikZu = 0$ on Γ, and formulated this as the boundary integral equation

$$-Hv + k^2 ZS(Zv) - \mathrm{i}kD'(Zv) - \mathrm{i}kZDv = g_k := -2\frac{\partial u_I}{\partial n} - 2\mathrm{i}kZu_I,$$

where $v = u|_\Gamma$, the restriction of u to Γ. The Galerkin discretisation of this integral equation yields a symmetric stiffness matrix and has no spurious frequencies provided $\Re Z > 0$. The authors argued (partly referring to earlier results [Gir82]) that, due to the oscillatory solution of (1.1), in general the conventional boundary element approximation v_h of this equation, using step-size h and polynomial degree p, would satisfy an error estimate $\|v - v_h\|_{L^2(\Gamma)} \leq C(k)(hk)^{p+1}$. Ignoring the unknown factor $C(k)$ this shows that in order to preserve accuracy as $k \to \infty$, we would require $h \sim k^{-1}$ and so, for integral equations on surfaces in $\mathbb{R}^d$, the number of degrees of freedom N would have to grow at least with $\mathcal{O}(k^{d-1})$. To remedy this, [ANZ94, ANZ95] suggested taking $M = 1$ and $\gamma_1(x) = x \cdot \hat{a}$ in (2.1), yielding

$$v(x,k) = k \exp(\mathrm{i}kx \cdot \hat{a})\, V(x,k), \tag{2.2}$$

and then approximating the unknown "slow variable" $V(\cdot, k)$ using conventional finite element methods. This may be thought of as a numerical implementation of the "Geometric Optics" or "Kirchhoff" approximation which assumes the phase of the scattered wave u^S in (1.1) to be the same as the phase of the incoming wave u^I. (The scaling factor k in (2.2) arises from the differentiation appearing in $v = \partial u/\partial n$.) The function $V(x,k)$ is known to be completely non-oscillatory only in certain regimes (for example when Γ is smooth and convex and x is in the illuminated zone and is bounded away from the "shadow boundary" which divides illuminated parts of Γ from the parts which are in "shadow"). However for general smooth convex Γ, $V(x,k)$ can be expected to be less oscillatory than $v(x,k)$ and it was argued in [ANZ94, ANZ95] that $\|V(\cdot,k)\|_{H^{n+1}(\Gamma)} \leq Ck^{(n+1)/3}$. (More details of how this estimate can be made rigorous are in subsection 2.1). The formal argument sketched above then suggests that using a finite element space of dimension $N = O(k^{(d-1)/3})$ to approximate V in a numerical method based on (2.2) should preserve accuracy as $k \to \infty$. This, while not being fully "robust" as $k \to \infty$, is a considerable improvement on the estimate for conventional methods sketched in subsection 1 although we emphsise that this is not fully rigorous since it ignores the unknown behaviour of $C(k)$.

In more recent work [BGMR04, BG07], Bruno et. al. tackled the breakdown of the geometric optics ansatz by employing a more careful discretisation scheme. Focussing on the formulation (1.4), using also the ansatz (2.2), and multiplying each side of the result by $\exp(-\mathrm{i}kx\cdot\hat{a})$, one obtains

$$V + \widetilde{D}'V - \mathrm{i}\eta\widetilde{S}V \;=\; 2\mathrm{i}(k\hat{a}\cdot\hat{n} - \eta). \tag{2.3}$$

The integral operators $\widetilde{D}'$ and $\widetilde{S}$ are analogues of D' and S with the additional factor $\exp(\mathrm{i}k(y-x)\cdot\hat{a})$ in their kernels, leading to kernel functions with easily identified phase. For example,

$$\begin{aligned}\widetilde{S}V(x) &= 2\int_\Gamma \Phi(x,y)\exp(\mathrm{i}k(y-x)\cdot\hat{a})V(y)\mathrm{d}y\\ &= \int_\Gamma \exp(\mathrm{i}k(|x-y| + (y-x)\cdot\hat{a}))M_k(x,y)V(y)\mathrm{d}y, \end{aligned}\tag{2.4}$$

where the factor M_k is weakly singular at $y = x$ but is not oscillatory for large k.

The approach taken in [BGMR04, BG07] is now to apply a Nyström method to (2.3) based on a suitable quadrature rule for oscillatory integrals of the form (2.4). This involves the approximation of V on a coarse k-independent grid. Since (as we shall see more precisely in the following section), the geometric optics approximation breaks down in a boundary layer of width $\mathcal{O}(k^{-1/3})$ around the shadow boundary, the mesh is graded in $\mathcal{O}(k^{-1/3})$ neighbourhoods of the shadow boundaries. Based on sampling V at points in this coarse mesh, integration for operators such as (2.4) are employed based on partitions of unity and (exponentially convergent) trapezoidal rules. The partition of unity is designed to localise around special points (with respect to the observation point x) namely (i) the singular point $y = x$; (ii) the stationary points where the gradient of the phase of (2.4) vanishes; (iii) shadow boundary points $n(x) \cdot \hat{a} = 0$. As $k \to \infty$ integration regions become more localised around these points. This is a high-frequency variant of the matrix-free Nyström method of [BK01]. Since this method is not based on a Galerkin formulation, the analysis of its k-robustness is a challenging open problem. We shall return to methods for oscillatory integrals arising in scattering problems in subsection 4.

A very interesting extension of the method in [BGMR04, BG07] to non-convex scattering is given in [BGR05]. There it is explained how the integral equation (1.4) may be solved by a Neumann series approach, where each term in the Neumann series corresponds to the scattering by a single obstacle of an incident field consisting of the incident wave combined with previously scattered waves. Each of these single-obstacle scattering problems can be solved by a method similar to the methods described above, except that now the ansatz (2.1) becomes somewhat more complicated: the phase $x \cdot \hat{a}$ appearing in (2.2) has to be replaced by a function reflecting the optical distance travelled by rays through all the previous reflections. Preliminary numerical tests were provided in [BGR05] which demonstrated the potential for the method. The theory of this method was substantially advanced in the subsequent work [Ece05, ER06]. There the implementation of the Neumann series was shown to correspond to a sum over increasing period of a sequence of periodic orbits. Each orbit corresponds to reflections off a fixed set of scatterers, and this allows the convergence rate of the Neumann series to be estimated, for sufficiently high frequency and permits the formulation of methods for accelerating its convergence. The most recent work in this direction [ABER06] extended the analysis to the three dimensional case, where additional considerations on the relative orientation of the scattering bodies come into play.

In [Ece05, ER06, ABER06] the emphasis is on the convergence of the Neumann series, and assumes the robust solution of the integral equations arising at each iteration. Thus the proof of the k-robustness of the overall algorithm remains a challenging open problem.

One of the substantial challenges which will arise in the rigorous numerical analysis of non-convex scattering problems is that the k-dependence of the constant C in (1.7) is likely to be considerably more complicated than it is in the convex case. We make this statement because C contains in some sense a bound on the inverse operator A^{-1}, where A appears in (1.5) (see subsection 3.1). While this inverse is uniformly bounded with respect to k in the convex case (see subsection 3.2), this is not true in the non-convex case. An explicit counter-example is given in [CGLL07]. Finally, it is important to point out that, while there has been some progress on aspects of algorithms for the 3D problem (e.g. [BGR05, GLS07]), the theoretical numerical analysis for this case is limited to date. Moreover the underlying asymptotic theory is much more challenging (e.g. [BGS05] gives a numerical approach to a 3D "canonical problem" and contains extensive references to the asymptotic theory).

In the next two subsections we describe recent work on 2D problems where more precise rigorous estimates are available, namely scattering by smooth convex obstacles and by convex polygons. We note that, in very recent work [LMC08], numerical experiments have been carried out which suggest that the algorithms for these two cases can be successfully combined to compute high frequency scattering by curvilinear convex polygons.

2.1 The case of smooth Γ *in 2D*

In this subsection we assume that Γ is a C^∞ strictly convex contour. Under plane wave illumination Γ is naturally divided by the two tangency points T_1 and T_2 into an "illuminated zone" (**I**) and a "shadow zone" (**S**), as depicted in Fig. 1. Letting $\boldsymbol{\gamma} : [0, 2\pi] \to \Gamma$ be a 2π-periodic parametrization of Γ we define $t_i \in [0, 2\pi)$ to be the preimages of the T_i: $\boldsymbol{\gamma}(t_i) = T_i$, for $i = 1, 2$. The "Geometric Optics" ansatz (2.2) may then be written (writing $v(\boldsymbol{\gamma}(s), k)$ as $v(s, k)$ for convenience)

$$v(s, k) = k \exp(\mathrm{i}k\boldsymbol{\gamma}(s) \cdot \hat{a})V(s, k). \qquad (2.5)$$

In [DGS07], the parameter space $[0, 2\pi)$ is covered by four intervals Λ_i, $i = 1, \ldots, 4$, with Λ_1 and Λ_2 being suitably small neighbourhoods of the tangency points t_1 and t_2, and Λ_3 and Λ_4 contained in the illuminated and shadow zones respectively. Letting χ_i denote a corresponding partition of unity, the p-version of the Galerkin method is then applied to (1.8), i.e. we use in (1.10) the approximating space of dimension N,

$$\mathcal{S}_{N,k} := \mathrm{span}\{k \exp(\mathrm{i}k\gamma(s) \cdot \widehat{a})\chi_j(s)P(s) : P \in \mathbb{P}_{p_j}, \, j = 1, 2, 3\}, \qquad (2.6)$$

where $\mathbb{P}_p$ denotes the algebraic polynomials of degree p. That is, we use possibly different degree polynomial approximations to V in each of the illuminated and transition zones, and zero approximation to V in the shadow zone.

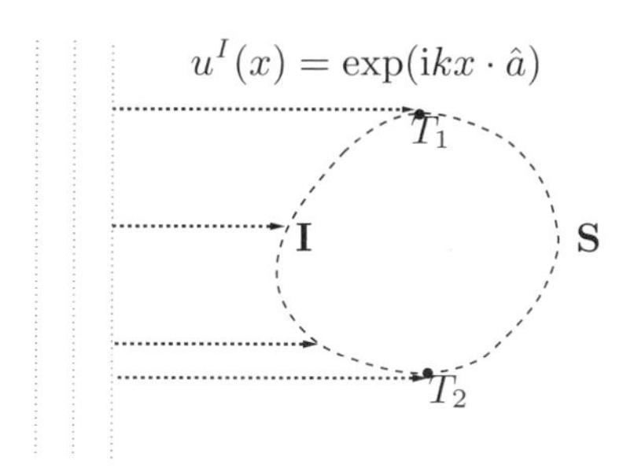

Fig. 1. Physical domain: **I** denotes the illuminated zone, **S** the shadow zone and T_1, T_2 the tangency points.

Considering the numerical analysis of this method we recall the two key questions **Q1** and **Q2** highlighted in subsection 1. To answer **Q1** we need estimates for the derivatives of $V(s,k)$ with respect to s which are explicit in k in the illuminated and shadow zones. Moreover we need estimates for the exponential decay of V in the deep shadow zone Λ_1. This requires a substantial study of the theory of the "geometric optics" approximation (2.2). In [DGS07] the following result is presented.

Theorem 1 *For all $L, M \in \mathbb{N} \cup \{0\}$, the function $V(s,k)$ admits a decomposition of the form:*

$$V(s,k) = \left[\sum_{\ell,m=0}^{L,M} k^{-1/3-2\ell/3-m} b_{\ell,m}(s) \Psi^{(\ell)}(k^{1/3} Z(s)) \right] + R_{L,M}(s,k), \tag{2.7}$$

for $s \in [0, 2\pi]$, where the remainder term has its nth derivative bounded, for $n \in \mathbb{N} \cup \{0\}$, by

$$|D_s^n R_{L,M}(s,k)| \leq C_{L,M,n}(1+k)^{\mu+n/3}, \tag{2.8}$$

where $\mu := -\min\left\{\frac{2}{3}(L+1),\ (M+1)\right\}$ and $C_{L.M,n}$ is independent of k. The functions $b_{\ell,m}$ and Z are C^∞ 2π-periodic functions. Z has simple zeros at t_1 and t_2, is positive-valued on $[t_1, t_2]$ and negative-valued elsewhere on $[0, 2\pi]$. Moreover $\Psi : \mathbb{C} \to \mathbb{C}$ is an entire function which may be spcified explicitly by a certain contour integral involving the Airy function (often called "Fock's integral" - see [Foc65, §7, 12], a book originally published in the Russian literature in the late 1940's).

Theorem 1 is derived in [DGS07] using the often cited paper [MT85], combined with the technique of matched asymptotic expansions and also referring to results from the classical literature such as [Bus63, ML68, Cha73, Bus75, Zwo90, BB91].

The asymptotics of $\Psi(\tau)$ for large $|\tau|$ are of key importance for the behaviour of V. Since Z is positive in Λ_3 (inside the illuminated zone), the behaviour of $V(s,k)$ for $s \in \Lambda_3$ and for large k is determined by the asymptotics of $\Psi(\tau)$ and its derivatives as $\tau \to \infty$. Similarly, the behaviour of $V(s,k)$ for $s \in \Lambda_1$ (the deep shadow), depends on the asymptotics of $\Psi(\tau)$ for $\tau \to -\infty$. More complicated behaviour arises in the transition zones Λ_1, Λ_2. The required properties of Ψ are known. In particular, (see [MT85, Lemma 9.9]):

$$\Psi(\tau) = a_0\tau + a_1\tau^{-2} + a_2\tau^{-5} + \ldots + a_n\tau^{1-3n} + \mathcal{O}(\tau^{1-3(n+1)}), \quad \text{as } \tau \to \infty, \tag{2.9}$$

where $a_0 \neq 0$ and this expansion remains valid for all derivatives of Ψ by formally differentiating each term on the right hand side, including the error term. Moreover, there exists $\beta > 0$ and $c_0 \neq 0$ such that, for any $n \in \mathbb{N} \cup \{0\}$, as $\tau \to -\infty$,

$$D_\tau^n \Psi(\tau) = c_0 \; D_\tau^n \{\exp(-\mathrm{i}\tau^3/3 - \mathrm{i}\tau\alpha_1)\} \; (1 + \mathcal{O}(\exp(-|\tau|\beta))), \tag{2.10}$$

where $\alpha_1 = \exp(-2\pi\mathrm{i}/3)\nu_1$ and $\nu_1 < 0$ is the right-most root of the (all real and negative) roots of the Airy function Ai. Hence, when $\tau \to -\infty$ the function Ψ, as well as its derivatives decrease exponentially but in a very oscillating way. The asymptotics in (2.10) may be deduced by applying the theory of residues to the contour integral defining Ψ - see [BB91, p.393], [Bus63, Lemma 8]. More details are in [DGS07]. Combining these asymptotics with Theorem 1, the following estimates for the derivatives of V are proved in [DGS07].

Theorem 2 *For all $n \in \mathbb{N} \cup \{0\}$ there exist constants $C_n > 0$ independent of k and $s \in [0, 2\pi]$, such that for all k sufficiently large,*

$$|D_s^n V(s,k)| \le C_n \left\{ \begin{array}{ll} 1, & n = 0, 1, \\ k^{-1}(k^{-1/3} + |\omega(s)|)^{-n-2}, & n \ge 2, \end{array} \right. \tag{2.11}$$

where $\omega(s) := (s - t_1)(t_2 - s)$. These estimates are uniform in $s \in [0, 2\pi]$.

This statement follows from [DGS07, Theorem 5.4] but is in a somewhat simpler form than given there. The essential point which follows from this is that for s in the illuminated zone and bounded away from t_1, t_2 (and hence $|\omega(s)|$ is bounded away from zero), all derivatives of V

are bounded as $k \to \infty$. This shows the correctness of the Geometric Optics approximation in the interior of the illuminated zone. However, in a region of width $\mathcal{O}(k^{-1/3})$ around t_1 or t_2, $|\omega(s)| \lesssim k^{-1/3}$ and $D_s^n V(s,k)$ may blow up with $\mathcal{O}(k^{(n-1)/3})$. (Notations like $\lesssim$ indicate that there is a hidden constant independent of k). This corresponds to the estimates employed in the analysis in [ANZ94, ANZ95] and the motivation for the mesh grading scheme used in [BGMR04] which we have described above.

In [DGS07] we considered, for sufficiently large $k > 0$ and parameters $\varepsilon, \delta \in (0, 1/3]$, and $c_1 > 0$, $c_2 > 0$ (to be chosen), the transition zones:

$$\begin{aligned} \Lambda_1 &:= [t_1 - c_2 k^{-1/3+\delta},\ t_1 + c_1 k^{-1/3+\varepsilon}], \\ \Lambda_2 &:= [t_2 - c_1 k^{-1/3+\varepsilon},\ t_2 + c_2 k^{-1/3+\delta}], \end{aligned}$$

and the illuminated and shadow zones, respectively:

$$\begin{aligned} \Lambda_3 &:= [t_1 + c_1 k^{-1/3+\varepsilon},\ t_2 - c_1 k^{-1/3+\varepsilon}]\ , \\ \Lambda_4 &:= [t_2 - 2\pi + c_2 k^{-1/3+\delta},\ t_1 - c_2 k^{-1/3+\delta}]. \end{aligned}$$

The regions Λ_j touch only at their endpoints.

In each of the zones Λ_1, Λ_2 and Λ_3 the error in best approximation by polynomials can be estimated by standard methods. Using Theorem 2 and standard error estimates for polynomial approximation it turns out that there are two conflicting choices for ε. The best estimate in the illuminated zone is obtained with $\varepsilon = 1/3$ (i.e. the boundary of Λ_1 does not approach t_1 or t_2 as $k \to \infty$) and the best error in the transition zones is obtained with $\varepsilon = 0$ (these zones shrink as fast as possible with k). To balance the error the best choice turns out to be $\varepsilon = 1/9$. These estimates have to be combined with the following theorem on exponential decay in the deep shadow which is stated in [DGS07].

Theorem 3 *There exist positive constants c_0, c_0' such that for all k sufficiently large,*

$$\|v\|_{L^2(\Lambda_4)} \le c_0' \exp(-c_0 k^{\delta}). \tag{2.12}$$

This result can be formally inferred by using the asymptotics (2.10) in the first term of the right-hand side of (2.7), but this is not a rigorous proof since the remainder term in (2.7) enjoys only algebraically decaying estimates. A brief account of how the proof of Theorem 3 follows from the results in the literature is given in [DGS07]. In particular we refer to [Urs68, Fil76, ZF85, ZF86, Leb84, HL94, Pop87] for the highly non-trivial proofs. An interesting side remark is that the results on exponential decay (in two-dimensional problems) in [ZF85, ZF86] do not require the contour to be analytic but only C^∞. There are also extensions to arbitrary dimension, but these require analytic scattering surfaces and (as stated) are only valid in the "deep shadow" (i.e. a bounded distance away from the shadow boundary) – see, e.g. [Pop87, Thm 3] which uses the ideas of [Leb84].

Combining the estimate from Theorem 3 with the estimates for polynomial approximation in the illuminated and transition zones, we obtain (in the special case $p_j = p$ for each $j = 1, 2, 3$ - see [DGS07] for more general cases), that the Galerkin method solution, defined by (1.10), satisfies the error estimate

$$\|v - v_N\| \leq \left(\frac{B}{\alpha}\right) \inf_{\phi_N \in \mathcal{S}_{N,k}} \|v(\cdot, k) - \phi_N\|_{L^2(\Gamma)} \tag{2.13}$$

$$\leq C_n \left(\frac{B}{\alpha}\right) k \left\{ k^{-4/9} \left(\frac{k^{1/9}}{p}\right)^n + \exp(-c_0 k^\delta) \right\}. \tag{2.14}$$

Here (2.13) follows from (1.7) and that, as observed after (1.10), $C \leq B/\alpha$, where B and α are the continuity and coercivity constants from (1.9). Moreover (2.14) follows from the estimates for polynomial approximation and exponential decay described above and holds for all $6 \leq n \leq p + 1$ (so in particular, for fixed k and C^∞ data we have superalgebraic convergence, as is normal in the p-version of the boundary element method.)

To make the estimate (2.14) of rigorous use we have to estimate the constant B from above and α from below with respect to k. In [DGS07] it is shown that for sufficiently smooth contours Γ, and in the case $\eta = k$, we have $B \lesssim k^{1/2}$. Substantial generalisations of this in [CGLL07] are discussed in subsection 3. Estimates for α from below are much harder. In [DGS07] it was proved by Fourier analysis (on circular or spherical boundaries) that $\alpha \gtrsim 1$ (i.e. k-independent coercivity), again for $\eta = k$. However the extension of this result to general Γ is a challenging open problem. Neverthess there is hope for success, since in [CM08] it was proved that $\|A^{-1}\| \lesssim 1$ for Lipschitz star-shaped Γ and any $\eta \in \mathbb{R}\backslash\{0\}$ (where A is the integral operator in (1.5)) and this estimate is a necessary (although not sufficient) condition for k-independent coercivity of a defined in (1.8).

On the assumption that k-independent coercivity holds for a the result (2.14) shows that the error in the Galerkin approximation is of the order of a low power of k times $(k^{1/9}/p)^n$ (for $6 \leq n \leq p + 1$), plus a term which is exponentially small in k. Roughly speaking this shows that by choosing p to grow slightly faster than $k^{1/9}$ we preserve the accuracy of the method as k increases. Numerical results in [DGS07] support this conclusion.

Before leaving this discussion we mention that using the asymptotics (2.10) when s is near to but less than t_1 (i.e. in the shadow region but near the transition point), then the first term in (2.7) has the asymptotics (as $k \to \infty$)

$$k^{-1/3} b_{0,0} \exp(\mathrm{i}k|Z(s)|^3/3) \exp(\mathrm{i}\Re(\alpha_1) k^{1/3} |Z(s)|) \exp(-\Im(\alpha_1) k^{1/3} |Z(s)|). \tag{2.15}$$

Since α_1 is in the first quadrant of the complex plane (see (2.10)), (2.15) contains two oscillatory factors, one oscillating with scale k and one with scale $k^{1/3}$, damped by the exponentially decaying third term. These two scales were modelled in the basis functions used in the collocation method of Giladi and Keller, which took into account the existence of "creeping waves" behind the shadow boundary [GK04].

We now turn our attention to scattering by convex polygonal bodies.

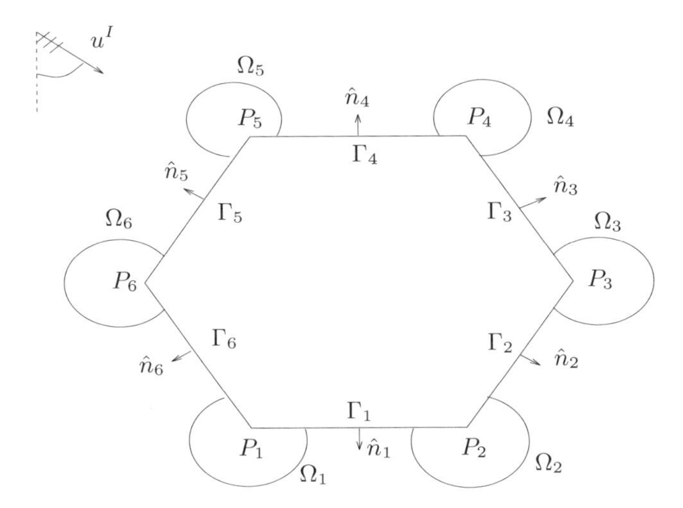

Fig. 2. Our notation for the polygon. The M corners are numbered anticlockwise, we denote the first corner by both P_1 and P_{M+1}, and Ω_m is the exterior corner angle at P_m.

2.2 The case of polygonal Γ

In separate work [ACL07, CL07], scattering by convex polygons, and more recently by curvilinear convex polygons [LMC08], has been considered. For the case of a polygon an appropriate specific form of (2.1) is shown in [CL07] to be

$$\begin{aligned} v(x,k) \approx\ & k \exp(\mathrm{i}kx\cdot\hat{a})\, V(x,k) \\ & + k\sum_{m=1}^{M} [\exp(\mathrm{i}kx\cdot\hat{a}_m) V_m^{+}(x,k) + \exp(-\mathrm{i}kx\cdot\hat{a}_m) V_m^{-}(x,k)], \end{aligned} \tag{2.16}$$

for $x \in \Gamma$, where M is the number of sides of the polygon, the unit vector $\hat{a}_m$ is parallel to the mth side (directed from corner P_m to corner P_{m+1} in Fig. 2), and the function $V_m^{\pm}$ is assumed non-zero only on side m. (Physically the terms in the summation are included to represent the parts of the field arising from diffraction at the corners.) In fact, it is shown in [CL07] that for a polygon it is adequate to take $V(x,k)$ to be constant on each side, precisely to define $V(x,k)$ so that the first term in the above expression is the high frequency Kirchhoff or physical optics approximation, i.e.

$$\frac{\partial u}{\partial n}(x) \approx \begin{cases} 2\frac{\partial u^I}{\partial n}(x) & \text{on illuminated sides,} \\ 0, & \text{on sides in shadow.} \end{cases}$$

This means to set, for $x \in \Gamma_m$,

$$V(x,k) := \begin{cases} 2\mathrm{i}\hat{n}_m\cdot\hat{a}, & \text{if } \hat{n}_m\cdot\hat{a} < 0, \\ 0, & \text{otherwise,} \end{cases} \tag{2.17}$$

where $\hat{n}_m$ is the outward unit normal on side m. Then, if $V_m^{\pm}$ are approximated by piecewise polynomials on carefully chosen graded meshes, uniformly accurate approximations are obtained as $k \to \infty$, provided the number of degrees of freedom grows like $O(\log^{3/2} k)$. The results [ACL07, CL07] are inspired by earlier results [CLR04, LC06] (indeed the method and analysis for the convex polygon case are outlined in [CLR04]). In these papers, as a prototype of developing and analysing boundary element methods for high frequency scattering, high frequency algorithms are proposed for the problem of 2D scattering by a flat surface with piecewise-constant impedance boundary condition, and a complete proof of k-independent accuracy and complexity is provided, the first such result for any scattering algorithm.

We give now a little more detail of the methods and results of [CL07]. A key component in the design of the ansatz (2.16) and the design of the graded meshes to approximate the functions $V_m^{\pm}(x,k)$ (both of which are essential to the overall low complexity) is understanding the high frequency behaviour of the solution. We have seen in subsection 2.1 that obtaining rigorous results on high frequency asymptotics is, in general, a complex business. But this case of a convex polygon is one in which rigorous high frequency asymptotics, at least a sufficient understanding of these for the purpose of designing an effective numerical scheme, can be achieved by elementary arguments.

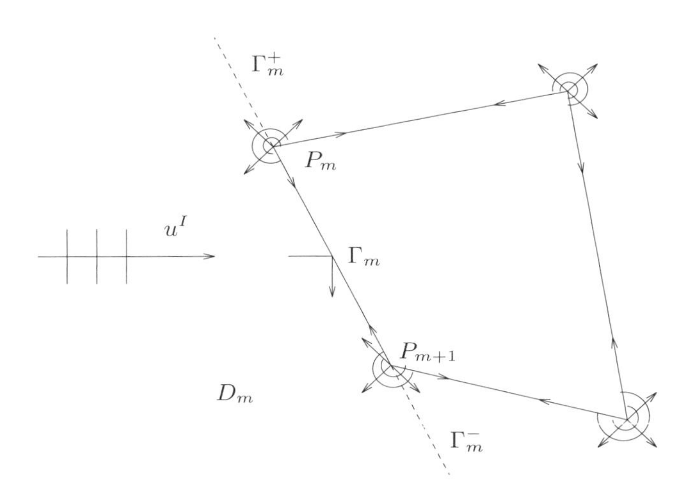

Fig. 3. Extension of Γ_m, for derivation of regularity estimates.

The trick (adapted from [CLR04]) is to observe that one can write down an explicit solution to the Dirichlet boundary value problem for the Helmholtz equation in a half-plane, since we (trivially) know by the method of images the Green's function for a half-plane. This observation leads to the following useful representation for the solution to the scattering problem. Let $D_m \subset D$ denote the half-plane whose boundary contains Γ_m, the mth side of the polygon (see Fig. 3), and let $G_m(x,y)$ be the Dirichlet Green's function for the half-plane D_m, i.e. $G_m(x,y) = \Phi(x,y) - \Phi(x,y_m')$, where y_m' denotes the image of y in the straight line ∂D_m. Then

$$u(x) = \epsilon_m(u^I(x) + u^R(x)) + \int_{\partial D_m} \frac{\partial G_m(x,y)}{\partial n(y)} u(y)\, \mathrm{d}s(y), \quad \text{for} \quad x \in D_m.$$

Here, $\epsilon_m = 0$ if side Γ_m is in shadow, $\epsilon_m = 1$ if it is illuminated, and u^R is the plane wave that would be reflected if u^I were incident on the straight line ∂D_m, on which a Dirichlet boundary condition holds (so $u^R(x) = R_m \exp(\mathrm{i}kx \cdot \hat{a}_m^R)$, where $\hat{a}_m^R = \hat{a} \cdot \hat{a}_m \, \hat{a}_m - \hat{a} \cdot \hat{n}_m \, \hat{n}_m$ and the constant R_m is chosen so that $u^I(x) + u^R(x) = 0$ on Γ_m; see [CL07] for details). It is an easy calculation that $\frac{\partial G_m(x,y)}{\partial n(y)} = 2\frac{\partial \Phi(x,y)}{\partial n(y)}$, and we recall that $u = 0$ on Γ. Thus, and taking the normal derivative in the above equation, we see that

$$\frac{\partial u}{\partial n}(x) = 2\epsilon_m \frac{\partial u^I}{\partial n}(x) + 2\int_{\partial D_m \setminus \Gamma_m} \frac{\partial^2 \Phi(x,y)}{\partial n(x)\partial n(y)} u(y)\,\mathrm{d}s(y), \quad x \in \Gamma_m. \tag{2.18}$$

It is this equation which justifies the ansatz (2.16), with the explicit formula (2.17) for $V(x,k)$. The first term in the above equation is the Kirchhoff or physical optics approximation and the integral is thus the correction to this approximation on the side Γ_m. To understand the behaviour of this correction we write $\partial D_m \setminus \Gamma_m$ as the union of the two disjoint half-lines Γ_m^+ and Γ_m^- (see Fig. 3). Then we note that, explicitly, for $x \in \Gamma_m$ and $y \in \Gamma_m^\pm$,

$$\begin{aligned} \frac{\partial^2 \Phi(x,y)}{\partial n(x)\partial n(y)} &= \frac{\mathrm{i}kH_1^{(1)}(k|x-y|)}{4|x-y|} \\ &= \frac{\mathrm{i}k^2}{4}\exp(\pm \mathrm{i}kx \cdot \hat{a}_m)\exp(\mp \mathrm{i}ky \cdot \hat{a}_m)\mu(k|x-y|), \end{aligned} \tag{2.19}$$

where $\mu(z) := \mathrm{e}^{-\mathrm{i}z} H_1^{(1)}(z)/z$, for $z > 0$. Thus, we see that, for $x \in \Gamma_m$,

$$\begin{aligned} v(x,k) &= k\exp(\mathrm{i}kx \cdot \hat{a})\,V(x,k) \\ &+ k[\exp(\mathrm{i}kx \cdot \hat{a}_m)V_m^+(x,k) + \exp(-\mathrm{i}kx \cdot \hat{a}_m)V_m^-(x,k)], \end{aligned} \tag{2.20}$$

with $V(x,k)$ given by (2.17) and

$$V_m^\pm(x,k) := \frac{\mathrm{i}k}{2}\int_{\Gamma_m^\pm} \exp(\mp \mathrm{i}ky \cdot \hat{a}_m)\mu(k|x-y|)u(y)\,\mathrm{d}s(y). \tag{2.21}$$

The point here is that, while we cannot evaluate the integrals $V_m^\pm(x,k)$, because these integrals involve the unknown u on $\Gamma_m^\pm$, we can show, in a precise quantitative way, that these functions are not oscillatory on Γ_m. This is the case since $\mu(k|x-y|)$ is very smooth, except when $k|x-y|$ is small, as the function $\mu(z)$, while singular at $z = 0$, is increasingly slowly varying as $z \to \infty$, as quantified in the following lemma from [CL07].

Lemma 4 *For every $\epsilon > 0$,*

$$|\mu^{(m)}(z)| \leq (1 + \epsilon^{-1/2})\,(m+1)!\,z^{-3/2-m},$$

for $z \geq \epsilon$ and $m = 0, 1, \ldots$.

Applying this lemma leads to bounds on the tangential derivatives on Γ_m of $V_m^{\pm}(x,k)$. In this theorem and subsequently it is is convenient to use the abbreviation

$$u_M := \sup_{x \in D} |u(x)|,$$

to let L_m denote the length of Γ_m, and to let $\partial/\partial s$ denote the derivative in the tangential direction on Γ.

Theorem 5 *[CL07, Theorem 3.2] For $x \in \Gamma_+$, let s denote the distance of x from the corner P_m. Then, for $n = 0, 1, \ldots$, it holds that*

$$\left|\frac{\partial^n}{\partial s^n} V_m^+(x,k)\right| \leq 2(1 + \epsilon^{-1/2})\ u_M\ n!\,k^{-1/2} s^{-1/2-n}, \quad \textit{for } \epsilon \leq ks \leq kL_m.$$

The same bound applies to the tangential derivatives of $V_m^-(x,k)$ but with s denoting distance along Γ_m from the corner P_{m+1}.

This bound captures the behaviour of $V_m^+(x,k)$ on Γ_m except near the corner P_m. But this can be understood by standard elliptic regularity estimates for behaviour of solutions near corners of the domain (e.g. [Gri85]) or more explicitly by writing down a representation for u near the corner P_m based on separation of variables in polar coordinates centred on P_m [CL07, theorem 2.3]. The resulting bound is the following, showing that $V_m^+(x,k)$ has the classic corner singularity behaviour near P_m where the exterior corner angle is Ω_m.

Theorem 6 *[CL07, Corollary 3.4] Suppose that each side of the polygon has length at least $\lambda/8$, and, for $x \in \Gamma_+$, let s denote distance of x from the corner P_m. Then, for $n = 0, 1, \ldots$, it holds that*

$$\left|\frac{\partial^n}{\partial s^n} V_m^+(x,k)\right| \leq C_n\, u_M\, k^{-\alpha_m}\, s^{-\alpha_m - n}, \qquad \textit{for } 0 < ks \leq \pi/12,$$

where $\alpha_m := 1 - \pi/\Omega_m \in (0, 1/2)$, and the value of the constant $C_n > 0$ depends only on n. The same bound applies to the tangential derivatives of $V_m^-(x,k)$ but with s denoting distance along Γ_m from the corner P_{m+1}, and with α_m replaced by α_{m+1}.

The detailed information in the above bounds enables us to construct finite element spaces which are well-adapted to approximating the functions $V_m^\pm$. One possibility, used in [CL07], adapted from [CLR04], is to use a discontinuous piecewise polynomial approximation for each of V_m^+ and V_m^- on Γ_m. Thus V_m^+ can be approximated using piecewise polynomials of some degree p, on a mesh which has a classical grading near the corner P_m where V_m^+ is singular, as quantified in Theorem 6, and then has a geometric grading over the rest of Γ_m, with the two meshes joined in a smooth manner. In detail, in the case when each side of the polygon has length $L_m \geq \lambda$, defining $q := (2p+3)/(1-2\alpha_m)$, it is shown in [CL07] that a mesh appropriate for approximating $V_m^+(x,k)$, given the bounds in the above theorems, consists of the points

$$s_i = \lambda \left(\frac{i}{N}\right)^q, \ i = 0, \ldots, N, \text{ and } s_{N+j} := \lambda \left(\frac{L_m}{\lambda}\right)^{j/N_m^+}, \ j = 1, \ldots, N_m^+. \tag{2.22}$$

Here N, the number of subintervals in the interval of length λ adjacent to the corner P_m, is the parameter controlling the mesh refinement, and N_m^+ is the smallest integer greater than or equal to $-\log(L_m/\lambda)/(q\log(1-1/N))$. Based on this mesh, an appropriate piecewise polynomial approximation space for $V_m^+(x,k)$ is

$$\mathcal{S}_m^+ = \text{span}\left\{\chi_j(s)P(s) : P \in \mathbb{P}_p, \ j = 0, 1, \ldots, N + N_m^+\right\}, \tag{2.23}$$

where $\chi_j(s)$ is the characteristic function of the interval (s_{j-1}, s_j). From the mean value theorem applied to $\log(1-1/N)$ it follows that the number of subintervals in the geometrically graded part of the mesh satisfies

$$N_m^+ < q^{-1} N \log(L_m/\lambda) + 1. \tag{2.24}$$

The equations defining a suitable approximation space $\mathcal{S}_m^-$ and mesh for approximating $V_m^-(x,k)$ are identical to (2.22) and (2.23), except that q is defined replacing α_m by α_{m+1} and s_i is now the distance of the mesh point from the corner P_{m+1}. Since the value of q is different, there is a different number N_m^- of mesh points in the geometrically graded part of the mesh.

Recalling (2.20), we see that an appropriate approximation space for $\partial u/\partial n$ on Γ_m is

$$k\exp(\mathrm{i}kx \cdot \hat{a})V(x,k) + \exp(\mathrm{i}kx \cdot \hat{a}_m)\mathcal{S}_m^+ + \exp(-\mathrm{i}kx \cdot \hat{a}_m)\mathcal{S}_m^-, \tag{2.25}$$

with $V(x,k)$ given by (2.17). The full approximation space used in

[ACL07, CL07] is the affine space $\tilde{\mathcal{S}}_{N,k} := k\exp(\mathrm{i}kx\cdot\hat{a})V(x,k) + \mathcal{S}_{N,k}$, where $\mathcal{S}_{N,k}$ is the linear space of functions whose restriction to side Γ_m is in the set $\exp(\mathrm{i}kx\cdot\hat{a}_m)\mathcal{S}_m^+ + \exp(-\mathrm{i}kx\cdot\hat{a}_m)\mathcal{S}_m^-$, for $m = 1,\ldots,M$. The dimension of $\mathcal{S}_{N,k}$, i.e. the number of degrees of freedom, is

$$\begin{aligned} D_N &= (p+1)\textstyle\sum_{m=1}^{M}(2N + N_m^+ + N_m^-) \\ &< 2MN\left((p+1)(1+N^{-1}) + \tfrac{1}{2}\log(\bar{L}/\lambda)\right), \end{aligned} \tag{2.26}$$

by (2.24), where $\bar{L} := (L_1 \ldots L_M)^{1/M}$.

The main numerical analysis result of [CL07] is the following best approximation estimate:

Theorem 7 *If each side of the polygon has length at least λ, then, for some positive constant C_p, depending only on p and on the corner angles Ω_1, Ω_2, ..., Ω_M, it holds that*

$$k^{-1/2} \inf_{\phi_N \in \tilde{\mathcal{S}}_{N,k}} \|v(\cdot,k) - \phi_N\|_{L^2(\Gamma)} \le C_p u_M \frac{(M[1+\log(\bar{L}/\lambda)])^{p+3/2}}{D_N^{p+1}}.$$

This theorem shows that, to maintain a given bound on the left hand side of the above inequality, aiming to keep (1.15) fixed as k increases, it is necessary to increase the number of degrees of freedom D_N only in proportion to $(\log k)^{3/2}$. If one is happy to maintain control instead on $k^{-1}\inf_{\phi\in\mathcal{S}_{N,k}}\|v(\cdot,k)-\phi\|_{L^2(\Gamma)}$, aiming to keep (1.13) fixed, then one can even decrease D_N as k increases.

In [CL07] the approximation space $\mathcal{S}_{N,k}$ is used as the basis of a Galerkin method for the integral equation (1.4). Precisely, cf. (1.10), the approximation v_N to v is defined by: Seek $v_N \in \tilde{\mathcal{S}}_{N,k}$ such that

$$a(v_N, w_N) = (f, w_N)_{L^2(\Gamma)}, \quad \text{for all} \quad w_N \in \mathcal{S}_{N,k}.$$

As discussed in subsection 2.1, in the case that Γ is a circle it is shown in [DGS07] that the sesquilinear form a is coercive, with a coercivity constant α independent of k. For general domains it is not yet clear whether a is coercive, and in [CL07] the stability analysis is approached by the classical methods for analysis of second kind equations discussed before equation (1.8) and in subsection 3.1 below. Using these standard methods it is shown in [CL07] that, for each fixed k, there exists a value for the stability constant $C > 0$ and an integer N_0 such that the Galerkin solution is well-defined and (1.7) holds for $N \ge N_0$. Combining this with Theorem 7 gives the estimate

$$k^{-1/2}\|v(\cdot,k) - v_N\|_{L^2(\Gamma)} \le C\,C_p u_M \frac{(M[1+\log(\bar{L}/\lambda)])^{p+3/2}}{D_N^{p+1}}, \tag{2.27}$$

for $N \geq N_0$, but, as discussed in section 1 and subsection 3.1 below, this classical analysis does not give any information about the dependence of C and N_0 upon k. An approximation u_N to u can be computed by replacing $v = \partial u/\partial n$ by its the approximation v_N in (1.3). This is shown in [CL07, Theorem 5.4], via the inequality (1.14), to satisfy the error bound

$$\frac{\sup_{x \in D} |u(x) - u_N(x)|}{\sup_{x \in D} |u(x)|} \leq C\, C'_p \frac{(M[1 + \log(\bar{L}/\lambda)])^{p+2}}{D_N^{p+1}}, \tag{2.28}$$

for $N \geq N_0$, where C'_p is a positive constant depending only on p and the corner angles $\Omega_1, \Omega_2, \ldots, \Omega_M$. Numerical results supporting these error estimates are shown in [CL07].

A collocation method based on the identical approximation space $\mathcal{S}_{N,k}$ and the identical integral equation formulation is implemented in [ACL07]. No stability analysis (even one based on classical second-kind theory) is made in [ACL07], but the numerical results support the conclusion that there is little difference in accuracy between the Galerkin and the (rather easier to implement) collocation method.

3 Stability and conditioning

3.1 General considerations

In section 1 we have split the numerical analysis of high frequency boundary element methods into research on three related questions. We turn in this section to research related to the second of these questions, namely the problem of estimating the value of the stability constant C in (1.7). We note that, while the emphasis of this review is on boundary integral equation methods specifically adapted to high frequency scattering, the results of this section are equally applicable to stability analysis and conditioning for conventional piecewise polynomial boundary element methods at high frequency.

We have noted already that, in the case that the sesquilinear form a is coercive, an upper bound on the stability constant C in the case when v_N is defined by the Galerkin method (i.e. by (1.10)) is

$$C \leq \frac{B}{\alpha} \tag{3.1}$$

where B and α are the continuity and coercivity constants in (1.9). These constants are closely related to the norms of A' and its inverse as operators on $L^2(\Gamma)$. Indeed, by Cauchy-Schwarz, for $v, w \in L^2(\Gamma)$ (and with $\|\cdot\|$ denoting throughout the norm of a bounded linear operator on $L^2(\Gamma)$),

$$\begin{aligned}|a(v,w)| = |(A'v,w)_{L^2(\Gamma)}| &\leq \|A'v\|_{L^2(\Gamma)}\,\|w\|_{L^2(\Gamma)} \\ &\leq \|A'\|\,\|v\|_{L^2(\Gamma)}\,\|w\|_{L^2(\Gamma)}.\end{aligned}$$

Thus $\|A'\|$ is a possible value for the constant B in (1.9). In fact (as follows from setting $w = A'v$ in the above inequality), $\|A'\|$ is the smallest possible value for the constant B for which (1.9) holds. Similarly, from the second of the inequalities (1.9),

$$\|A'v\|_{L^2(\Gamma)}\,\|v\|_{L^2(\Gamma)} \geq |(A'v,v)_{L^2(\Gamma)}| = |a(v,v)| \geq \alpha\|v\|^2_{L^2(\Gamma)},$$

so that

$$\|A'^{-1}\| \leq \alpha^{-1}.$$

Thus, the ratio B/α is bounded below by the *condition number* of the operator A':

$$\frac{B}{\alpha} \geq \operatorname{cond} A' := \|A'\|\,\|A'^{-1}\|. \tag{3.2}$$

This gives one motivation for studying the condition number of A' and its dependence on k, which will be a main topic of this section. Another motivation is the following. The inequality (3.1) is only useful if a is coercive, which we will see below is known to be the case if Γ is a circle or sphere, but not, so far, more generally. Whether or not a is coercive, it is known that the Galerkin method (1.10) is well-defined if and only if a satisfies the *discrete inf-sup condition*: that, for some $\gamma_N > 0$ (the discrete inf-sup constant),

$$\sup_{0\neq w\in\mathcal{S}_{N,k}} \frac{|a(v,w)|}{\|w\|_{L^2(\Gamma)}} \geq \gamma_N\,\|v\|_{L^2(\Gamma)}, \quad \text{for } v\in\mathcal{S}_{N,k}. \tag{3.3}$$

If (3.3) holds (which it does, for example, if a is coercive, with $\gamma_N = \alpha$), then a standard upper bound for the stability constant C is

$$C \leq 1 + \frac{B}{\gamma_N}. \tag{3.4}$$

We do not know of any explicit bounds on the discrete inf-sup constant for high frequency boundary element methods, but we note that $\|A'^{-1}\| = \gamma^{-1}$, where γ is the corresponding *continuous inf-sup constant*, i.e.

$$\gamma := \inf_{0\neq v\in\mathcal{S}_{N,k}}\ \sup_{0\neq w\in\mathcal{S}_{N,k}} \frac{|a(v,w)|}{\|v\|_{L^2(\Gamma)}\,\|w\|_{L^2(\Gamma)}}. \tag{3.5}$$

Thus, if B is the smallest value for which the left-hand inequality in (1.9) holds, then

$$\text{cond } A' = \frac{B}{\gamma}.$$

One can hope that studying cond A' sheds some light on the behaviour, e.g. as a function of k and the coupling parameter η, of B/γ_N and hence of the upper bound (3.4).

As mentioned in subsection 1, stability can also be studied by classical second-kind integral equation methods [Atk97]: these have the attraction that they also apply to classes of collocation methods, indeed to projection methods in general. Focusing on the Galerkin case, introducing the operator P_N of orthogonal projection from $L^2(\Gamma)$ to $\mathcal{S}_{N,k}$, and writing A' as $A' = I + K$, so that $K = D' - \mathrm{i}\eta S$, it can be shown (e.g. [Atk97]) that the formulation (1.10) is equivalent to the operator equation

$$(I - P_N K)v_N = P_N f,$$

and that the Galerkin method is well-defined if and only if $I - P_N K$ is invertible. Moreoever, if the Galerkin method is well-defined, then

$$v - v_N = (I - P_N K)^{-1}(v - P_N v). \tag{3.6}$$

Thus a possible value for the stability constant C in (1.7) is

$$C = \|(I - P_N K)^{-1}\|.$$

In the case that Γ is C^1, so that K is compact [FJR78, CK83], the classical results tell us that, as long as the spaces $\mathcal{S}_{N,k}$ have the standard approximation property that $\inf_{\phi_N \in \mathcal{S}_{N,k}} \|\phi - \phi_N\|_{L^2(\Gamma)} \to 0$ as $N \to \infty$, for every $\phi \in L^2(\Gamma)$, then $\|P_N K - K\| \to 0$ as $N \to \infty$. Thus

$$C = \|(I - P_N K)^{-1}\| \to \|(I - K)^{-1}\| = \|A'^{-1}\|$$

as $N \to \infty$ (with k fixed). Alternatively, one can write (3.6) as

$$v - v_N = (I + L_N)(v - P_N v), \quad \text{where} \quad L_N := (I - P_N K)^{-1} P_N K (I - P_N), \tag{3.7}$$

from which equation it is clear that (1.7) holds with

$$C = 1 + \|L_N\|.$$

When K is compact, $\|L_N\| \to 0$ holds as $N \to \infty$ (since $\|K - KP_n\| \to 0$). Thus we see that, at least in the case that Γ is C^1, the bound (1.7) holds for every $C > 1$ provided N is sufficiently large ($N \geq N_0$).

However, as we have emphasised already in section 1, it is not clear, for a fixed value of $C > 1$, how large N_0 needs to be for (1.7) to hold, and how this value N_0 depends on k. More fundamentally, for Galerkin methods based on hybrid approximation spaces, such as we have discussed in section 2, it is the aim to keep N fixed or almost fixed as $k \to \infty$, so that it is not clear that we will ever be in the regime where KP_N is a good approximation to K in operator norm so that we can prove that $\|L_N\| \approx 0$. On the other hand, it is reasonable to hope that $\|(I - P_N K)^{-1}\|$ will be well-approximated by $\|(I - K)^{-1}\| = \|A'^{-1}\|$ before $\|P_N K - K\|$ is small, in which case (1.7) will hold with

$$C \approx \|A'^{-1}\|.$$

We have summarised what the known variational-based and classical second-kind integral equation techniques can tell us about the stability constant C. For a recent analysis which is, roughly speaking, intermediate between the two techniques, and its application to the error analysis of conventional boundary integral equation methods at high frequency, see [BS07].

In the next subsection we discuss what is known about the coercivity constant α, the continuity constant B (the smallest choice for which is $\|A'\|$), $\|A'^{-1}\|$, and cond $A' = \|A'\| \, \|A'^{-1}\|$, and their dependence on the wavenumber k and the coupling parameter η in (1.5).

3.2 Coercivity and condition numbers

Studies of the conditioning and spectral properties of integral operators (and their discretisations) in acoustic and electromagnetic scattering date back to Kress and Spassov [KS83] (and see [Kre85, Ami90, Ami93, Gie97, WC99, WC01, WC04, BS06, DGS07, BS07, CM08, CGLL07]). Most studies have focussed on the special case when Γ is a circle or sphere in which case a very complete theory is possible due to the fact that all the integral operators S, D, D' and H, defined in section 1, operate diagonally in the basis of trigonometric polynomials ($d = 2$) or spherical harmonics ($d = 3$). The analysis is further simplified by the fact that $D = D'$ and so $A = A'$ when Γ is a circle/sphere.

Suppose Γ is the unit circle, with parametrisation $\gamma(s) = (\cos s, \sin s)$. With this parametrisation $L^2(\Gamma)$ is isometrically isomorphic to $L^2[0, 2\pi]$.

We can write any $w \in L^2[0,2\pi] = L^2(\Gamma)$ as

$$w(s) = \frac{1}{2\pi}\sum_{m\in\mathbb{Z}} \widehat{w}_m \exp(\mathrm{i}ms), \quad \text{where} \quad \widehat{w}_m := \int_0^{2\pi} \varphi(s)\exp(-\mathrm{i}ms)\,\mathrm{d}s,$$

in which case the L^2-inner product and norm are given by $(v,w)_{L^2(\Gamma)} = \frac{1}{2\pi}\sum_{m\in\mathbb{Z}} \widehat{v}_m \overline{\widehat{w}_m}$ and $\|w\|^2_{L^2(\Gamma)} = \frac{1}{2\pi}\sum_{m\in\mathbb{Z}} |\widehat{w}_m|^2$. Then (see [Kre85, equation (4.4)] or [DGS07, Lemma 4.1]), we have the Fourier representation:

$$\begin{array}{lll} A'w(s) & = & \frac{1}{2\pi}\sum_{m\in\mathbb{Z}} \lambda_m \widehat{w}_m \exp(\mathrm{i}ms) \\ \text{with} \quad \lambda_m & = & \pi H^{(1)}_{|m|}(k)\left[\mathrm{i}kJ'_{|m|}(k) + \eta J_{|m|}(k)\right]. \end{array} \tag{3.8}$$

Note that λ_m is the eigenvalue of $A' = A$ corresponding to the eigenfunction $\exp(\pm \mathrm{i}ms)$. As argued in [Kre85], since the eigenfunctions $\exp(\mathrm{i}ms)$, $m \in \mathbb{Z}$, are a complete orthonormal system in $L^2[0,2\pi] = L^2(\Gamma)$, it holds that

$$\|A'\| = \sup_{m\in\mathbb{N}\cup\{0\}} |\lambda_m|, \quad \|A'^{-1}\| = \left(\inf_{m\in\mathbb{N}\cup\{0\}} |\lambda_m|\right)^{-1}, \tag{3.9}$$

so that

$$\mathrm{cond}\, A' = \frac{\sup_{m\in\mathbb{N}\cup\{0\}} |\lambda_m|}{\inf_{m\in\mathbb{N}\cup\{0\}} |\lambda_m|}. \tag{3.10}$$

Further, for $w \in L^2(\Gamma)$ we have that

$$\begin{aligned} |a(w,w)| = |(A'w,w)_{L^2(\Gamma)}| &\geq \Re(A'w,w)_{L^2(\Gamma)} \\ &= \frac{1}{2\pi}\sum_{m\in\mathbb{Z}} \Re(\lambda_m)|\widehat{w}_m|^2 \geq \alpha\|w\|^2_{L^2(\Gamma)}, \end{aligned}$$

where

$$\alpha = \inf_{m\in\mathbb{N}\cup\{0\}} \Re(\lambda_m). \tag{3.11}$$

For the case $d = 3$, when Γ is a sphere of unit radius, a similar analysis applies, based on the fact that the integral operators on the sphere are diagonal operators in the space of spherical harmonics. The corresponding expression for the symbol λ_m is

$$\lambda_m = \mathrm{i}kh^{(1)}_m(k)\left(kj'_m(k) + \mathrm{i}\eta j_m(k)\right), \tag{3.12}$$

where j_m and $h^{(1)}_m$ are the spherical Bessel and Hankel functions respectively. This formula can be found, for example, in [Kre85, Gie97] – see also [BS06]. The formulae (3.9), (3.10), and (3.11) hold also in the 3D case [Kre85, DGS07], with λ_m given by (3.12).

We see that, for the case of a circle/sphere, studying the coercivity and conditioning of A' reduces to the study of the behaviour of explicitly known eigenvalues λ_m, as a function of m, the wavenumber k, and the coupling parameter η. The early papers [KS83, Kre85] carried out a precise theoretical and numerical study of this issue for the case of small k, with emphasis on choosing η so as to minimise the condition number. The thesis of Giebermann [Gie97] made a similar careful study of the large k case, which is our main focus here, with a mixture of rigorous analysis, indicative asymptotics, and numerical calculation. The substantial gaps in the analysis in [Gie97] (in particular, the estimates of the coercivity constant α in [Gie97] were suggestive rather than rigorous: the issue is to obtain sufficiently sharp bounds on the relevant combinations of Bessel functions that are uniform in argument and order) were filled recently in [DGS07], for the explicit choice $\eta = k$ (previously proposed as optimal for conditioning for the unit circle when $k \geq 1$ in e.g. [KS83, Ami90, Ami93]). Further, one of the bounds in [DGS07] was refined recently in [BS07] (an improved upper-bound on the norm of D, the double-layer operator part of A'). Additionally, there are techniques and results that we shall mention below, described in [CM08, CGLL07], that apply to general boundaries, and so to the circle/sphere in particular. The upshot (see [CGLL07] for more historical detail) is that the following results are now known for the circle/sphere. In all these bounds $c \geq 1$ denotes some absolute constant, not necessarily the same at each occurrence.

Coercivity for the circle/sphere ([DGS07, CGLL07]). With the choice of coupling parameter $\eta = k$, $A' = A$ is coercive for all sufficiently large k, with α bounded above and below by constants independent of k. Indeed, for the circle,

$$\alpha = 1 \text{ for all sufficiently large } k.$$

Conditioning for the circle/sphere ([DGS07, CGLL07]).

$$1 \leq \|A'^{-1}\| = \|A^{-1}\| \leq c\left(1 + \frac{1+k}{\eta}\right); \tag{3.13}$$

indeed, for a circle and $\eta = k$ it holds that $\|A'^{-1}\| = \|A^{-1}\| = 1$ for all sufficiently large k. For a sphere,

$$1 \leq \|A'\| = \|A\| \leq c\left(1 + \eta(1+k)^{-2/3}\right); \tag{3.14}$$

the same bound holds for a circle in the case $\eta = k$. Thus, for a sphere,

$$\text{cond } A' = \text{cond } A \le c\left(1 + k^{1/3}\right), \tag{3.15}$$

if $\eta = 1 + k^p$, for some $p \in [2/3, 1]$. The same bound holds for a circle, for $k \ge 1$, with the choice $\eta = 1 + k$.

We note that the above bounds (3.13)–(3.15) suggest that taking $\eta = 1 + k^p$, for some $p \in [2/3, 1]$ will be approximately optimal in terms of minimising the condition numbers of A' and A. In fact, for a sphere of radius R_0, based on low frequency calculations and analysis, the specific choice

$$\eta = \max\left(\frac{1}{2R_0}, k\right) \tag{3.16}$$

was made in [Kre85], and there is further evidence supporting this choice for higher frequencies in [Ami90, Ami93]. Recently, Banjai and Sauter [BS07] have pointed out that, as is clear from (3.15), choosing $\eta = k^{2/3}$ gives the same growth rate as $k \to \infty$ as the choice $\eta = k$, and calculations for the case of a circle confirm almost identical values of condition number for $\eta = k/2$ and $\eta = k^{2/3}$ at high wavenumbers.

In recent work by the authors and their collaborators [DGS07, CM08, CGLL07, CGLL08] rigorous upper and lower bounds on $\|A\| = \|A'\|$ and $\|A^{-1}\| = \|A'^{-1}\|$ have been obtained for rather general classes of scatterers, which results show that: (i) the detail of the geometry of Γ plays a strong role in determining the dependence of these norms on k; (ii) the growth of the condition number with k can be much faster than the mild growth (3.15) for a circle/sphere. We briefly summarise the techniques that have been used and the results that have been obtained.

A first, simple, observation [CGLL07, Lemma 4.1] is that both $\|A'\|$ and $\|A'^{-1}\|$ are bounded below by the value 1, as a consequence of being perturbations of the identity, if some part of Γ is at least C^1 smooth. To obtain upper bounds on $\|A'\|$ rather crude methods are used in [CGLL07] which ignore the oscillation in the kernels of the integral operators D' and S (whose norms are bounded separately, and then $\|A'\|$ is bounded using the triangle inequality). For example, we bound $\|S\|$ using the estimate

$$\|S\| \le 2 \sup_{x\in\Gamma} \int_\Gamma |\Phi(x,y)|\, \mathrm{d}s(y) \le k^{(d-3)/2}(2\pi)^{(1-d)/2} \int_\Gamma \frac{\mathrm{d}s(y)}{|x-y|^{(d-1)/2}}, \tag{3.17}$$

the last inequality in fact an equality in the 3D case ($d = 3$). Our resulting bound on the norm of A' is the following:

Theorem 8 *[CGLL07, Theorem 3.6] For every Lipschitz Γ, there exist positive constants c_1 and c_2, dependent only on Γ, such that*

$$\|A\| = \|A'\| \leq 1 + c_1 k^{(d-1)/2} + c_2 \eta k^{(d-3)/2},$$

for all $k > 0$.

In 2D ($d = 2$), for the case Γ simply-connected and smooth, this bound was shown previously, for all sufficiently large k, in [DGS07].

We note that these bounds predict, for the usual choice $\eta = k$, a faster growth than (3.14) for a circle/sphere as k increases, namely proportional to $k^{1/2}$ in 2D, k in 3D. Perhaps surprisingly, although the techniques used to obtain the above bounds ignore the increasing oscillation in the kernels of the integral operators as k increases, it is shown in [CGLL07] that in 2D (nothing is known yet about the 3D case) the above bounds are sharp, in the sense that there exist Lipschitz boundaries Γ for which $\|S\|$ grows proportional to $k^{-1/2}$ and $\|D'\|$ arbitrarily close to $k^{1/2}$. In particular:

Lemma 9 *[CGLL07, Theorem 4.2] In the 2D case, if Γ contains a straight line section of length a, then*

$$\|S\| \geq \sqrt{\frac{a}{\pi k}} + O(k^{-1})$$

as $k \to \infty$ and

$$\|A\| = \|A'\| \geq \eta \sqrt{\frac{a}{\pi k}} - 1 + O(\eta k^{-1})$$

as $k \to \infty$, uniformly in $\eta > 0$.

The quantitative information in the above lemma is pretty sharp. Indeed if Γ is a straight line of length a then the formula (3.17) tells us that

$$\|S\| \leq 2\sqrt{\frac{a}{\pi k}}.$$

The technique used to obtain Lemma 9 is to construct a $\phi_k \in L^2(\Gamma)$, dependent on the wavenumber k, so as to approximately maximise

$$\frac{\|S\phi_k\|_{L^2(\Gamma)}}{\|\phi_k\|_{L^2(\Gamma)}}.$$

For the proof of Lemma 9 the choice $\phi_k(x) = \exp(\mathrm{i}kx \cdot \hat{c})$ on the straight line part of Γ, and zero elsewhere on Γ, where the unit vector $\hat{c}$ is parallel to the straight line section of Γ, does the trick.

The same technique can be used to construct lower bounds that explore the subtle interaction between the geometry and the size of $\|S\|$ and $\|D\| = \|D'\|$. For example, one result from [CGLL07] is (cf. the bound (3.14) for the case of a circle/sphere):

Lemma 10 *[CGLL07, Corollary 4.5] Suppose (in the 2D case) that Γ is locally C^2 in a neighbourhood of some point x^* on the boundary and let R be the radius of curvature at x^*. If $R < \infty$, then, as $k \to \infty$,*

$$\|S\| \geq \frac{1}{2}\left(\frac{R}{\pi}\right)^{1/3} (2k)^{-2/3}(1 + o(1)). \tag{3.18}$$

If also $\eta k^{-2/3} \to \infty$ as $k \to \infty$, then also

$$\|A'\| = \|A\| \geq \frac{\eta}{2}\left(\frac{R}{\pi}\right)^{1/3} (2k)^{-2/3}(1 + o(1)),$$

as $k \to \infty$.

Other results in [CGLL07] explore what happens if the radius of curvature vanishes (and other higher order smoothness conditions) and under what conditions $\|D'\| = \|D\|$ can be large. The lower bounds in the above lemmas meet the upper bounds in Theorem 8 in some cases. For example, if Γ is a polygon (as in subsection 2.2) and the usual choice $\eta = k$ is made then, for some constants c_1 and c_2,

$$c_1 k^{1/2} \leq \|A'\| = \|A\| \leq c_2 k^{1/2},$$

for all sufficiently large k. In other cases, for example for an ellipse or some other smooth, strictly convex obstacle, there is a gap between our upper and lower bounds: e.g. for $\eta = k$ our upper bound (Theorem 8) gives a growth rate of $k^{1/2}$ while our lower bound (Lemma 10) has a growth rate of $k^{1/3}$. We suspect, from the case of the circle (3.14), and from the evidence of numerical simulations in [CGLL08], that it is our lower bounds that are sharp.

One technique that has not been employed yet to obtain upper bounds, which is standard in the harmonic analysis literature [Ste93], and which could be the tool to close the gap, is the observation that, e.g.

$$\|D'\| = \|D\| = \|DD^*\|^{1/2}.$$

Here D^* is the Hilbert space adjoint of D (whose kernel is the complex conjugate of the kernel of D'). The point is that DD^* is itself an integral operator whose norm can be estimated by the (relatively crude) methods we use to prove Theorem 8, and that the kernel of the integral operator DD^* is given as an oscillatory integral involving the wavenumber k, whose values may be estimated by standard oscillatory integral techniques [Ste93].

To provide upper bounds on $\|A'^{-1}\| = \|A^{-1}\|$ a completely different technique has been used, namely a priori bounds derived from Rellich-type identities and subtle properties of radiating solutions of the Helmholtz equation [CM08]. These upper bounds apply for a general class of geometries, namely whenever the scatterer Ω is simply-connected, piecewise smooth, starlike, and Lipschitz. For the rest of this section we assume, without loss of generality, that the origin lies in Ω ($0 \in \Omega$). Then the class of domains studied in [CM08] are those satisfying the following assumption (Assumption 3 in [CM08]):

Assumption 11 *Γ is Lipschitz and is C^2 in a neighbourhood of almost every $x \in \Gamma$. Further*

$$\delta_- := \operatorname*{ess\,inf}_{x\in\Gamma} \; x \cdot n(x) > 0.$$

Note that Assumption 11 holds, for example, if Ω is a convex polyhedron (and $0 \in \Omega$), with δ_- the distance from the origin to the nearest side of Γ.

Define

$$R_0 := \sup_{x\in\Gamma} |x|, \quad \delta_+ := \operatorname*{ess\,sup}_{x\in\Gamma} \; x \cdot n(x), \quad \delta^* := \operatorname*{ess\,sup}_{x\in\Gamma} |x - (x \cdot n(x))n(x)|.$$

Then a main result in [CM08] is the following:

Theorem 12 *Suppose that Assumption 11 holds and $\eta > 0$. Then*

$$\|A'^{-1}\| = \|A^{-1}\| \le B \tag{3.19}$$

where B is given by the formula:

$$\frac{1}{2} + \left[\left(\frac{\delta_+}{\delta_-} + \frac{4\delta^{*2}}{\delta_-^2}\right)\left[\frac{\delta_+}{\delta_-}\left(\frac{k^2}{\eta^2} + 1\right) + \frac{d-2}{\delta_-\eta} + \frac{\delta^{*2}}{\delta_-^2}\right] + \frac{(1+2kR_0)^2}{2\delta_-^2\eta^2}\right]^{1/2}.$$

To understand this expression for B, suppose first that Γ is a circle or sphere, i.e. $\Gamma = \{x : |x| = R_0\}$. Then $\delta_- = \delta_+ = R_0$ and $\delta^* = 0$ so

$$B = B_0 := \frac{1}{2} + \left[1 + \frac{k^2}{\eta^2} + \frac{d-2}{R_0\eta} + \frac{(1+2kR_0)^2}{2R_0^2\eta^2}\right]^{1/2}. \tag{3.20}$$

In the general case, since $\delta_- \leq \delta_+ \leq R_0$ and $0 \leq \delta_* \leq R_0$, it holds that $B \geq B_0$. Note that the expression B blows up if $k/\eta \to \infty$ or if $\delta_+/\delta_- \to \infty$, or if $\delta_-\eta \to 0$, uniformly with respect to the values of other variables.

An important implication of Theorem 12 is that, whenever Γ is starlike in the sense of Assumption 11, if η is chosen so that

$$\max(l_1 R_0^{-1}, l_2 k) \leq \eta \leq \max(u_1 R_0^{-1}, u_2 k), \tag{3.21}$$

for some positive constants l_1, l_2, u_1, and u_2, then, for some constant $c > 0$, $\|A'^{-1}\| = \|A^{-1}\| \leq c$, for all $k > 0$. For example, choosing

$$\eta = R_0^{-1} + k, \tag{3.22}$$

which satisfies (3.21) with $l_1 = l_2 = 1$ and $u_1 = u_2 = 2$, defining $\theta := R_0/\delta_-$, and noting that $\delta_+/\delta_- \leq \theta$, $\delta^*/\delta_- \leq \theta$, we see that Theorem 12 implies that

$$\|A'^{-1}\| = \|A^{-1}\| \leq B \leq \frac{1}{2} + \theta[2 + (1 + 4\theta)(d + \theta)]. \tag{3.23}$$

Based on computational experience, Bruno and Kunyansky [BK01], [Bru07] recommend the choice $\eta = \max(6T^{-1}, k/\pi)$, where T is the diameter of the obstacle, which satisfies (3.21), this formula chosen on the basis of minimising the number of GMRES iterations in an iterative solver. Another choice of η satisfying (3.21) is (3.16), recommended as optimal for a sphere for low frequency in [Kre85].

Putting together the bounds of Theorems 8 and 12, we see that, in the case when Γ is piecewise C^2, Lipschitz and starlike, satisfying Assumption 11, it holds, for some constant $c \geq 1$ depending on Γ, that

$$1 \leq \text{cond } A' = \text{cond } A \leq c\left(1 + k^{(d-1)/2} + \eta k^{(d-3)/2}\right)\left(1 + \frac{1+k}{\eta}\right). \tag{3.24}$$

Thus, for some constant $c' \geq 1$,

$$1 \leq \text{cond } A' = \text{cond } A \leq c'\left(1 + k^{(d-1)/2}\right), \tag{3.25}$$

if η is chosen to satisfy (3.21), e.g. given specifically by (3.16) or (3.22).

If Γ is not starlike then $\|A'^{-1}\| = \|A^{-1}\|$ can grow as k increases. In particular, a 2D example is presented in [CGLL07, CGLL08] in which Γ is a trapping-type obstacle, with two straight parallel sides separated by the medium of propagation. It is shown in [CGLL07], by combining arguments from [CM08] with methods of estimating multi-dimensional oscillatory integrals from [IN06], that, for some constant $c > 0$,

$$\|A'^{-1}\| = \|A^{-1}\| \geq ck^{9/10}(1 + \eta/k)^{-1}$$

and that

$$\text{cond } A' = \text{cond } A \geq c(1 + k^{14/10})$$

for the usual choice of η satisfying (3.21).

4 Implementation

This paper has concentrated on the theory of integral equation formulations for the Helmholtz equation and their numerical solution by Galerkin and collocation methods in the high-frequency case. A hugely important question, which we only have space to deal with briefly, is whether these methods can be realised with computation times which are reasonable as $k \to \infty$, in particular, do the computation times reflect the theoretical estimates which we have given above? We describe briefly in this section, work on two different issues which are related to this question.

Computation of Oscillatory Stiffness Matrix Entries.
The Galerkin and collocation methods described above require work on the assembly of stiffness matrices, the entries of which are given as oscillatory integrals. The Nyström approach of Bruno et. al. [BGMR04, BG07] involves a direct approach to the integration problem, without the intermediate step of considering it as part of an expansion method for the integral equation. In any case oscillatory integrals defined on (subsets of) obstacle boundaries with, in addition, weakly singular integrands and complicated phase functions must be computed.

In particular, the hybrid Galerkin discretisation (with (2.1)) of the representative integral operator $v(x) \mapsto \int_\Gamma \Phi(x,y)v(y)ds(y)$ taken from (1.4), leads to double integrals of the form

$$\int_{S_{n'}} \int_{S_n} \left\{ \Phi(x,y) \exp(ik[\gamma_m(y) - \gamma_{m'}(x)]) \right\} P_n(y) P_{n'}(x) ds(y) ds(x), \tag{4.1}$$

where the P_n are (piecewise) polynomial basis functions with supports S_n. Because the phase of the fundamental solution Φ is known, the kernel (in the braces) in (4.1) may also be written as $\exp(ik[|x-y| + \gamma_m(y) - \gamma_{m'}(x)])K(x,y)$, with K (weakly) singular but non-oscillatory, revealing an oscillatory double integral with a complicated geometry-dependent phase. Collocation methods lead to the simpler (but still oscillatory) single integrals:

$$\int_{S_n} \left\{ \Phi(x,y) \exp(\mathrm{i}k[\gamma_m(y) - \gamma_{m'}(x)]) \right\} P_n(y) ds(y), \qquad (4.2)$$

to be evaluated at collocation points x.

There has been considerable recent activity on problems of oscillatory integration in general (e.g. [IN04], [IN06]), which has provided new insight and analysis for classical methods such as Filon's rule and Levin's method, and, by particularly exploiting asymptotic theory, has also generated new classes of methods. We refer to the separate review [HO08] in this volume for more detail.

Building on the progress on oscillatory integration in general, Huybrechs and Vandewalle [HV07] described a general method for computing integrals of the form (4.2) using a numerical variant of the method of steepest descent, in which the integral over S_n (parametrised by a real interval) is computed via an integral over a path in the complex plane over which the integrand is not oscillatory. A very nice observation which then follows is that if the collocation point x is not in S_n, if the phase has no stationary points, and if P_n has sufficiently many vanishing derivatives then (4.2) vanishes rapidly as k increases. Hence if local basis functions are used in the collocation method, then the collocation matrix can be replaced by a sparse matrix for large k. The only non-zero entries of the sparse matrix correspond to points x and supports S_n where either $x \in S_n$ (a "singular point"), or the phase has a stationary point. Although there is no stability analysis of the method in [HV07], the numerical results suggest this idea produces a powerful novel algorithm. This idea was further developed in [HTH06, Tre07] in the context of the partition of unity boundary integral method with plane wave basis functions, applied to general Helmholtz problems (as distinct from the plane wave scattering considered here). In this case the oscillatory integrals can contain very complicated distributions of stationary points.

An application of the method of stationary phase to the computation of collocation matrices arising in boundary integral methods with global basis functions on domains which are diffeomorphic to the sphere is presented in [GLS07]. There the chief difficulty is the problem of locating the stationary points for general geometries.

When choosing quadrature rules for implementing boundary integral methods in the high frequency case, one should bear in mind error estimates for the solution of the integral equation such as (2.14) and Theorem 7. It is clear that one requires sharp quadrature error estimates as $N \to \infty$ with explicit dependence on k in the asymptotic constant. When we apply quadrature to approximate the stiffness matrix for these methods, we only need to ensure that the resulting perturbations satisfy the same kind of error estimates and can then apply classical "Strang Lemma" arguments to obtain error estimates for the whole practical method. Progress on this issue for the 2D Galerkin case on a polygon in the context of *hp*-Galerkin methods has been made by Melenk and Langdon [ML07]. This approach employs a change of variable of the form $\tau = |x - y| + \gamma_m(y) - \gamma_{m'}(x)$ for either fixed x or y. This ensures that the oscillation in (4.1) is in one variable only, and then applies Filon quadrature. However much work remains to prove rigorous error estimates and extend the results to 3D.

In an intriguing different approach [Dar02] computes integrals such as (4.1) on a subgrid which resolves the oscillations and does this efficiently using a multipole expansion of the kernel factor $\Phi(x, y)$ for (x, y) in each block of a tree-based decomposition of $\Gamma \times \Gamma$. This is a practical alternative to difficult stationary-phase based methods.

Fast methods for dense systems.

Matrix compression and fast solvers for non-local operator equations are a major development in numerical analysis in the last 25 years. Typical solvers usually consist of a (preconditioned) Krylov iterative method coupled with a fast matrix-vector multiplication based on kernel approximation (e.g. multipole or panel clustering, or more recently $\mathcal{H}$-matrices). The fast multiplication algorithms work by approximating the (weakly singular) kernel $K(x, y)$ by combinations of separable functions of the form $a_i(x)b_j(y)$ when x, y are sufficiently separated. Blocks of the dense stiffness matrix are thus replaced by low rank matrices, with the choice of blocking and approximation controlled by a tree-based hierarchical algorithm. This allows matrix-vector multiplications with the $N \times N$ dense boundary element stiffness matrix in close to $O(N)$ time. This method has been extended to high frequency Helmholtz problems approximated by *conventional* boundary elements (e.g. [CSCVHH04, DH04]), but the extension to *hybrid* approximations is an open and fascinating problem. This is important, especially in 3D, since then, even using the hybrid approximation spaces proposed above, N may still be large. The results of [Dar02] show that replacing the Helmholtz kernel with a separable expansion in the far field can still yield low rank approximations even in the hybrid case, but much work remains to be done to yield a solver for which the cost is close to $O(N)$ with a k-independent constant.

Acknowledgement We would like to thank Valery Smyshlyaev for guiding us through the substantial literature in this field from the former Soviet Union.

Bibliography

[ANZ94] T. Abboud, J.-C. Nédélec & B. Zhou (1994). Méthode des équations intégrales pour les hautes fréquencies. *C.R. Acad. Sci. Paris.* **318** Série I, 165–170.

[ANZ95] T. Abboud, J.-C. Nédélec & B. Zhou (1995). Improvement of the integral equation method for high-frequency problems, in *Proceedings of 3rd International Conference on Mathematical Aspects of Wave Propagation Problems* (SIAM, Phildelphia).

[Ami90] Amini, S. (1990). On the choice of the coupling parameter in boundary integral formulations of the exterior acoustics problem. *Appl. Anal.* **35**, 75–92.

[Ami93] S. Amini (1993). Boundary integral solution of the exterior acoustic problem, *Comput. Mech.* **13**, 2–11.

[ABER06] A. Anand, Y. Boubendir, F. Ecevit & F. Reitich (2006). Analysis of multiple scattering iterations for high-frequency scattering problems II: The three-dimensional scalar case, report 147, Max-Planck-Institut für Mathematik in den Naturwissenschaften, Leipzig.

[ACL07] S. Arden, S. N. Chandler-Wilde & S. Langdon (2007). A collocation method for high frequency scattering by convex polygons. *J. Comp. Appl. Math.* **204**, 334–343.

[Atk97] K. E. Atkinson (1997). *The Numerical Solution of Integral Equations of the Second Kind*, (Cambridge University Press, Cambridge).

[BB91] V. M. Babich & V. S. Buldyrev (1991). *Short-wavelength Diffraction Theory* (Springer-Verlag, Berlin).

[BS07] L. Banjai & S. Sauter (2007). A refined Galerkin error and stability analysis for highly indefinite variational problems, *SIAM J. Numer. Anal.* **45** 37–53.

[BGS05] B. D. Bonner, I. G. Graham & V. P. Smyshlyaev (2005). The computation of conical diffraction coefficients in high frequency acoustic wave scattering, *SIAM J. Numer. Anal.* **43**, 1202–1230.

[BW65] H. Brakhage & P. Werner (1965). Über das Dirichletsche Außenraumproblem für die Helmholtzsche Schwingungsgleichung, *Arch. Math.* **16**, 325–329.

[BG07] O. P. Bruno & C. A. Geuzaine (2007). An $\mathcal{O}(1)$ integration scheme for three-dimensional surface scattering problems, *J. Comp. Appl. Math.* **204**, 463–476.

[BGMR04] O. P. Bruno, C. A. Geuzaine, J. A. Monro & F. Reitich (2004). Prescribed error tolerances within fixed computational times for scattering problems of arbitrarily high frequency: the convex case. *Phil. Trans. R. Soc. Lond. A.* **362**, 629–645.

[BGR05] O. P. Bruno, C. A. Geuzaine & F. Reitich (2005). On the $\mathcal{O}(1)$ solution of multiple-scattering problems, *IEEE Trans. Magn.* **41**, 1488–1491.

[BK01] O. P. Bruno & L. Kunyansky (2001). Surface scattering in three dimensions: an accelerated high-order solver, *Proc. R. Soc. Lond.* **A 457**, 2921-2934.

[Bru07] O. P. Bruno (2007). Private communication.

[BS06] A. Buffa & S. Sauter, S. (2006). On the acoustic single layer potential: Stabilisation and Fourier analysis, *SIAM J. Sci. Comput.* **28**, 1974–1999.

[Bus63] V. S. Buslaev (1964). Short-wave asymptotic behaviour in the problem of diffraction by smooth convex contours (in Russian), *Trudy Mat. Inst. Steklov.* **73**, 14–117. Abbreviated English summary: On the shortwave asymptotic limit in the problem of diffraction by convex bodies, *Soviet Physics Doklady* **7**, 685–687. (1963)

[Bus75] V. S. Buslaev (1975). The asymptotic behavior of the spectral characteristics of exterior problems for the Schrödinger operator (in Russian). *Izv. Akad. Nauk SSSR Ser. Mat.* **39**, 149–235; English translation (1975). *Math. USSR–Izv.* **9**, 139-223.

[CGLL07] S. N. Chandler-Wilde, I. G. Graham, S. Langdon & M. Lindner (2007). Condition number estimates for combined potential boundary integral operators in acoustic scattering, Isaac Newton Institute for Mathematical Sciences Preprint NI07067-HOP, Isaac Newton Institute.

[CGLL08] S. N. Chandler-Wilde, I. G. Graham, S. Langdon & M. Lindner (2008). Condition number estimates for combined potential integral operators in acoustics and their boundary element discretisation, in preparation.

[CL07] S. N. Chandler-Wilde & S Langdon (2007). A Galerkin boundary element method for high frequency scattering by convex polygons, *SIAM J. Numer. Anal.* **45**, 610–640.

[CLR04] S. N. Chandler-Wilde, S. Langdon & L. Ritter (2004). A high-wave number boundary-element method for an acoustic scattering problem, *Phil. Trans. R. Soc. Lond. A.* **362**, 647–671.

[CM08] S. N. Chandler-Wilde & P. Monk (2008). Wave-number-explicit bounds in time-harmonic scattering, *SIAM J. Math. Anal.* **39**, 1428-1455.

[Cha73] J. Chazarain (1973). Construction de la paramétrix du problème mixte hyperbolique pour l'equation des ondes, *C. R. Acad. Sc. Paris* **276**, 1213–1215.

[CSCVHH04] W. C. Chew, J. M. Song, T. J. Cui, S. Velamparambil, L. Hastriter & B. Hu (2004). Review of large scale computing in electromagnetics with fast integral equation solvers, *Computer Modeling in Engineering and Sciences*, **5**, 361-372.

[CK83] D. Colton & R. Kress (1983). *Integral Equation Methods in Scattering Theory* (Wiley, New York).

[Dar02] E. Darrigrand (2002). Coupling of fast multipole method and microlocal discretization for the 3-D Helmholtz equation, *J. Comput. Phys.* **181**, 126–154.

[DH04] E. Darve & P. Havé (2004). A fast multipole method for Maxwell equations stable at all frequencies, *Phil. Trans. R. Soc. Lond. A*, **362**, 603–628.

[DGS07] V. Dominguez, I. G. Graham & V. P. Smyshlyaev, V.P. (2007). A hybrid numerical-asymptotic boundary integral method for high-frequency acoustic scattering, *Numer. Math.* **106**, 471–510.

[Ece05] F. Ecevit (2005). Integral equation formulations of electromagnetic and acoustic scattering problems: convergence of multiple scattering iterations and high-frequency asymptotic expansions. PhD Thesis, University of Minnesota.

[ER06] F. Ecevit & F. Reitich (2006). Analysis of multiple scattering iterations for high-frequency scattering problems. I: The two dimensional case, report 137, Max-Planck-Institut für Mathematik in den Naturwissenschaften, Leipzig.

[FJR78] E. B. Fabes, M. Jodeit & N. M. Riviere (1978). Potential techniques for boundary value problems on C^1 domains, *Acta Math.*, **141**, 165–186.

[Fil76] V. B. Filippov (1976). Rigorous justification of the shortwave asymptotic theory of diffraction in the shadow zone. *J. Sov. Math.* **6**, 577-626.

[Foc65] V. A. Fock (1965) *Electromagnetic Diffraction and Propagation Problems* (Pergamon Press, New York).

[GLS07] M. Ganesh, S. Langdon & I. H. Sloan. (2007). Efficient evaluation of highly oscillatory acoustic scattering surface integrals. *J. Comp. Appl. Math.* **204**, 363–374.

[Gie97] K. Giebermann (1997). Schnelle Summationsverfahren zur numerischen Lösung von Integralgleichungen für Streuprobleme im $\mathbb{R}^3$. PhD Thesis, University of Karlsruhe.

[GK04] E. Giladi & J. B. Keller (2004). An asymptotically derived boundary element method for the Helmholtz equation, in *Proceedings of the 20th Annual Review of Progress in Applied Computational Electromagnetics*, (Syracuse, New York).

[Gil07] E. Giladi (2007). Asymptotically derived boundary elements for the Helmholtz equation in high frequencies, *J. Comp. Appl. Math.*, **198**, 52 - 74.

[Gir82] J. Giroire (1982). Integral equation methods for the Helmholtz equation, *Integral Equations and Operator Theory* **5**, 506-517.

[Gri85] P. Grisvard (1985). *Elliptic Problems in Nonsmooth Domains* (Pitman, Boston).

[HL94] T. Hargé & G. Lebeau (1994). Diffraction par un convexe. *Invent. Math.* **118**, 161-196 .

[HTH06] M. E. Honnor, J. Trevelyan & D. Huybrechs (2006). Numerical evaluation of 2D partition of unity boundary integrals for Helmholtz problems, Preprint.

[HV07] D. Huybrechs & S. Vandewalle (2007). A sparse discretisation for integral equation formulations of high-frequency scattering problems, *SIAM J. Sci. Comput.*, **29**, 2305–2328.

[HO08] D. Huybrechs & S. Olver (2008). Highly oscillatory quadrature, this volume.

[IN04] A. Iserles & S. P. Nørsett (2004). On quadrature methods for highly oscillatory integrals and their implementation, *BIT* **44**, 755-772.

[IN06] A. Iserles & S. P. Nørsett (2006). Quadrature methods for multivariate highly oscillatory integrals using derivatives", *Math. Comp.* **75**, 1233-1258.

[Kre85] R. Kress (1985). Minimizing the condition number of boundary integral operators in acoustic and electromagnetic scattering. *Q. Jl. Mech. appl. Math.*, **38**, 323-341.

[KS83] R. Kress & W. T. Spassov (1983). On the condition number of boundary integral operators for the exterior Dirichlet problem for the Helmholtz equation, *Numer. Math.* **42**, 77-85.

[LC06] S. Langdon & S. N. Chandler-Wilde (2006). A wavenumber independent boundary element method for an acoustic scattering problem, *SIAM J. Numer. Anal.* **43**, 2450–2477.

[LMC08] S. Langdon, M. Mokgolele & S. N. Chandler-Wilde (2008). High frequency scattering by convex curvilinear polygons, Isaac Newton Institute for Mathematical Sciences Preprint NI08012-HOP.

[Leb84] G. Lebeau (1984). Régularité Gevrey 3 pour la diffraction, *Comm. in Partial Differential Equations* **9**, 1437–1494.

[Lei65] R. Leis (1965). Zur Dirichtletschen Randwertaufgabe des Aussenraums der Schwingungsgleichung, *Math. Z.* **90**, 205–211.

[ML07] J. M. Melenk & S. Langdon (2007). An *hp*-BEM for high frequency scattering by convex polygons, Proceedings of WAVES2007, University of Reading.

[MT85] E. B. Melrose & M. E. Taylor (1985). Near peak scattering and the corrected Kirchhoff approximation for a convex obstacle, *Adv. in Math.* **55**, 242-315.

[ML68] C. S. Morawetz & D. Ludwig (1968). An inequality for the reduced wave equation and the justification of geometrical optics, *Comm. Pure Appl. Math.* **21**, 187-203.

[MR08] M. Motamed & O. Runborg (2008). Approximation of high frequency wave propagation problems, this volume.

[Pan65] O. I. Panič (1965). On the question of the solvability of the exterior boundary-value problems for the wave eqaution and Maxwell's equations, *Usp. Mat. Nauk* **20A**, 221–226.

[PLBT04] E. Perrey-Debain, O. Lagrouche, P. Bettess & J. Trevelyan (2004). Plane-wave basis finite elements and boundary elements for three-dimensional wave scattering, *Phil. Trans. R. Soc. Lond. A*, **362**, 561–577.

[Pop87] G. Popov (1987). Some estimates of Green's functions in the shadow, *Osaka J. Math.* **24**, 1-12.

[Ste93] E. M. Stein (1993). *Harmonic Analysis: Real-Variable Methods, Orthogonality, and Oscillatory Integrals* (Princeton University Press).

[Tre07] J. Trevelyan (2007). Numerical steepest descent evaluation of 2D partition of unity boundary integrals for Helmholtz problems, *Oberwolfach Reports* **5**, 354–356.

[Urs68] F. Ursell (1968). Creeping modes in a shadow, *Proc. Camb. Phil. Soc.* **68**, 171–191.

[WC99] K. F. Warnick & W. C. Chew (1999). Convergence of moment-method solutions of the electric field integral equation for a 2-D open cavity, *Microwave Optical Tech. Letters* **23**, 212–218.

[WC01] K. F. Warnick & W. C. Chew (2001). On the spectrum of the electric field integral equation and the convergence of the moment method, *Int. J. Numer. Meth. Engng.* **51**, 31–56.

[WC04] K. F. Warnick & W. C. Chew (2004). Error analysis of the moment method, *IEEE Ant. Prop. Mag.* **46**, 38–53.

[ZF85] A. B. Zayaev & V. P. Filippov (1985). Rigorous justification of the asymptotic solutions of "sliding-wave" type. *J. Sov. Math.* **30**, 2395–2406.

[ZF86] A. B. Zayaev & V. P. Filippov (1986). Rigorous justification of the Friedlander-Keller formulas. *J. Sov. Math.* **32**, 134–143.

[Zwo90] M. Zworski (1990). High frequency scatering by a convex obstacle. *Duke Math. J.* **61**, 545–634.

8

Novel analytical and numerical methods for elliptic boundary value problems

Athanasios S. Fokas
Department of Applied Mathematics and Theoretical Physics
University of Cambridge
Wilberforce Rd
Cambridge CB3 0WA
United Kingdom
Email: T.Fokas@damtp.cam.ac.uk

Euan A. Spence
Department of Applied Mathematics and Theoretical Physics
University of Cambridge
Wilberforce Rd
Cambridge CB3 0WA
United Kingdom
Email: E.A.Spence@damtp.cam.ac.uk

Abstract

A new method for solving boundary value problems has recently been introduced by the first author. Although this method was first developed for non-linear integrable PDEs (using the crucial notion of a *Lax pair*), it has also given rise to new analytical and numerical techniques for linear PDEs. Here we review the application of the new method to linear elliptic PDEs, using the modified Helmholtz equation as an illustrative example.

1 Introduction

Almost forty years ago an ingenious new method was discovered for the solution of the initial value problem of the Korteweg–de Vries (KdV) equation [GGKM67]. This new method, which was later called the inverse scattering transform (IST) method was based on the mysterious fact that the KdV equation is equivalent to two linear eigenvalue equations called a Lax pair (in honor of Peter Lax, [Lax68] who first understood that the IST method was the consequence of this remarkable property). The KdV equation belongs to a large class of nonlinear equations which are called *integrable*. Although there exist several types of integrable equations, which include PDEs, ODEs, singular-integrodifferential equations, difference equations and cellular automata, the existence of an associated Lax pair provides a common feature of all these equations.

After several attempts to extend the inverse scattering transform method from initial value problems to boundary value problems, a unified method for solving boundary value problems for linear and integrable nonlinear PDEs was introduced by the first author in [Fok97] and reviewed in [Fok08]. This method is based on two novel ideas:

1. Perform the spectral analysis of both parts of the Lax pair of the given integrable PDE *simultaneously* (this is to be contrasted with the case of the inverse scattering transform where one performs the spectral analysis of only the t-independent part of the Lax pair).
2. Analyse a certain *global relation* which couples the given boundary data and the unknown boundary values.

Lax pairs for linear evolution PDEs were first introduced in [FG94]. Actually it is straightforward to construct Lax pairs for a large class of linear PDEs which include PDEs with constant coefficients, such as the following PDE

$$\nabla^2 q(\boldsymbol{x}) + \lambda q(\boldsymbol{x}) = 0, \tag{1.1}$$

where λ is a real constant; for $\lambda = 0$ this is Laplace's equation, $\lambda > 0$ the Helmholtz equation, and $\lambda < 0$ the Modified Helmholtz equation.

In this paper we will review the application of the new method to such second order linear elliptic PDEs, using the modified Helmholtz equation as a specific example. We begin by recalling the classic theory for the solution of (1.1).

1.1 The classic theory

Let (1.1) be valid in Ω which is a piecewise smooth domain. Green's theorem gives the following integral representation

$$q(\boldsymbol{x}) = \int_{\partial\Omega} \left(q(\boldsymbol{\xi}) \frac{\partial E}{\partial n}(\boldsymbol{\xi}, \boldsymbol{x}) - E(\boldsymbol{\xi}, \boldsymbol{x}) \frac{\partial q}{\partial n}(\boldsymbol{\xi}) \right) dS(\boldsymbol{\xi}), \quad \boldsymbol{x} \in \Omega \tag{1.2}$$

where E is the fundamental solution (or free space Green's function) satisfying

$$\left(\nabla^2_{\boldsymbol{\xi}} + \lambda \right) E(\boldsymbol{\xi}, \boldsymbol{x}) = \delta(\boldsymbol{\xi} - \boldsymbol{x}), \quad \boldsymbol{\xi} \in \Omega. \tag{1.3}$$

E is given for the Laplace, Helmholtz and modified Helmholtz equations respectively by the following expressions:

$$E = \frac{1}{2\pi}\log|\boldsymbol{\xi} - \boldsymbol{x}|$$
$$E = -\frac{i}{4}H_0^{(1)}(\sqrt{\lambda}|\boldsymbol{\xi} - \boldsymbol{x}|)$$
$$E = -\frac{1}{2\pi}K_0(\sqrt{-\lambda}|\boldsymbol{\xi} - \boldsymbol{x}|)$$

where $H_0^{(1)}$ is a Hankel function and K_0 a modified Bessel function. We note that a drawback for both Helmholtz and modified Helmholtz is that there does not exist an explicit representation for the fundamental solution.

The integral representation (1.2) involves both q and its normal derivative on the surface, that is both the *Dirichlet* and *Neumann* boundary values respectively. However for a BVP problem to be well-posed either the Dirichlet, or the Neumann, or a combination of these boundary values is given as boundary conditions. Actually, in many applications it is precisely the unknown boundary values that are required. If Dirichlet boundary conditions are given, the determination of the unknown Neumann boundary values is achieved by finding the *Dirichlet to Neumann map*.

For certain simple domains the method of images or separation of variables can be used to find the appropriate Green's function eliminating the dependence of (1.2) on the unknown boundary values.

For more general domains, a linear integral equation can be derived for the unknown boundary values by taking the limit of (1.2) as $\boldsymbol{x}$ tends to the boundary of the domain. The solution of this *singular integral equation* is known as the *boundary integral method*, and its numerical implementation the *boundary element method.*

1.2 From Green to Lax

The starting point of the classic theory is to write the PDE (1.1) in *divergence form.* This can be done by utilizing the formal adjoint. For example, let q satisfy the modified Helmholtz equation,

$$q_{xx} + q_{yy} - 4\beta^2 q = 0, \quad (x, y) \in \Omega, \tag{1.4}$$

Denote by $\tilde{q}$ a solution of the adjoint PDE, which in this case coincides with equation (1.4). The equations for q and $\tilde{q}$ imply

$$(\tilde{q}q_x - \tilde{q}_x q)_x - (q\tilde{q}_y - q_y\tilde{q})_y = 0, \quad (x,y) \in \Omega. \tag{1.5}$$

Choosing instead of $\tilde{q}$ the fundamental solution $E(x', y'; x, y)$ and employing Green's theorem we find

$$\int_{\partial\Omega} [(Eq_{y'} - E_{y'}q)\, dx' - (Eq_{x'} - E_{x'}q)\, dy'] = \begin{cases} q(x,y), & (x,y) \in \Omega, \\ 0, & (x,y) \notin \Omega. \end{cases} \tag{1.6}$$

where $\partial\Omega$ denotes the boundary of Ω. The first of these equations is (1.2).

Alternatively, separation of variables implies that we can choose $\tilde{q} = \exp[k_1 x + k_2 y]$, where $k_1^2 + k_2^2 = 4\beta^2$. Hence, introducing the potential M, equation (1.5) implies

$$M_y = e^{k_1 x + k_2 y}(q_x - k_1 q), \quad M_x = e^{k_1 x + k_2 y}(k_2 q - q_y), \tag{1.7}$$

i.e. M satisfies

$$dM = e^{k_1 x + k_2 y}\left[(k_2 q - q_y)dx + (q_x - k_1 q)dy\right]. \tag{1.8}$$

Using the parameterization $k_1 = 2\beta \sin\lambda$, $k_2 = 2\beta\cos\lambda$ and letting $\exp[i\lambda] = k$, $M = \mu \exp[k_1 x + k_2 y]$, equations (1.7) yield the following *Lax pair*

$$\mu_y + \beta\left(\tfrac{1}{k} + k\right)\mu = q_x + i\beta\left(k - \tfrac{1}{k}\right)q,$$

$$\mu_x + i\beta\left(\tfrac{1}{k} - k\right)\mu = -q_y + \beta\left(k + \tfrac{1}{k}\right)q, \quad (x,y) \in \Omega, \quad k \in \mathbf{C}.$$

These two equations are compatible (i.e. $\mu_{xy} = \mu_{yx}$) if and only if q satisfies (1.4). These equations are equivalent (compare with (1.8)) with

$$d\left[\mu e^{i\beta(\frac{1}{k} - k)x + \beta(\frac{1}{k} + k)y}\right] = e^{i\beta(\frac{1}{k} - k)x + \beta(\frac{1}{k} + k)y}\left\{\left[q_x + i\beta\left(k - \frac{1}{k}\right)q\right]dy\right.$$
$$\left. + \left[-q_y + \beta\left(k + \frac{1}{k}\right)q\right]dx\right\}, \quad (x,y) \in \Omega, \quad k \in \mathbf{C}. \tag{1.9}$$

Applying Green's theorem in the domain Ω, equation (1.9) yields

$$\int_{\partial\Omega} e^{i\beta(\frac{1}{k} - k)x + \beta(\frac{1}{k} + k)y}\left\{\left[q_x + i\beta\left(k - \frac{1}{k}\right)q\right]dy\right.$$
$$\left. + \left[-q_y + \beta\left(k + \frac{1}{k}\right)q\right]dx\right\} = 0, \quad k \in \mathbf{C}. \tag{1.10}$$

We call this equation the *global relation*. It characterizes the *Dirichlet to Neumann correspondence* and plays a crucial role in the new method. Note that a second global relation can be obtained from (1.10) by letting $k \mapsto 1/k$.

1.3 The new method

The new method involves two ingredients, namely:

1. An integral representation expressing the solution q in terms of transforms of the Dirichlet and Neumann boundary values.
2. One or more global relations, i.e. equations coupling the transforms of the Dirichlet and Neumann boundary values.

It is shown in §2 that the integral representation and global relations for the modified Helmholtz equation in the interior of a convex polygon are given as follows:

Proposition 1 *Let Ω be the interior of a convex bounded polygon in the complex z-plane, with corners $z_1, \dots, z_n, z_{n+1} = z_1$, see Fig. 1. Assume that there exists a solution $q(z, \bar{z})$ of the modified Helmholtz equation*

$$q_{z\bar{z}} - \beta^2 q = 0, \quad z \in \Omega, \quad \beta > 0, \tag{1.11}$$

valid in the interior of Ω and suppose that this solution has sufficient smoothness all the way to the boundary of the polygon. Then q can be expressed in the form

$$q(z, \bar{z}) = \frac{1}{4i\pi} \sum_{j=1}^{n} \int_{l_j} e^{i\beta(kz - \frac{\bar{z}}{k})} \hat{q}_j(k) \frac{dk}{k}, \quad z \in \Omega, \tag{1.12}$$

where the functions $\{\hat{q}_j(k)\}_1^n$ are defined by

$$\hat{q}_j(k) = \int_{z_j}^{z_{j+1}} e^{-i\beta(kz - \frac{\bar{z}}{k})} \left[(q_z + ik\beta q)\frac{dz}{ds} - \left(q_{\bar{z}} + \frac{\beta}{ik} q \right) \frac{d\bar{z}}{ds} \right] ds, \quad k \in \mathbf{C},$$
$$j = 1, \dots, n, \quad z_{n+1} = z_1, \tag{1.13}$$

$z(s)$ is a parametrization of the side (z_j, z_{j+1}) and $\{l_j\}_1^n$ are the rays on the complex k-plane oriented toward infinity and defined by

$$l_j = \{k \in \mathbf{C} : \arg(k) = -\arg(z_{j+1} - z_j)\}, j = 1, \dots, n, z_{n+1} = z_1. \tag{1.14}$$

Equation (1.13) can be written in the form

$$\hat{q}_j(k) = \int_{z_j}^{z_{j+1}} e^{-i\beta(kz - \frac{\bar{z}}{k})} \left[iq_n + i\beta \left(\frac{1}{k}\frac{d\bar{z}}{ds} + k\frac{dz}{ds} \right) q \right] ds, \quad k \in \mathbf{C},$$

$$j = 1, \ldots, n, \quad z_{n+1} = z_1, \tag{1.15}$$

where q_n denotes the derivative of q normal to the boundary of Ω. Furthermore, the following *global relations* are valid

$$\sum_{j=1}^{n} \hat{q}_j(k) = 0, \quad \sum_{j=1}^{n} \tilde{q}_j(k) = 0, \quad k \in \mathbf{C}, \tag{1.16}$$

where $\{\tilde{q}_j(k)\}_1^n$ are defined by

$$\tilde{q}_j(k) = \int_{z_j}^{z_{j+1}} e^{i\beta(k\bar{z} - \frac{z}{k})} \left[iq_n + i\beta \left(\frac{1}{k}\frac{dz}{ds} + k\frac{d\bar{z}}{ds} \right) q \right] ds, \quad k \in \mathbf{C},$$

$$j = 1, \ldots, n, z_{n+1} = z_1. \tag{1.17}$$

The global relations couple the functions q and q_n on the boundary, i.e. they couple the Dirichlet and Neumann boundary values, and thus they characterize the Dirichlet to Neumann correspondence.

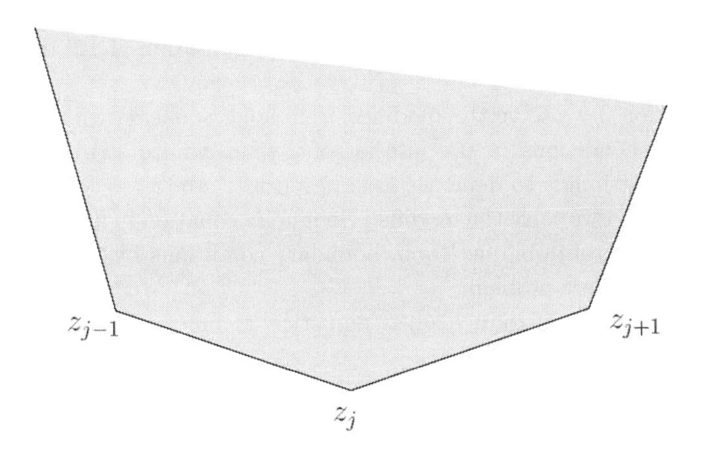

Fig. 1. Part of the convex polygon.

The key idea is to use the global relations to either eliminate the contributions of the unknown boundary values from the integral representation, or determine these contributions directly. This is possible due to the fact that the global relations (1.16) are valid with an *arbitrary complex number* k. So far there have been three different approaches for doing this:

1. **Use the *invariant properties* of the global relations to eliminate the contributions of the unknown boundary values from the integral representation using only algebraic manipulations.** This is possible for certain simple domains and boundary conditions (including those solvable by the method of images or separation of variables). Also, whereas the classic techniques yields solutions in the form of infinite series, the new method yields solutions in the form of integrals which can be deformed so their integrands decay exponentially. This has both analytical and computational advantages over the classical solutions.
2. For more complicated boundary conditions, **formulate a Riemann–Hilbert problem** [AF03] **for the unknown boundary values.**
3. For more general domains, **solve the global relations numerically to determine the unknown boundary values.**

Outline of the paper In section 2 we outline the different methods which can be used to derive the integral representation (1.12) of Proposition 1.

In section 3 we solve the modified Helmholtz equation with Dirichlet boundary conditions in the interior of a semi-infinite strip and of an equilateral triangle to demonstrate approach (i) above.

In section 4 we solve the modified Helmholtz equation in a semi-infinite strip with certain oblique Robin boundary conditions by formulating a Riemann–Hilbert problem.

In section 5 we introduce a new numerical technique, which solves the global relations numerically.

2 The derivations of integral representations

There exist three different ways of obtaining the integral representations:

1. Spectral analysis of the differential form (1.9)

2. Substituting a spectral representation of the fundamental solution into Green's integral representation
3. Applying the global relation in a subdomain of Ω.

Among these approaches only the first one can be generalised to non-linear PDEs [Fok08].

2.1 Spectral analysis of the differential form

It is convenient to change from cartesian to complex co-ordinates

$$z = x + iy, \quad \bar{z} = x - iy \tag{2.1}$$

so that

$$\partial_x = \partial_z + \partial_{\bar{z}}, \quad \partial_y = i\left(\partial_z - \partial_{\bar{z}}\right).$$

Then equations (1.9) and (1.10) become

$$\begin{aligned} &d\left[e^{-i\beta\left(kz - \frac{\bar{z}}{k}\right)}\mu(z,\bar{z},k)\right] \\ &= -ie^{-i\beta\left(kz-\frac{\bar{z}}{k}\right)}\left[(q_z + ik\beta q)dz - \left(q_{\bar{z}} + \frac{\beta}{ik}q\right)d\bar{z}\right], \qquad k \in \mathbf{C}, \end{aligned} \tag{2.2}$$

and

$$\int_{\partial\Omega} e^{-i\beta\left(kz-\frac{\bar{z}}{k}\right)}\left[(q_z + ik\beta q)dz - \left(q_{\bar{z}} + \frac{\beta}{ik}q\right)d\bar{z}\right] = 0, \quad k \in \mathbf{C}. \tag{2.3}$$

By performing the spectral analysis of the differential form (2.2) we will derive the result of Proposition 1. This result was first obtained in [Fok01].

For $z \in \Omega$ define

$$\mu_j(z,\bar{z},k) = \int_{z_j}^{z} e^{i\beta[k(z-\zeta)-\frac{1}{k}(\bar{z}-\bar{\zeta})]}\left[(q_\zeta + ik\beta q)d\zeta - \left(q_{\bar{\zeta}} + \frac{\beta}{ik}q\right)d\bar{\zeta}\right]. \tag{2.4}$$

Now μ_j involves the two exponentials

$$e^{i\beta k(z-\zeta)}, \quad e^{-\frac{i\beta}{k}(\bar{z}-\bar{\zeta})} = e^{-\frac{i\beta\bar{k}}{|k|^2}(\bar{z}-\bar{\zeta})}.$$

The real part of these two exponentials have the same sign, thus the exponentials have identical domains of boundness as k and $1/k$ tend to infinity. The term $\exp[i\beta k(z-\zeta)]$ is bounded as $k \to \infty$ for

$$0 \le \arg k + \arg(z-\zeta) \le \pi. \tag{2.5}$$

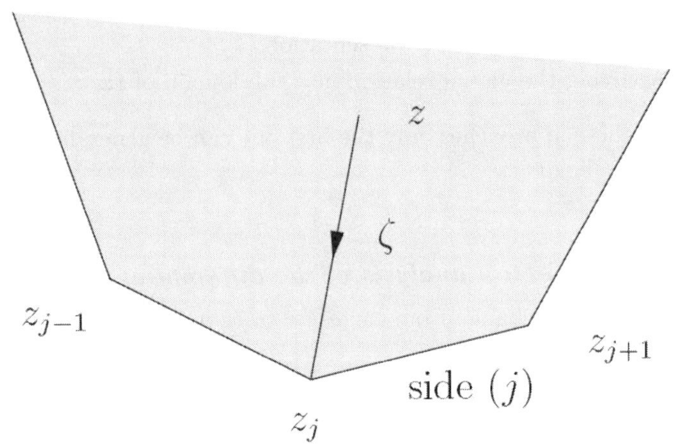

Fig. 2. The contour of integration for μ_j.

If z is inside the polygon and ζ is on a curve from z to z_j, see Fig. 2, then

$$\arg(z_{j+1} - z_j) \leq \arg(z - \zeta) \leq \arg(z_{j-1} - z_j), \quad j = 1, \ldots, n.$$

Hence, the inequalities (2.5) are satisfied provided that

$$-\arg(z_{j+1} - z_j) \leq \arg k \leq \pi - \arg(z_{j-1} - z_j).$$

Hence, the function μ_j is an entire function of k which is bounded as $k \to \infty$ and $k \to 0$ in the sector Σ_j defined by

$$\Sigma_j = \{k \in \mathbf{C}, \quad \arg k \in [-\arg(z_{j+1} - z_j), \pi - \arg(z_{j-1} - z_j)]\}, \quad j = 1, \ldots, n. \tag{2.6}$$

The angle of the sector Σ_j, which we denote by Ψ_j, equals

$$\Psi_j = \pi - \arg(z_{j-1} - z_j) + \arg(z_{j+1} - z_j) = \pi - \phi_j,$$

where ϕ_j is the angle at the corner z_j. Hence

$$\sum_{j=1}^{n} \Psi_j = n\pi - \sum_{j=1}^{n} \phi_j = n\pi - \pi(n-2) = 2\pi,$$

thus the sectors $\{\Sigma_j\}_1^n$ precisely cover the complex k-plane. Hence, the function

$$\mu = \mu_j, \quad z \in \Omega, \quad k \in \Sigma_j, \quad j = 1, \ldots, n,$$

defines a sectionally analytic function in the complex k-plane.

The differential form (2.2) is equivalent to the following Lax pair,

$$\mu_z - i\beta k\mu = q_z + i\beta kq, \quad \mu_{\bar{z}} + \frac{i\beta}{k}\mu = -\left(q_{\bar{z}} + \frac{\beta}{ik}q\right). \tag{2.7}$$

The first of these equations suggests that

$$\mu = -q + O\left(\frac{1}{k}\right), \quad k \to \infty. \tag{2.8}$$

This can be verified using equation (2.4) with $k \in \Sigma_j$ and integration by parts. Also subtracting equation (2.4) and the analogous equation for μ_{j+1} we find

$$\mu_j - \mu_{j+1} = e^{i\beta\left(kz - \frac{\bar{z}}{k}\right)}\hat{q}_j(k), \quad k \in l_j, \tag{2.9}$$

where $\{\hat{q}_j\}_1^n$ are defined by equation (1.13). This follows from the fact that (2.2) is an exact differential.

The ray l_j is the ray of overlap of the sectors Σ_j and Σ_{j+1}. Using the identity

$$\pi - \arg(z_j - z_{j+1}) = -\arg(z_{j+1} - z_j) \mod(2\pi),$$

it follows that l_j is defined by equation (1.14). Furthermore, Σ_j is to the left of Σ_{j+1}, see Fig. 3.

To show that the expression for $\hat{q}_j$ becomes the one of equation (1.15) we use the following identities

$$q_z dz = \frac{1}{2}(\dot{q} + iq_n)ds, \quad q_{\bar{z}} d\bar{z} = \frac{1}{2}(\dot{q} - iq_n)ds,$$

where $\dot{q}$ is the derivative along the side, i.e. $\dot{q} = dq(z(s))/ds$ and q_n is the derivative normal to the side in the outward direction.

The solution of the RH problem defined by (2.8) and (2.9) is given for all $k \in \mathbf{C} \setminus \sum_1^n l_j$ by

$$\mu = -q + \frac{1}{2i\pi}\sum_{j=1}^{n}\int_{l_j} e^{i\beta\left(lz - \frac{\bar{z}}{l}\right)}\hat{q}_j(l)\frac{dl}{l-k}, \quad z \in \Omega.$$

Substituting this expression in the second of equations (2.7) we find equation (1.12).

Using the definitions of $\{\hat{q}_j\}_1^n$ and $\{\tilde{q}_j\}_1^n$ (i.e. equations (1.13) and (1.17)) in equations (2.3) and the equation obtained by letting $k \mapsto 1/k$ in (2.3), we find the global relations (1.16). □

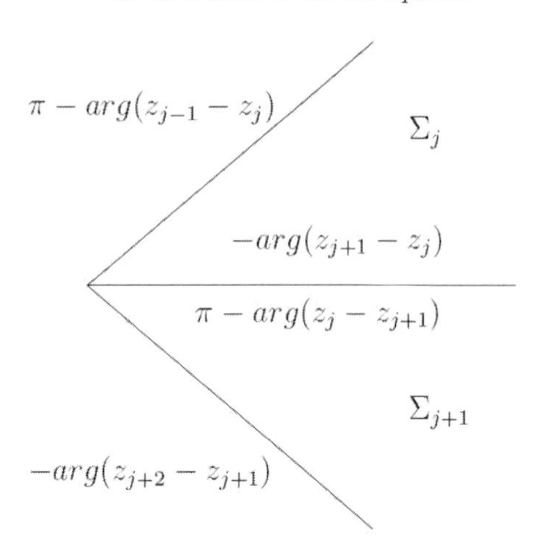

Fig. 3. The sectors Σ_j and Σ_{j+1}.

Remark 1 It is shown in (Fokas 01) that a similar result is valid for an unbounded polygon provided that q vanishes at infinity. In this case, if $z_1 = z_n = \infty$, then we define only $n-1$ functions $\hat{q}_j$. Furthermore, $\hat{q}_1$ and $\hat{q}_{n-1}$ are *not* defined for all complex k but only for k in certain sectors of the complex k-plane. The derivation is similar to the above, where the proof that $\{\Sigma_j\}_1^{n-1}$ cover the complex k-plane follows from the identity

$$\sum_{j=2}^{n-1} \phi_j = \pi(n-3) + \phi_\infty.$$

2.2 From the physical to the spectral

There exists a beautiful connection between the new method, which is formulated in the spectral (or Fourier) plane, and the classic technique of the method of images, formulated in the physical plane.

First, it is possible to obtain the integral representation (1.12) starting from Green's integral representation (1.2); this was implemented in [FZ02]. Consider for example the modified Helmholtz equation. Representing the product of δ functions in terms of Fourier integrals,

$$\delta(x-x')\delta(y-y') = \frac{1}{4\pi^2} \int\int_{\mathbf{R}^2} e^{ik_1(x-x')+ik_2(y-y')} dk_1 dk_2,$$

it follows that the fundamental solution of equation (1.4) is given by

$$E = -\frac{1}{4\pi^2}\iint_{\mathbf{R}^2} \frac{e^{ik_1(x-x')+ik_2(y-y')}}{k_1^2+k_2^2+4\beta^2} dk_1 dk_2.$$

Substituting this expression in the left-hand side of the first of equations (1.6) and comparing the resulting equation with the formula for q given by equation (1.12) (the representation in the spectral plane) we observe that the former equation involves integrations with respect to both dk_1 and dk_2, whereas the latter equation involves integration only with respect to dk. It is shown in [FZ02] that after performing a change of variables from (k_1, k_2) to (k_T, k_N) where k_T and k_N are tangent and normal to $\partial\Omega$, the convexity of Ω means it is possible to compute explicitly the integral with respect to dk_N and then the integral representation in the physical plane yields the integral representation in the spectral plane.

Second, let us refer to equations (1.6) as the integral representation in the physical plane and the global relation in the physical plane respectively; it can be shown that using the *invariant* properites of the global relation to eliminate the unknown boundary values in the integral representation, is equivalent to the method of images [Fok08].

2.3 Applying the global relation in a subdomain

By applying the global relation in a subdomain of Ω it is possible to find an integral transform of the solution q, in this case the usual Fourier transform. This can be inverted, and then the integral representation (1.12) can be obtained by deforming contours [Spe08]. This method has the advantage that it involves the least mathematical machinery: only the Fourier transform and Cauchy's theorem. Integral representations for linear evolution PDEs were first obtained using this method in [Fok02].

3 The elimination of the unknown boundary values using algebraic manipulations

For certain simple domains and boundary conditions (including those solvable by the method of images or separation of variables) it is possible to use the invariant properties of the global relations to eliminate the contributions of the unknown boundary values from the integral representation using *algebraic manipulations*. Moreover,

- In constrast to the classic techniques, which yield solutions in the form of *infinite series*, the new method yields solutions in the form of *integrals* which can be deformed so their integrands decay exponentially. This has both analytical and computational advantages over the classical solutions.
- This method yields explicit solutions to certain problems which it appears cannot be obtained using classical techniques. An example of this is the solution of the modified Helmholtz equation in the interior of an equilateral triangle with certain oblique Robin boundary conditions, see Remark 5.

Notation

- The given Dirichlet or Neumann boundary data will be denoted by d and n respectively and their transforms by D and N. Other type of given boundary data will be denoted by g (for given) and their transforms by G. The transforms of the unknown boundary values will be denoted by U (for unknown).
- If q is real, then the second global relation for the modified Helmholtz equation (i.e. the second of equations (1.16)) can be obtained from the first global relation (i.e. the first of equations (1.16)) by complex conjugation and then by replacing $\bar{k}$ with k. If q is complex then the second global relation can be obtained from the first by complex conjugation of every term *except* of those terms involving q and then by replacing $\bar{k}$ with k. We will refer to this latter procedure as *Schwarz conjugation.*

3.1 The modified Helmholtz equation in a semi-infinite strip

Let Ω be the interior of the semi-infinite strip with finite corners at the origin and at the point $z_2 = il$ where l is a finite positive constant, see Fig. 4.

For concreteness we solve the Dirichlet problem. Several other types of boundary value problems can be solved similarly (see Remark 2).

Proposition 2 *Let the complex-valued function $q(x,y)$ satisfy the modified Helmholtz equation* (1.11) *in the semi-infinite strip defined above with the Dirichlet boundary conditions*

$$q(x,l) = d_1(x),\ q(x,0) = d_3(x),\ 0 < x < \infty;\ q(0,y) = d_2(y),\ 0 < y < l. \tag{3.1}$$

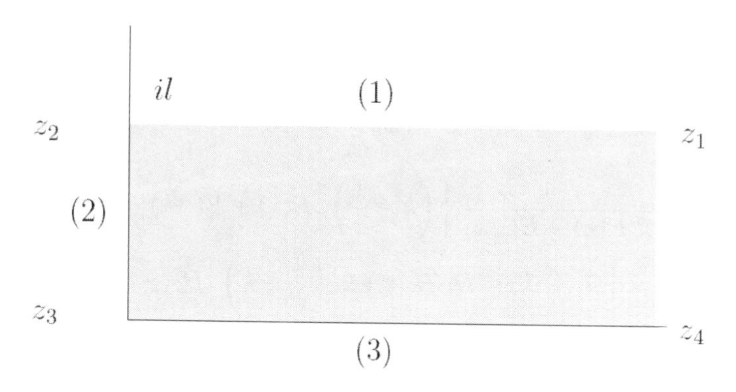

Fig. 4. A semi-infinite strip.

Let us assume that the complex-valued functions $\{d_j\}_1^3$ *have appropriate smoothness and are compatible at the corners* $(0,0)$, $(0,l)$ *and also that the functions* d_1 *and* d_3 *have appropriate decay for large* x.

Define the following transforms of the given data

$$D_1(k) = -\int_0^\infty e^{\beta(k+\frac{1}{k})x} d_1(x)dx, \quad \mathrm{Re}\ k \le 0,$$

$$D_2(k) = -\int_0^l e^{\beta(k+\frac{1}{k})y} d_2(y)dy, \quad k \in \mathbf{C},$$

$$D_3(k) = \int_0^\infty e^{\beta(k+\frac{1}{k})x} d_3(x)dx, \quad \mathrm{Re}\ k \le 0. \tag{3.2}$$

The solution $q(x,y)$ *is given by equation* (1.12) *with* $n=3$ *and with the rays* $\{l_j\}_1^3$ *depicted in Fig. 5, where* $\{\hat{q}_j\}_1^3$ *are defined in terms of* $\{D_j\}_1^3$ *as follows:*

$$\hat{q}_1(k) = i\beta E(k)\left[\left(k+\frac{1}{k}\right)D_1(-ik) + F_1(k)\right], \quad k \in \mathbf{R}^-, \tag{3.3}$$

$$\hat{q}_2(k) = -i\beta\left[\left(k+\frac{1}{k}\right)E(k)D_1(ik) + 2i\left(k-\frac{1}{k}\right)D_2(k) + \left(k+\frac{1}{k}\right)D_3(ik)\right], \quad k \in i\mathbf{R}^+, \tag{3.4}$$

$$\hat{q}_3(k) = i\beta\left[\left(k+\frac{1}{k}\right)D_3(-ik) + F_3(k)\right], \quad k \in \mathbf{R}^+, \tag{3.5}$$

where

$$E(k) = e^{\beta\left(k+\frac{1}{k}\right)l}, \quad k \in \mathbf{C} \tag{3.6}$$

and F_1, F_3 *are defined for* $k \in \mathbf{R}$, *by*

$$\begin{aligned} F_1(k) &= -\frac{1}{E(k)-E(-k)}\left\{\left(k+\frac{1}{k}\right)\left[E(k)+E(-k)\right]\right. \\ &\quad \times \left[D_1(-ik)-D_1(ik)\right] + 2i\left(\frac{1}{k}-k\right)\left[D_2(-k)-D_2(k)\right] \\ &\quad \left. + 2\left(k+\frac{1}{k}\right)\left[D_3(-ik)-D_3(ik)\right]\right\}, \end{aligned} \tag{3.7}$$

$$\begin{aligned} F_3(k) &= \frac{1}{E(k)-E(-k)}\left\{2\left(k+\frac{1}{k}\right)\left[D_1(-ik)-D_1(ik)\right]\right. \\ &\quad + 2i\left(\frac{1}{k}-k\right)\left[E(k)D_2(-k)-E(-k)D_2(k)\right] \\ &\quad \left. + \left(k+\frac{1}{k}\right)\left[E(k)+E(-k)\right]\left[D_3(-ik)-D_3(ik)\right]\right\}. \end{aligned} \tag{3.8}$$

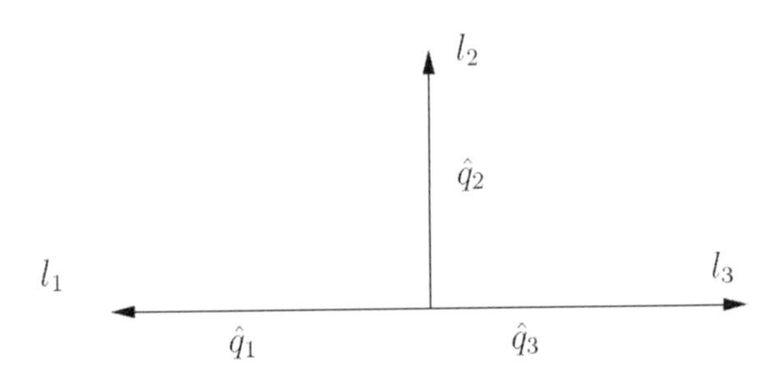

Fig. 5. The rays for the semi-infinite strip.

Proof Using $z = x + il$, $z = iy$, $z = x$ for the sides (1), (2), (3) respectively equation (1.13) yields

$$\hat{q}_1(k) = e^{\beta\left(k+\frac{1}{k}\right)l}\int_0^{\infty} e^{-i\beta\left(k-\frac{1}{k}\right)x}\left[iq_y(x,l) - i\beta\left(k+\frac{1}{k}\right)q(x,l)\right]dx, \quad \operatorname{Im} k \le 0, \tag{3.9}$$

$$\hat{q}_2(k) = -\int_0^l e^{\beta\left(k+\frac{1}{k}\right)y}\left[iq_x(0,y)+\beta\left(\frac{1}{k}-k\right)q(0,y)\right]dy,$$
$$k \in \mathbf{C}, \tag{3.10}$$

$$\hat{q}_3(k) = \int_0^\infty e^{-i\beta\left(k-\frac{1}{k}\right)x}\left[-iq_y(x,0)+i\beta\left(k+\frac{1}{k}\right)q(x,0)\right]dx,$$
$$\mathrm{Im}\, k \leq 0. \tag{3.11}$$

Replacing in the expressions for $\{\hat{q}_j\}_1^3$ the Dirichlet boundary values by the given data we find

$$\hat{q}_1(k) = E(k)\left[iU_1(-ik)+i\beta\left(k+\frac{1}{k}\right)D_1(-ik)\right], \quad \mathrm{Im}\, k \leq 0,$$

$$\hat{q}_2(k) = iU_2(k)+\beta\left(\frac{1}{k}-k\right)D_2(k), \quad k \in \mathbf{C},$$

$$\hat{q}_3(k) = iU_3(-ik)+i\beta\left(k+\frac{1}{k}\right)D_3(-ik), \quad \mathrm{Im}\, k \leq 0, \tag{3.12}$$

where the unknown functions $\{U_j\}_1^3$ denote the transforms of the unknown Neumann boundary values, i.e.

$$U_1(k) = \int_0^\infty e^{\beta\left(k+\frac{1}{k}\right)x} q_y(x,l)dx, \quad \mathrm{Re}\, k \leq 0,$$

$$U_2(k) = -\int_0^l e^{\beta\left(k+\frac{1}{k}\right)y} q_x(0,y)dy, \quad k \in \mathbf{C},$$

$$U_3(k) = -\int_0^\infty e^{\beta\left(k+\frac{1}{k}\right)x} q_y(x,0)dx, \quad \mathrm{Re}\, k \leq 0. \tag{3.13}$$

The first of global relations (1.16) yields

$$E(k)U_1(-ik)+U_2(k)+U_3(-ik) = J(k), \quad \mathrm{Im}\, k \leq 0, \tag{3.14}$$

where the known function $J(k)$ is defined by

$$J(k) = -\beta\left(k+\frac{1}{k}\right)E(k)D_1(-ik)+i\beta\left(\frac{1}{k}-k\right)D_2(k)$$
$$-\beta\left(k+\frac{1}{k}\right)D_3(-ik), \quad \mathrm{Im}\, k \leq 0. \tag{3.15}$$

The Schwarz conjugate of equation (3.14) yields

$$E(k)U_1(ik)+U_2(k)+U_3(ik) = \overline{J(\bar{k})}, \quad \mathrm{Im}\, k \geq 0, \tag{3.16}$$

where $\overline{J(\bar{k})}$ denotes the function obtained from $J(k)$ by taking the complex conjugate of each term of $J(k)$ *except* of the terms $\{d_j\}_1^3$. The integral representation of q involves integrals in $\mathbf{C}^+$, thus we will express the unknown functions $U_1(-ik)$, $U_2(k)$, $U_3(-ik)$ appearing in equations (3.12) in terms of the functions $U_1(ik)$ and $U_3(ik)$ which are analytic in $\mathbf{C}^+$. In this respect we note that equation (3.16) immediately implies $U_2(k)$ in terms of $U_1(ik)$ and $U_3(ik)$. In order to express $U_1(-ik)$ and $U_3(-ik)$ in terms of $U_1(ik)$ and $U_3(ik)$ we subtract equations (3.14), (3.16)

$$E(k)\left[U_1(-ik) - U_1(ik)\right] + \left[U_3(-ik) - U_3(ik)\right] = J(k) - \overline{J(\bar{k})}, \quad k \in \mathbf{R}. \tag{3.17}$$

Replacing in this equation k by $-k$ we obtain a second equation involving the same brackets that appear in equation (3.17). Solving this equation and equation (3.17) for these two unknown brackets we find

$$U_1(-ik) = U_1(ik) + \beta F_1(k), \quad k \in \mathbf{R},$$

$$U_3(-ik) = U_3(ik) + \beta F_3(k), \quad k \in \mathbf{R}, \tag{3.18}$$

where the known functions F_1 and F_3 and defined in (3.7) and(3.8). Substituting in equations (3.12) the expressions for $U_1(-ik)$, $U_2(k)$, $U_3(-ik)$ we find

$$\hat{q}_1(k) = i\beta\left(k + \frac{1}{k}\right)E(k)D_1(-ik) + i\beta E(k)F_1(k) + iE(k)U_1(ik), \; k \in \mathbf{R},$$

$$\hat{q}_2(k) = \beta\left(\frac{1}{k} - k\right)D_2(k) + i\overline{J(\bar{k})} - iE(k)U_1(ik) - iU_3(k), \quad \mathrm{Im}\, k \geq 0,$$

$$\hat{q}_3(k) = i\beta\left(k + \frac{1}{k}\right)D_3(-ik) + i\beta F_3(k) + iU_3(ik), \quad k \in \mathbf{R}. \tag{3.19}$$

The first two terms in the right-hand side of equations (3.19) yield equations (3.3)–(3.5), whereas the remaining terms, which are shown in Fig. 6, yield a zero contribution. Indeed the integral along $-l_3 \cup l_2$ involves the function

$\exp[i\beta(kz - \bar{z}/k)]/k$ which is bounded as $k \to 0$ and as $k \to \infty$ in the first quadrant of the complex k-plane, and the function $U_3(ik)$ which is analytic and of $O(1/k)$ as $k \to \infty$ and of $O(k)$ as $k \to 0$; hence this integral vanishes. Similar considerations are valid for the integral along $-l_3 \cup l_1$, since the relevant exponential satisfies

$$\mathrm{Re}\left[e^{i\beta\left(kz - \frac{\bar{z}}{k}\right) + \beta\left(k + \frac{1}{k}\right)l}\right] = e^{\beta\left(1 + \frac{1}{|k|^2}\right)\left[-k_I x + k_R(l-y)\right]}.$$

$\square$

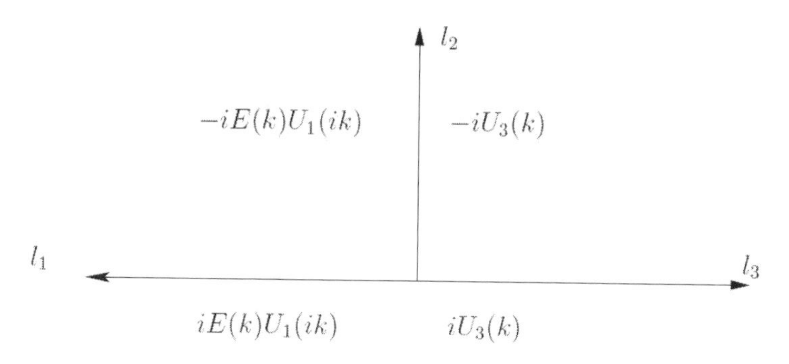

Fig. 6. The terms involving $U_3(k)$ and $U_1(ik)$.

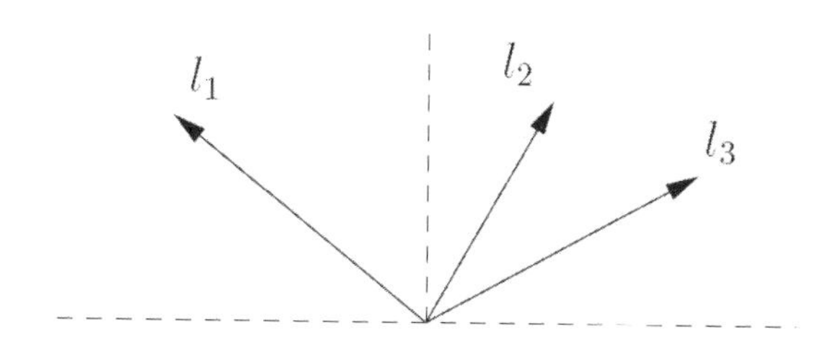

Fig. 7. The contours $\{l_i\}_1^3$ for Example 1.

Contour deformations

By considering the term $E(k)$ appearing in $\hat{q}_1(k)$ together with the term $\exp[i\beta(kz - \bar{z}/k)]$, it follows that the representation for q contains the following functions on the contours $(0, -\infty)$, $(0, i\infty)$ and $(0, \infty)$ respectively

$$-\frac{1}{|k|}e^{i\beta\left(\frac{1}{|k|}-|k|\right)x-\beta\left(|k|+\frac{1}{|k|}\right)(l-y)}, \quad \frac{1}{i|k|}e^{-\beta\left(\frac{1}{|k|}+|k|\right)x+i\beta\left(\frac{1}{|k|}-|k|\right)y},$$

$$\frac{1}{|k|}e^{-i\beta\left(\frac{1}{|k|}-|k|\right)x-\beta\left(|k|+\frac{1}{|k|}\right)y}.$$

Each of these terms contains a function which decays exponentially as $|k| \to \infty$ or $|k| \to 0$. In addition, the representation for q contains the functions

$$i\beta\left(k+\frac{1}{k}\right)D_1(-ik)+i\beta F_1(k), \quad \hat{q}_2(k), \quad \hat{q}_3(k)$$

which either are bounded or they decay. Actually, using appropriate contour deformations it is possible to obtain decaying instead of oscillatory exponentials. This is illustrated in the following example.

Example 1 Let

$$d_1 = d_2 = 0, \quad d_3 = xe^{-ax}, \quad 0 < x < \infty, \quad a > 0.$$

Then

$$D_1 = D_2 = 0, \quad D_3(ik) = \frac{1}{\left[i\beta\left(k-\frac{1}{k}\right)-a\right]^2}.$$

Hence

$$\begin{aligned}
\hat{q}_1(k) &= E(k)\tilde{q}_1(k), \quad \tilde{q}_1(k) = -\frac{2i\beta\left(k+\frac{1}{k}\right)}{E(k)-E(-k)}\Big[D_3(-ik)-D_3(ik)\Big],\\
\hat{q}_2(k) &= i\beta\left(k+\frac{1}{k}\right)D_3(ik),\\
\hat{q}_3(k) &= i\beta\left(k+\frac{1}{k}\right)\Bigg\{D_3(-ik) \qquad\qquad (3.20)\\
&\quad + \frac{\Big[E(k)+E(-k)\Big]}{E(k)-E(-k)}\Big[D_3(-ik)-D_3(ik)\Big]\Bigg\}.
\end{aligned}$$

Thus

$$\begin{aligned}
q(x,y) &= \frac{1}{2\pi}\left\{\int_{L_1} e^{i\beta\left(kz-\frac{\bar{z}}{k}\right)+l\beta\left(k+\frac{1}{k}\right)}\tilde{q}_1(k)\frac{dk}{k}+\int_{L_2} e^{i\beta\left(kz-\frac{\bar{z}}{k}\right)}\hat{q}_2(k)\frac{dk}{k}\right.\\
&\quad \left.+\int_{L_3} e^{i\beta\left(kz-\frac{\bar{z}}{k}\right)}\hat{q}_3(k)\frac{dk}{k}\right\}, \qquad (3.21)
\end{aligned}$$

where the contours $\{L_j\}_1^3$, depicted in Fig. 7, are defined by the requirement that $\arg k$ is in the following open intervals

$$\left(\frac{\pi}{2},\pi\right), \quad \left(0,\frac{\pi}{2}\right), \quad \left(0,\frac{\pi}{2}\right). \qquad (3.22)$$

The numerical evaluation of the right-hand side of equation (3.21) is straightforward; analogous computations for evolutionary PDEs are performed in [FF08].

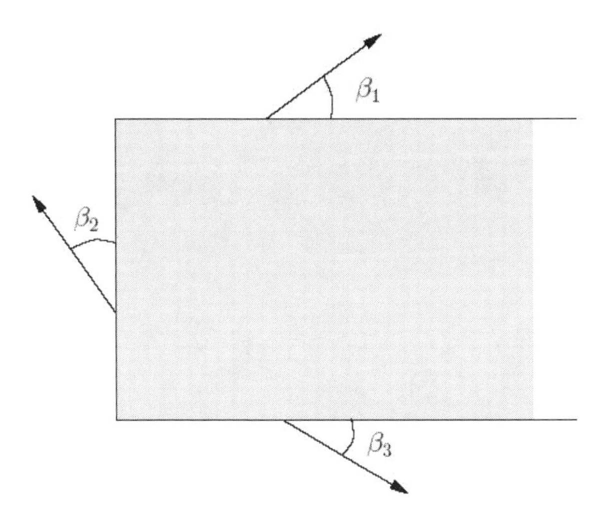

Fig. 8. The angles of the oblique Robin boundary conditions in a semi-infinite strip.

Remark 2 The modified Helmholtz equation with the oblique Robin boundary conditions defined by

$$q_x \cos\beta_1 + q_y \sin\beta_1 + \gamma_1 q = g_1(x), \quad 0 < x < \infty, \quad y = l, \tag{3.23a}$$
$$q_y \cos\beta_2 - q_x \sin\beta_2 + \gamma_2 q = g_2(y), \quad x = 0, \quad 0 < y < l, \tag{3.23b}$$
$$q_x \cos\beta_3 - q_y \sin\beta_3 + \gamma_3 q = g_3(x), \quad 0 < x < \infty, \quad y = 0, \tag{3.23c}$$

is investigated in [AF04] where it is shown that the new transform method yields an explicit solution provided that the real constants $\{\beta_j\}_1^3$ and $\{\gamma_j\}_1^3$ satisfy the following conditions

$$\begin{aligned} e^{4i(\beta_2-\beta_1)} = e^{4i(\beta_2+\beta_3)} &= 1, \\ (2\beta^2 - \gamma_2^2)\sin 2\beta_1 + (2\beta^2 - \gamma_1^2)\sin 2\beta_2 &= 0, \\ (2\beta^2 - \gamma_2^2)\sin 2\beta_3 - (2\beta^2 - \gamma_3^2)\sin 2\beta_2 &= 0. \end{aligned} \tag{3.24}$$

Requiring that $0 < \beta_j < \pi$ and $\gamma_j \geq 0$, $j = 1, 2, 3$, equations (3.24) yield

$$\begin{aligned} &\beta_3 + \beta_2 = \frac{\pi}{2}m, \quad 2\beta^2 - \gamma_2^2 = (2\beta^2 - \gamma_3^2)(-1)^{m-1}, \quad m = 1, 2, 3, \\ &\beta_2 - \beta_1 = \frac{\pi}{2}n, \quad 2\beta^2 - \gamma_2^2 = (2\beta^2 - \gamma_1^2)(-1)^{n-1}, \quad n = -1, 0, 1. \end{aligned} \tag{3.25}$$

3.2 The modified Helmholtz equation in an equilateral triangle

Let Ω be the interior of an equilateral triangle with corners at the following points

$$z_1 = \frac{l}{\sqrt{3}} e^{-\frac{i\pi}{3}}, \quad z_2 = \bar{z}_1, \quad z_3 = -\frac{l}{\sqrt{3}}.$$

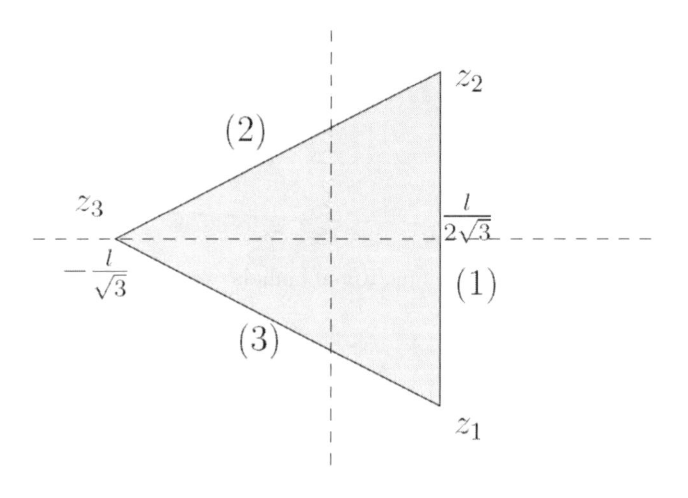

Fig. 9. An equilateral triangle.

Proposition 3 *Let the complex-valued function $q(x, y)$ satisfy the modified Helmholtz equation (1.11) in the interior of the equilateral triangle described above, see Fig. 9. Assume that the same smooth complex-valued function $d(s)$ is prescribed on each side as the Dirichlet boundary condition, i.e.*

$$q^{(j)}(s) = d(s), \quad s \in \left(-\frac{l}{2}, \frac{l}{2}\right), \quad j = 1, 2, 3, \tag{3.26}$$

where each side of the triangle is parameterized in terms of s by

$$z(s) = \frac{l}{2\sqrt{3}} + is, \quad z(s) = \left(\frac{l}{2\sqrt{3}} + is\right)\alpha, \quad z(s) = \left(\frac{l}{2\sqrt{3}} + is\right)\bar{\alpha}, \tag{3.27}$$

where

$$\alpha = e^{\frac{2\pi i}{3}}.$$

The solution $q(x,y)$ is given by

$$q = \frac{1}{4i\pi}\int_{l_1}\left\{A(k,z,\bar{z})E(-ik)\left[\beta\left(\frac{1}{k}-k\right)D(k)+\beta\frac{G(k)}{\Delta(\alpha k)}\right]\right\}\frac{dk}{k}$$
$$+\frac{1}{4\pi i}\int_{l_1'}A(k,z,\bar{z})E^2(i\alpha k)\beta\frac{G(k)}{\Delta(\alpha k)\Delta(k)}\frac{dk}{k}, \tag{3.28}$$

where

$$A = e^{i\beta(kz-\frac{\bar{z}}{k})}+e^{i\beta(\bar{\alpha}k-\frac{\bar{z}}{\bar{\alpha}k})}+e^{i\beta(\alpha k-\frac{\bar{z}}{\alpha k})}, \tag{3.29}$$

$$E(k) = e^{\beta(k+\frac{1}{k})\frac{l}{2\sqrt{3}}},\quad D(k)=\int_{-\frac{l}{2}}^{\frac{l}{2}}e^{\beta(k+\frac{1}{k})s}d(s)ds,$$

$$G(k) = \left[\Delta^{+}(\bar{\alpha}k)\left(\frac{1}{k}-k\right)D(k)+2\left(\frac{1}{\bar{\alpha}k}-\bar{\alpha}k\right)D(\bar{\alpha}k)\right.$$
$$\left.+\Delta^{+}(k)\left(\frac{1}{\alpha k}-\alpha k\right)D(\alpha k)\right], \tag{3.30}$$

$$\Delta(k) = e(k)-e(-k),\quad \Delta^{+}(k)=e(k)+e(-k),\quad e(k)=e^{\frac{\beta l}{2}(k+\frac{1}{k})}, \tag{3.31}$$

and l_1' is a ray directed toward infinity such that $-\pi/2<\arg k<-\pi/6$.

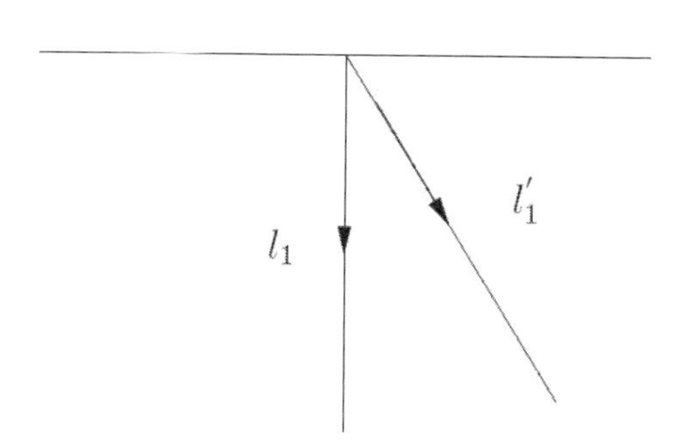

Fig. 10. The contours of integration for the modified Helmholtz equation in the interior on an equilateral triangle.

Proof Using the parametrizations (3.27), equation (1.15) implies that for all $k\in\mathbf{C}$,

$$\hat{q}_1(k) = e^{\frac{l}{2\sqrt{3}}(-ik-\frac{1}{ik})}\int_{-\frac{l}{2}}^{\frac{l}{2}}e^{\beta(k+\frac{1}{k})s}\left[iq_n^{(1)}+\beta\left(\frac{1}{k}-k\right)q^{(1)}\right]ds, \tag{3.32}$$

$$\hat{q}_2(k) = e^{\frac{l}{2\sqrt{3}}\left(-i\alpha k - \frac{1}{i\alpha k}\right)} \int_{-\frac{l}{2}}^{\frac{l}{2}} e^{\beta\left(\alpha k + \frac{1}{\alpha k}\right)s} \left[iq_n^{(2)} + \beta\left(\frac{1}{\alpha k} - \alpha k\right) q^{(2)}\right] ds, \tag{3.33}$$

$$\hat{q}_3(k) = e^{\frac{l}{2\sqrt{3}}\left(-i\bar{\alpha} k - \frac{1}{i\bar{\alpha} k}\right)} \int_{-\frac{l}{2}}^{\frac{l}{2}} e^{\beta\left(\bar{\alpha} k + \frac{1}{\bar{\alpha} k}\right)s} \left[iq_n^{(3)} + \beta\left(\frac{1}{\bar{\alpha} k} - \bar{\alpha} k\right) q^{(3)}\right] ds. \tag{3.34}$$

Hence

$$\hat{q}_1(k) = \hat{q}(k), \quad \hat{q}_2(k) = \hat{q}(\alpha k), \quad \hat{q}_3(k) = \hat{q}(\bar{\alpha} k), \tag{3.35}$$

with

$$\hat{q}(k) = E(-ik)\left[iU(k) + \beta\left(\frac{1}{k} - k\right) D(k)\right]. \tag{3.36}$$

Substituting these expressions in the first of the global relations, i.e. the first of equations (1.13), multiplying the resulting equation by $E(i\bar{\alpha}k)$ and using the identities

$$i(\bar{\alpha} - \alpha) = \sqrt{3}, \quad i(\alpha - 1) = \sqrt{3}\bar{\alpha}, \tag{3.37}$$

the first of the global relations becomes

$$e(-\alpha k)U(k) + e(k)U(\alpha k) + U(\bar{\alpha} k) = i\beta J(k), \quad k \in \mathbf{C}, \tag{3.38}$$

where $e(k) = e^{\frac{\beta l}{2}\left(k + \frac{1}{k}\right)}$ and

$$J(k) = \left(\frac{1}{k} - k\right) e(-\alpha k)D(k) + e(k)\left(\frac{1}{\alpha k} - \alpha k\right) D(\alpha k) +$$

$$\left(\frac{1}{\bar{\alpha} k} - \bar{\alpha} k\right) D(\bar{\alpha} k). \tag{3.39}$$

Taking the Schwarz conjugate of the global relation (3.38) and multiplying the resulting equation by $e(-k)$ we find

$$e(\alpha k)U(k) + e(-k)U(\alpha k) + U(\bar{\alpha} k) = -i\beta e(-k)\overline{J(\bar{k})}, k \in \mathbf{C}, \tag{3.40}$$

where we have used the identity

$$1 + \alpha + \bar{\alpha} = 0 \tag{3.41}$$

and $\overline{J(\bar{k})}$ denote the function obtained from $J(k)$ by taking the complex conjugate of each term of $J(k)$ except of $d(s)$. Subtracting equations (3.38), (3.40) we find the following equation which is valid for all $k \in \mathbf{C}$,

$$[e(\alpha k) - e(-\alpha k)]\, U(k) = [e(k) - e(-k)]\, U(\alpha k) - i\beta\left[J(k) + e(-k)\overline{J(\bar{k})}\right]. \tag{3.42}$$

Replacing in the equation for $\hat{q}$, i.e. (3.36), the expression for $U(k)$ obtained from equation (3.42) and using the identity

$$E(-ik)[e(k) - e(-k)] = E^2(i\bar{\alpha}k) - E^2(i\alpha k),$$

we find

$$\begin{aligned}\hat{q}(k) &= \beta\left(\frac{1}{k} - k\right) E(-ik)D(k) + \frac{\beta G(k)E(-ik)}{\Delta(\alpha k)} \\ &\quad + i\left[E^2(i\bar{\alpha}k) - E^2(i\alpha k)\right]\frac{U(\alpha k)}{\Delta(\alpha k)},\end{aligned} \tag{3.43}$$

where $\Delta(k)$ and $G(k)$ are defined, for all $k \in \mathbf{C}$, by

$$\Delta(k) = e(k) - e(-k), \quad G(k) = [J(k) + e(-k)\overline{J(\bar{k})}]. \tag{3.44}$$

The functions $\hat{q}_2(k)$ and $\hat{q}_3(k)$ can be obtained from the right-hand side of equation (3.43) by replacing k with αk and $\bar{\alpha}k$.

The vector $z_2 - z_1$ makes an angle $\pi/2$ with the positive axis, the vector $z_3 - z_2$ an angle $-5\pi/6$ and the vector $z_1 - z_3$ an angle $-\pi/6$ and the rays $\{l_j\}_1^3$ are depicted in Fig. 11.

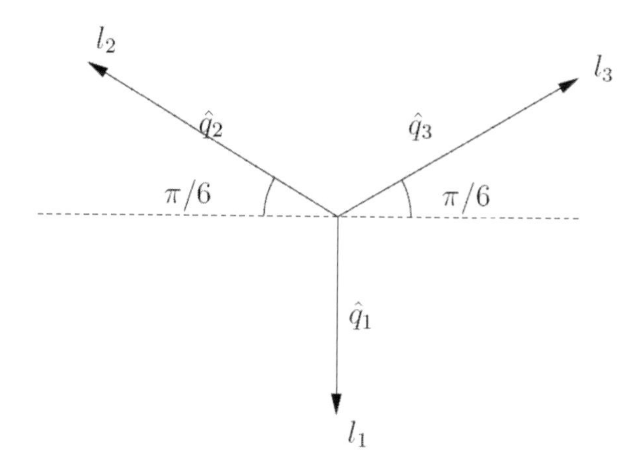

Fig. 11. The rays $\{l_j\}_1^3$ for the equilateral triangle.

In what follows we will show that the contribution to the solution q of the unknown functions $U(\alpha k)$, $U(\bar{\alpha}k)$, $U(k)$ can be computed in terms of the given boundary conditions. In this respect we will use the following facts:

(a) The zeros of the functions $\Delta(k)$, $\Delta(\alpha k)$, $\Delta(\bar{\alpha}k)$ occur on the following lines respectively in the complex k-plane

$$i\mathbf{R}, \quad e^{\frac{5i\pi}{6}}\mathbf{R}, \quad e^{\frac{i\pi}{6}}\mathbf{R}. \tag{3.45}$$

Indeed, the zeros of $\Delta(k)$ occur on the imaginary axis and then the zeros of $\Delta(\alpha k)$ and $\Delta(\bar{\alpha}k)$ can be obtained by appropriate rotations.

(b) The functions

$$e^{i\beta\left(kz-\frac{\bar{z}}{k}\right)}E^2(i\alpha k), \quad e^{i\beta\left(kz-\frac{\bar{z}}{k}\right)}E^2(ik), \quad e^{i\beta\left(kz-\frac{\bar{z}}{k}\right)}E^2(i\bar{\alpha}k), \tag{3.46}$$

with z in the interior of the triangle, are bounded as $k \to 0$ and $k \to \infty$, for $\arg k$ in

$$\left[-\frac{\pi}{2}, \frac{\pi}{6}\right], \quad \left[\frac{\pi}{6}, \frac{5\pi}{6}\right], \quad \left[\frac{5\pi}{6}, \frac{3\pi}{2}\right] \tag{3.47}$$

respectively, see Fig. 12.

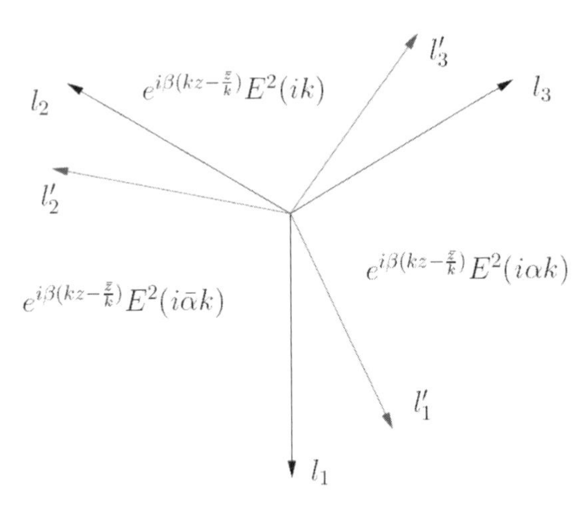

Fig. 12. The domains of boundedness of the function defined in (3.46).

Indeed, let us consider the first exponential in (3.46). Using $z_1 = -l\alpha/\sqrt{3}$, this exponential can be written as

$$e^{i\beta k(z-z_1)+\frac{\beta(\bar{z}-\bar{z}_1)}{ik}}.$$

If z is in the interior of the triangle then

$$\frac{\pi}{2} \leq \arg(z - z_1) \leq \frac{5\pi}{6},$$

thus, if

$$-\frac{\pi}{2} \leq \arg k \leq \frac{\pi}{6},$$

it follows that

$$0 \leq \arg\left[k(z-z_1)\right] \leq \pi.$$

Hence, the exponentials

$$e^{i\beta k(z-z_1)} \quad \text{and} \quad e^{\frac{\beta(\bar{z}-\bar{z}_1)}{ik}}$$

are bounded as $|k| \to \infty$ and $|k| \to 0$ respectively. The analogous results for the second and third exponentials in (3.46) can be obtained in a similar way.

(c) The functions

$$\frac{U(k)}{\Delta(k)}, \quad \frac{U(\alpha k)}{\Delta(\alpha k)}, \quad \frac{U(\bar{\alpha} k)}{\Delta(\bar{\alpha} k)},$$

are bounded in the entire complex k-plane except on the lines defined in (3.45) where the functions $\Delta(k)$, $\Delta(\alpha k)$, $\Delta(\bar{\alpha} k)$, have simple zeros.

Indeed, regarding $U(k)/\Delta(k)$ we note that $\Delta(k)$ is dominated by $e(k)$ for Re $k > 0$ and by $e(-k)$ for Re $k < 0$, hence

$$\frac{U(k)}{\Delta(k)} \sim \begin{cases} e(-k)U(k), & \text{Re } k > 0 \\ -e(k)U(k), & \text{Re } k < 0. \end{cases}$$

Furthermore, $e(-k)U(k)$ involves $k(s-l/2)$ which is bounded for Re $k \geq 0$, whereas $e(k)U(k)$ involves $k(s+l/2)$ which is bounded for Re $k \leq 0$. Similarly for the terms involving $1/k$.

The unknown function $U(\alpha k)$ in the expression for $\hat{q}(k)$, see equation (3.43), yields the following contribution $C_1(x,y)$ to the solution q,

$$C_1 = \frac{1}{4\pi}\int_{l_1} P[E^2(i\bar{\alpha}k) - E^2(i\alpha k)]\frac{U(\alpha k)}{\Delta(\alpha k)}\frac{dk}{k},$$

where P denotes the exponential

$$P = e^{i\beta\left(kz-\frac{\bar{z}}{k}\right)}.$$

The integral of the second term in the right-hand side of C_1 can be deformed from l_1 to l_1', where l_1' is a ray with $-\pi/2 < \arg k < -\pi/6$. Hence,

$$C_1 = \frac{1}{4\pi}\int_{l_1} PE^2(i\bar{\alpha}k)\frac{U(\alpha k)}{\Delta(\alpha k)}\frac{dk}{k} - \frac{1}{4\pi}\int_{l_1'} PE^2(i\alpha k)\frac{U(\alpha k)}{\Delta(\alpha k)}\frac{dk}{k}.$$

In the second integral of the right-hand side of this equation we use equation (3.42), i.e. the equation

$$\Delta(\alpha k)U(k) = \Delta(k)U(\alpha k) - i\beta G(k),$$

to replace $U(\alpha k)$. Hence, $C_1 = \tilde{C}_1 + \tilde{U}_1$, where

$$\tilde{U}_1 = \frac{1}{4\pi}\int_{l_1} PE^2(i\bar{\alpha}k)\frac{U(\alpha k)}{\Delta(\alpha k)}\frac{dk}{k} - \frac{1}{4\pi}\int_{l_1'} PE^2(i\alpha k)\frac{U(k)}{\Delta(k)}\frac{dk}{k} \tag{3.48}$$

and

$$\tilde{C}_1 = \frac{\beta}{4\pi i}\int_{l_1'} PE^2(i\alpha k)\frac{G(k)}{\Delta(k)\Delta(\alpha k)}\frac{dk}{k}. \tag{3.49}$$

In summary, the term $\hat{q}(k)$ gives rise to the contribution $F_1 + \tilde{U}_1$, where $\tilde{U}_1$ is defined in (3.48) and F_1 is defined by

$$\begin{aligned} F_1 &= \frac{1}{4\pi i}\int_{l_1} P\left[\beta\left(\frac{1}{k} - k\right)E(-ik)D(k) + \frac{\beta E(-ik)G(k)}{\Delta(\alpha k)}\right]\frac{dk}{k} \\ &\quad + \frac{\beta}{4\pi i}\int_{l_1'} PE^2(i\alpha k)\frac{G(k)}{\Delta(k)\Delta(\alpha k)}\frac{dk}{k}. \end{aligned} \tag{3.50}$$

The contributions of $\hat{q}_2$ and $\hat{q}_3$ can be obtained from F_1 and $\tilde{U}_1$ using the substitutions

$$l_1 \to l_2 \to l_3, \quad l_1' \to l_2' \to l_3', \quad k \to \alpha k \to \bar{\alpha}k. \tag{3.51}$$

We will now show that the contributions of $\tilde{U}_j$, $j = 1, 2, 3$ vanish. Indeed, the integrands

$$PE^2(i\bar{\alpha}k)\frac{U(\alpha k)}{k\Delta(\alpha k)}, \quad PE^2(ik)\frac{U(\bar{\alpha}k)}{k\Delta(\bar{\alpha}k)}, \quad PE^2(i\alpha k)\frac{U(k)}{k\Delta(k)}$$

occur on $l_1 \cup l_2'$, $l_2 \cup l_3'$, $l_3 \cup l_1'$ and in the domains bounded by these contours the above functions are bounded and analytic, see Fig. 12.

Hence,

$$q = F_1 + F_2 + F_3, \tag{3.52}$$

where F_2 and F_3 are obtained from F_1 using the substitutions (12.65).

We will now show that equation (3.52) is equivalent to equation (70a). We make the change of variables $k \to \bar{\alpha}k$ and $k \to \alpha k$ in the integrals defining F_2 and F_3 respectively. Regarding F_2 we note that under this transformation: (a) the fraction dk/k remains invariant; (b) the rays l_2 and l_2' are mapped to the rays l_1 and l_1' respectively; (c) the exponential $\exp[i\beta(kz - \bar{z}/k)]$ is mapped to $\exp[i\beta(\bar{\alpha}kz - \bar{z}/\bar{\alpha}k]$; (d) the remaining terms in the integrand of F_2 are identical to the corresponding terms of the integrand of F_2. Similar considerations are valid for F_3. □

Remark 3 The integrands appearing in the integrals along l_1 and l_1' defined in equation (3.28) contain terms which decay exponentially. Indeed, regarding the integral along l_1' we note that $G(k)/\Delta(k)\Delta(\alpha k)$ is bounded for k on l_1' and the function $A(k,z,\bar{z})E^2(i\alpha k)$ contains terms which decay exponentially since each of the three terms of this function contains exponentials with negative real parts. Regarding the integral along l_1 we note that the function $D(k)$ is bounded for k on l_1, $G(k)/\Delta(k)$ decays exponentially since $s \in (-l/2, l/2)$ and each of the real terms of the function $A(k,z,\bar{z})E(-ik)$ has an exponential with negative real part.

Example 1 Suppose that l and $d(s)$ are given by

$$l = \pi, \quad d(s) = \cos s. \tag{3.53}$$

Then the definitions of $D(k)$ and $G(k)$, see equations (67) and (70c), imply

$$D(s) = \frac{2}{1+\beta^2\left(k+\frac{1}{k}\right)^2}\cosh\left[\beta\left(k+\frac{1}{k}\right)\frac{\pi}{2}\right] \tag{3.54}$$

and

$$G(k) = 4\left[\frac{\frac{1}{k}-k}{1+\beta^2\left(k+\frac{1}{k}\right)^2} + \frac{\frac{1}{\alpha k}-\alpha k}{1+\beta^2\left(\alpha k+\frac{1}{\alpha k}\right)^2}\right]\cosh\left[\beta\left(k+\frac{1}{k}\right)\frac{\pi}{2}\right]$$

$$\cosh\left[\beta\left(\alpha k+\frac{1}{\alpha k}\right)\frac{\pi}{2}\right] + 4\frac{\frac{1}{\bar{\alpha}k}-\bar{\alpha}k}{1+\beta^2\left(\bar{\alpha}k+\frac{1}{\bar{\alpha}k}\right)^2}\cosh\left[\beta\left(\bar{\alpha}k+\frac{1}{\bar{\alpha}k}\right)\frac{\pi}{2}\right]. \tag{3.55}$$

Furthermore,

$$\Delta^+(k) = 2\cosh\left[\beta\left(k+\frac{1}{k}\right)\frac{\pi}{2}\right], \quad \Delta(k) = 2\sinh\left[\beta\left(k+\frac{1}{k}\right)\frac{\pi}{2}\right]. \tag{3.56}$$

Hence, the solution of the modified Helmholtz equation in the interior of the equilateral triangle with $l=\pi$ and with the Dirichlet boundary condition $d(s)=\cos(s)$ on each side of the triangle, is given by equation (3.28), where D, G, Δ, Δ^+ are given by equations (3.30) and (3.56) and the rays l and l' are defined by (see Fig. 10)

$$l = \left\{k\in\mathbf{C},\ \arg k = -\frac{\pi}{2}\right\},\ l' = \left\{k\in\mathbf{C},\ \arg k = \phi,\ -\frac{\pi}{2} < \phi < -\frac{\pi}{6}\right\}.$$

Remark 4 It can be verified directly that the integrands of the integrals appearing in (3.28) decay exponentially. Indeed, regarding the first integral, for which $\mathrm{Re}\, k = 0$, $\mathrm{Im}\, k < 0$, the following formulae are valid as $k \to 0$ or $k \to \infty$:

- $e^{i\beta\left(kz - \frac{\bar{z}}{k}\right)} E(-ik) \sim e^{\beta\left(ik + \frac{1}{ik}\right)\left(\mathrm{Re}\ (z) - \frac{\pi}{2\sqrt{3}}\right)} \sim e^{-\beta\left(t + \frac{1}{t}\right)\left(x - \frac{\pi}{2\sqrt{3}}\right)},$
$$t < 0, \quad x < \frac{\pi}{2\sqrt{3}};$$
- $D(k) \sim \dfrac{1}{k^2}$ as $k \to \infty$, $D(k) \sim k^2$ as $k \to 0$;
- $\dfrac{G(k)}{\Delta(k)} \sim e\left(-\dfrac{\sqrt{3}}{2} ik\right) \sim e^{\frac{\sqrt{3}}{2}\left(t + \frac{1}{t}\right)}, \quad t < 0.$

For the second integral, for which $\arg k \in \left(-\frac{\pi}{2}, -\frac{\pi}{6}\right)$, the following formula are valid as $k \to 0$ or $k \to \infty$:

- $e^{i\beta\left(kz - \frac{1}{k}\right)} E^2(iak) \sim \exp\left[\left(x - \dfrac{\pi}{2\sqrt{3}}\right)\cos\left(\phi + \dfrac{\pi}{2}\right) - \left(y + \dfrac{\pi}{2}\right)\sin\left(\phi + \dfrac{\pi}{2}\right)\right],$
where $x < \dfrac{\pi}{2\sqrt{3}}$, $y > -\dfrac{\pi}{2}$ and $\phi = \arg k$. Hence, since $\arg k \in \left(-\frac{\pi}{2}, -\frac{\pi}{6}\right)$, the argument of the exponential is negative.
- $\dfrac{G(k)}{\Delta(k)\Delta(ak)} \sim \dfrac{1}{k}.$

Similar consideration are valid for the other two terms of $A(k, z, \bar{z})$.

Remark 5 The Laplace, modified Helmholtz and Helmholtz equations with the following oblique Robin boundary conditions are investigated in [DF05],

$$\sin\beta_j q_n^{(j)}(s) + \cos\beta_j \frac{d}{ds} q^{(j)}(s) + \gamma_j q^{(j)}(s) = g_j(s),$$
$$s \in \left(-\frac{l}{2}, \frac{l}{2}\right), \quad j = 1, 2, 3, \tag{3.57}$$

where g_j are smooth functions and β_j, γ_j, $j = 1, 2, 3$ are real constants. It is shown that the unknown boundary values can be determined explicitly provided that the following conditions are satisfied

$$\beta_2 = \beta_1 + \frac{m\pi}{3}, \quad \beta_3 = \beta_1 + \frac{n\pi}{3}, \quad m, n \in \mathbf{Z},$$
$$\sin 3\beta_1 \left[\gamma_2(3\beta^2 - \gamma_2^2) - e^{im\pi}\gamma_1(3\beta^2 - \gamma_1^2)\right] = 0,$$
$$\sin 3\beta_1 \left[\gamma_3(3\beta^2 - \gamma_3^2) - e^{in\pi}\gamma_1(3\beta^2 - \gamma_1^2)\right] = 0. \tag{3.58}$$

4 Formulation of Riemann–Hilbert problems

For certain simple polygonal domains, the algebraic manipulation of the global relations and of the equations obtained from the global relations via certain transformations in the complex k-plane, immediately yield a Riemann–Hilbert problem for the characterization of $\{\hat{q}_j(k)\}^n$. In what follows we illustrate this approach for the simple case of the semi-infinite strip [AF04].

4.1 The modified Helmholtz equation in a semi-infinite strip

Proposition 4 *Let the complex-valued function $q(x,y)$ satisfy the modified Helmholtz equation in the semi-strip $\{0 < x < \infty,\ 0 < y < l\}$, with the oblique Robin boundary conditions (3.23), see Fig. 8, where the complex-valued functions $\{g_j\}_1^3$ have sufficient smoothness, the functions g_1, g_3 have sufficient decay and β_j, γ_j, are real constants with $\sin\beta_j \neq 0$, $j = 1, 2, 3$. Let $\{U_j(k)\}_1^3$ denote the following transforms of the unknown Dirichlet boundary values*

$$\begin{aligned} U_1(k) &= \frac{1}{\sin\beta_1}\int_0^\infty e^{\beta(k+\frac{1}{k})x} q(x,l)dx, \\ U_2(k) &= \frac{1}{\sin\beta_2}\int_0^l e^{\beta(k+\frac{1}{k})y} q(0,y)dy, \\ U_3(k) &= \frac{1}{\sin\beta_3}\int_0^\infty e^{\beta(k+\frac{1}{k})x} q(x,0)dx, \end{aligned} \tag{4.1}$$

where $U_1(k), U_3(k)$ are defined for $\mathrm{Re}\ k \le 0$, whereas $U_2(k)$ is defined for all $k \in \mathbf{C}$.

Let $\{G_j\}_1^3$ denote the following transforms of the given boundary conditions,

$$\begin{aligned} G_j(k) &= \frac{1}{\sin\beta_j}\int_0^\infty e^{\beta(k+\frac{1}{k})x} g_j(x)dx, \quad j = 1,3, \quad \mathrm{Re}\ k \le 0, \\ G_2(k) &= \frac{1}{\sin\beta_2}\int_0^l e^{\beta(k+\frac{1}{k})x} g_2(y)dy, \quad k \in \mathbf{C}. \end{aligned} \tag{4.2}$$

The functions U_1 and U_3 satisfy the following 2×2 matrix RH problem:

$$U_1(ik),\ \ U_3(ik)\ \text{ are analytic for }\ \mathrm{Im}\ k > 0, \tag{4.3a}$$

$$J(k)\begin{pmatrix} U_1(ik) \\ U_3(ik) \end{pmatrix} + \overline{J(\bar{k})}\begin{pmatrix} U_1(-ik) \\ U_3(-ik) \end{pmatrix} = \begin{pmatrix} \chi(k) \\ \chi(-k) \end{pmatrix}, \quad k \in \mathbf{R}, \tag{4.3b}$$

$$U_1(ik) = o(1), \quad U_3(ik) = o(1), \quad k \to \infty \text{ and } k \to 0, \tag{4.3c}$$

where $J_1(k)$ and $\chi(k)$ are defined by the following equations:

$$J(k) = \begin{pmatrix} E(ik)\overline{J_1(\bar{k})}J_2(k) & J_2(k)\overline{J_3(\bar{k})} \\ E(ik)J_1(-k)\overline{J_2(-\bar{k})} & \overline{J_2(-\bar{k})}J_3(-k) \end{pmatrix}, \quad k \in \mathbf{R} \tag{4.4}$$

$$\begin{aligned} E(k) &= e^{\beta l(k+\frac{1}{k})}, \\ J_1(k) &= \beta\left(e^{-i\beta_1}k - \frac{e^{i\beta_1}}{k}\right) - i\gamma_1, \\ J_2(k) &= \beta\left(e^{-i\beta_2}k + \frac{e^{i\beta_2}}{k}\right) - \gamma_2, \\ J_3(k) &= \beta\left(e^{i\beta_3}k - \frac{e^{-i\beta_3}}{k}\right) - i\gamma_3, \quad k \in \mathbf{C}, \end{aligned} \tag{4.5}$$

$$\begin{aligned} \chi(k) &= iE(ik)J_2(k)G_1(ik) - iE(-ik)\overline{J_2(\bar{k})}G_1(-ik) \\ &\quad + 2\sin\beta_2\left(k - \frac{1}{k}\right)G_2(k) + i\left[J_2(k)G_3(ik) - \overline{J_2(\bar{k})}G_3(ik)\right] \\ &\quad + i\delta_1\cot\beta_1\left[E(ik)J_2(k) - E(-ik)\overline{J_2(\bar{k})}\right] \\ &\quad + 2\beta\left(k - \frac{1}{k}\right)\left[\delta_2(\cos\beta_2 + \sin\beta_2\cot\beta_3) - \delta_1\cos\beta_2 E(k)\right], \\ \delta_1 &= q(0,l), \quad \delta_2 = q(0,0). \end{aligned} \tag{4.6}$$

Proof In order to formulate the global relation we must first use the boundary conditions to express the functions $\{\hat{q}_j\}_1^3$ defined by equations (1.13) in terms of $\hat{g}_j$ and U_j, $j = 1,2,3$. The function $\hat{q}_1$ involves $q(x,l)$ and $q_y(x,l)$. Solving equation (3.23a) for q_y, substituting the resulting expression in equation (3.9) and using integration by parts to eliminate $q_x(x,l)$ we find

$$\hat{q}_1(k) = E(-ik)\left[iG_1(-ik) + J_1(k)U_1(-ik) + i\cot\beta_1 q(0,l)\right], \quad \text{Im } k \le 0. \tag{4.7a}$$

Similarly, solving equations (3.23b) and (3.23c) for $q_x(0,y)$ and $g_y(x,0)$ and substituting the resulting expression in equations (3.10) and (3.11) we find

$$\hat{q}_2(k) = iG_2(k) + iJ_2(k)U_2(k) + i\cot\beta_2[q(0,0) - E(k)q(0,l)], \quad k \in \mathbf{C}, \tag{4.7b}$$

$$\hat{q}_3(k) = iG_3(-ik) + J_3(k)U_3(-ik) + i\cot\beta_3 q(0,0), \quad \text{Im } k \le 0. \tag{4.7c}$$

Substituting the expressions for $\{\hat{q}_j\}_1^3$ from equations (4.7) into the global relation, taking the Schwarz conjugate of the resulting equation and then eliminating the function $U_2(k)$ from these two equations we find the following equation which is valid for $k \in \mathbf{R}$,

$$[E(ik)\overline{J_1(\bar{k})}J_2(k)U_1(ik) + J_2(k)\overline{J_3(\bar{k})}U_3(ik)]$$

$$+[E(-ik)J_1(k)\overline{J_2(\bar{k})}U_1(-ik) + \overline{J_2(\bar{k})}J_3(k)U_3(-ik)] = \chi(k). \quad (4.8)$$

Replacing in this equation k with $-k$ and writing the resulting equation and equation (4.8) in matrix form we find equation (4.3b).

The definitions of $U_1(ik)$ and $U_3(ik)$ imply equations (4.3a) and (4.3c). □

Remark 6 The solvability of the RH problem (4.3), as well as the question of how to determine the values $q(0,0)$ and $q(0,l)$, are discussed in [AF04]. Here we only note that if *either* of the following conditions are satisfied,

$$e^{4i(\beta_2-\beta_1)} = 1 \quad \text{and} \quad (2\beta^2 - \gamma_2^2)\sin 2\beta_1 + (2\beta^2 - \gamma_1^2)\sin 2\beta_2 = 0, \quad (4.9)$$

or

$$e^{4i(\beta_2+\beta_3)} = 1 \quad \text{and} \quad (2\beta^2 - \gamma_2^2)\sin 2\beta_3 - (2\beta^2 - \gamma_3^2)\sin 2\beta_2 = 0, \quad (4.10)$$

then the RH problem defined by equations (4.3) becomes triangular and hence is reduced to a scalar RH problem that can be solved in closed form. Indeed, the (12) entry of the relevant jump matrix is proportional to $A(k) - \bar{A}(k)$, where

$$A(k) = (J_3(k)J_3(-k))\bar{J}_2(k)\bar{J}_2(-k).$$

Consider for economy of presentation the case that $\gamma_2 = \gamma_3 = 0$. Then

$$\begin{aligned}
J_3(k)J_3(-k) &= -\beta^2\left(e^{i\beta_3}k - \frac{e^{-i\beta_3}}{k}\right)^2, \\
\bar{J}_2(k)\bar{J}_2(-k) &= -\beta^2\left(e^{i\beta_2}k + \frac{e^{-i\beta_2}}{k}\right)^2, \\
A(k) - \bar{A}(k) &= \beta^4\left(k^4 - \frac{1}{k^4}\right)\left[e^{2i(\beta_2+\beta_3)} - e^{-2i(\beta_2+\beta_3)}\right] \quad (4.11) \\
&\quad + 2\beta^2\left(k^2 - \frac{1}{k^2}\right)\left[\left(e^{-2i\beta_2} - e^{2i\beta_2}\right) + \left(e^{2i\beta_3} - e^{-2i\beta_3}\right)\right].
\end{aligned}$$

The coefficient of $k^4 - 1/k^4$ vanishes if and only if the first of equations (4.10) is valid; the coefficient of $k^2 - 1/k^2$ vanishes as a consequence of the second of equations (4.10) (here $\gamma_2 = \gamma_3 = 0$).

5 A new numerical method

In what follows we present a simple technique for the numerical evaluation of the unknown boundary values. In a variety of physical applications, one is interested only in these boundary values instead of the solution inside the domain. For those problems that the solution is needed for z inside the polygon, an effective approach to computing the solution is to use the integral representation. Using this representation and appropriate contour deformations, it is possible to obtain integrals with exponentially decaying integrands which can be computed efficiently [SSF08b] (the analogous computations for evolution instead of elliptic PDEs are presented in [FF08].

The key to this new method is that the global relations (1.16) are valid **for all** $k \in \mathbb{C}$. Suppose we expand the n unknown functions (the unknown boundary value on each side) in some series (Fourier or Chebychev etc.) up to N terms, and then we evaluate either of the two global relations at nN points; this would yield nN equations for the nN unknowns, which in principle could be solved. We now face two questions:

1. How to choose the basis?
2. How to choose the points k?

Regarding the first question we recall that for sufficiently smooth unknown functions, a Chebychev or Legendre basis gives exponential convergence, and numerical experiments indicate that this is inherited by the method. However, the prescence of exponentials in the global relations mean that the choice of a Fourier basis leads to matrices with a low condition number. Regarding the second question we note that for a Fourier basis, the choice of points comes naturally from the basis elements. Because this novel approach involves enforcing the global relations to hold at a set of discrete points in the spectral plane, it has been called a spectral collocation method.

In what follows we give the details of the numerical method applied to the Dirichlet boundary value problem for the modified Helmholtz equation [SSF08a].

5.1 The Dirichlet to Neumann map

Proposition 5 *Let the complex-valued function $q(z, \bar{z})$ satisfy the modified Helmholtz equation in the interior of a convex polygon Ω with corners $\{z_j\}_1^n$ (indexed counterclockwise, modulo n), and let S_j denote the side (z_j, z_{j+1}), see Fig. 1. Let q satisfy Dirichlet boundary conditions on each side:*

$$q_j(s) = d_j(s), \quad j = 1, \ldots, n, \tag{5.1}$$

where s parametrizes the side S_j, q_j denotes q on this side and the given complex-valued functions $\{d_j\}_1^n$ have sufficient smoothness. Let $u_j(s)$ denote the unknown Neumann boundary data on S_j. Let $u_j(s)$ denote the unknown Neumann boundary data on S_j.

The n unknown complex-valued functions $\{u_j\}_1^n$ satisfy the following $2n$ equations for $l \in \mathbf{R}^+$ and $p = 1, \ldots, n$:

$$\int_{-\pi}^{\pi} e^{ils} u_p(s) ds = -\sum_{\substack{j=1 \\ j \neq p}}^{n} E_{jp}(k_p(l)) \int_{-\pi}^{\pi} e_j(k_p(l), s) u_j(s) ds - G_p(l), \tag{5.2a}$$

$$\int_{-\pi}^{\pi} e^{-ils} u_p(s) ds = -\sum_{\substack{j=1 \\ j \neq p}}^{n} \bar{E}_{jp}(\bar{k}_p(l)) \int_{-\pi}^{\pi} \bar{e}_j(\bar{k}_p(l), s) u_j(s) ds - \tilde{G}_p(l); \tag{5.2b}$$

where the known functions $E_{jp}(k)$, $e_j(k, s)$, $G_p(l)$ and $\tilde{G}_p(l)$ ($j = 1, \ldots, n$, $p = 1, \ldots, n$) are defined by

$$E_{jp}(k) = e^{-i\beta(m_j - m_p)k + \frac{i\beta}{k}(\bar{m}_j - \bar{m}_p)}, \quad e_j(k, s) = e^{-i\beta\left(kh_j - \frac{\bar{h}_j}{k}\right)s}, \quad k \in \mathbf{C}, \quad -\pi < s < \pi, \tag{5.3a}$$

$$G_p(l) = \sum_{j=1}^{n} E_{jp}(k_p(l)) \rho_j(k_p(l)) \int_{-\pi}^{\pi} e_j(k_p(l), s) d_j(s) ds, \quad l \in \mathbf{R}^+,$$

$$\tilde{G}_p(l) = \sum_{j=1}^{n} \bar{E}_{jp}(\bar{k}_p(l)) \bar{\rho}_j(\bar{k}_p(l)) \int_{-\pi}^{\pi} \bar{e}_j(\bar{k}_p(l), s) d_j(s) ds, \quad l \in \mathbf{R}^+,$$

with

$$h_j = \frac{1}{2\pi}(z_{j+1} - z_j), \quad m_j = \frac{1}{2}(z_{j+1} + z_j), \quad j = 1, \ldots, n, \tag{5.3b}$$

and

$$\rho_j(k) = \beta\left(\frac{\bar{h}_j}{k} + kh_j\right), \quad j = 1, \ldots, n, \quad k \in \mathbf{C}; \tag{5.3c}$$

and the function $k_p(l)$ is defined by

$$k_p(l) = -\frac{l + \sqrt{l^2 + 4\beta^2|h_p|^2}}{2h_p\beta}, \quad p = 1, \ldots, n, \quad l \in \mathbf{R}^+. \tag{5.3d}$$

Furthermore, each of the terms appearing in the four sums on the right-hand sides of equations (5.2) decays exponentially as $l \to \infty$, except for the terms with $j = p$ which oscillate and for those with $j = p \pm 1$ which decay linearly.

Proof Parametrizing the side S_j with respect to its midpoint m_j,

$$z(s) = m_j + sh_j, \quad -\pi < s < \pi \tag{5.4}$$

and using

$$q_z dz = \frac{1}{2}\left[\frac{dq_j}{ds}(s) + iu_j(s)\right] ds, \qquad q_{\bar{z}} d\bar{z} = \frac{1}{2}\left[\frac{dq_j}{ds}(s) - iu_j(s)\right] ds \tag{5.5}$$

yields

$$\hat{q}_j(k) = ie^{-i\beta\left(m_j k - \frac{\bar{m}_j}{k}\right)} \int_{-\pi}^{\pi} e_j(k,s)\left[u_j(s) + \rho_j(k)d_j(s)\right] ds, \tag{5.6}$$
$$j = 1, \ldots, n, \quad k \in \mathbb{C}.$$

Writing the first global relation (1.16a) in the form

$$\hat{q}_p(k) = -\sum_{\substack{j=1 \\ j \neq p}}^{n} \hat{q}_j(k), \quad p = 1, \ldots, n, \quad k \in \mathbb{C}, \tag{5.7}$$

substituting for $\{\hat{q}_j(k)\}_1^n$ from (5.6), multiplying by $-ie^{\beta(im_p k - \frac{i}{k}\bar{m}_p)}$, and evaluating the resulting equation at $k = k_p(l)$ yields (5.2a). Since the second global relation (1.16b) can be obtained from the first by taking the Schwartz conjugate of all terms except q, equation (5.2b) follows from (5.2a).

Convexity implies the estimate

$$[\arg(m_j - m_p) - \arg(h_p)] \in (0, \pi), \quad j \neq p, \tag{5.8}$$

see Fig. 13.

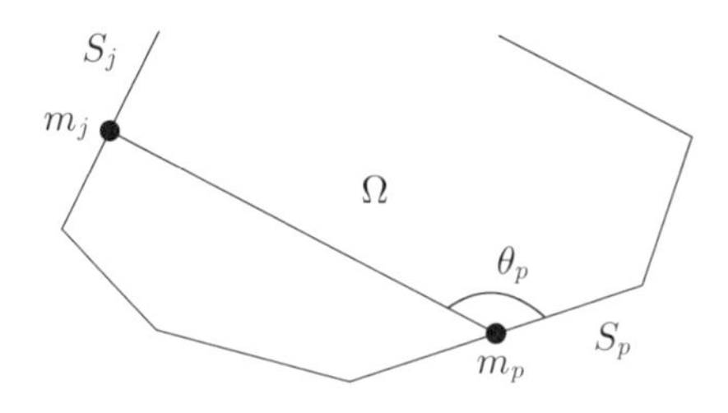

Fig. 13. The angle $\theta_p := \arg(m_j - m_p) - \arg(h_p)$

For $j = 1, \ldots, n$ and $p = 1, \ldots, n$,

$$E_{jp}(k_p(l))e_j(k_p(l), s) = e^{iL_p(l)\left(\frac{m_j - m_p + h_j s}{h_p}\right)} e^{-\frac{i\beta^2|h_p|^2}{L_p(l)}\left(\frac{\bar{m}_j - \bar{m}_p + \bar{h}_j s}{\bar{h}_p}\right)} \tag{5.9}$$

for all $l \in \mathbf{R}^+$, where $L_p(l) > 0$ is defined for $p = 1, \ldots, n$ by

$$L_p(l) = -h_p k_p(l) = \frac{l + \sqrt{l^2 + 4\beta^2|h_p|^2}}{2\beta}, \quad l \in \mathbf{R}^+. \tag{5.10}$$

The behaviour of (5.9) as $l \to \infty$ is governed only by the first exponential and so each of the terms appearing on the right-hand side of (5.2a) essentially involves

$$\int_{-\pi}^{\pi} e^{\frac{iL_p(l)}{h_p}(m_j + sh_j - m_p)} a_j(s) ds \quad \text{with} \quad a_j = u_j \quad \text{or} \quad a_j = d_j. \tag{5.11}$$

Convexity again implies that for $s \in (0, \pi)$,

$$0 < [\arg(m_j + sh_j - m_p) - \arg h_p] < \pi \quad \text{if} \quad j \neq p, p \pm 1.$$

If $j = p - 1$, then

$$m_p - \pi h_p = z_p = m_j + \pi h_j,$$

and hence

$$\arg(m_j + \pi h_j - m_p) - \arg h_p = \pi.$$

If $j = p + 1$, then

$$m_p + \pi h_p = z_{p+1} = m_j - \pi h_j,$$

and hence

$$\arg(m_j - \pi h_j - m_p) - \arg h_p = 0.$$

Integration by parts implies that the left-hand side of equation (5.11) equals

$$\frac{h_p}{iL_p(l)h_j}\left[e^{\frac{iL_p(l)}{h_p}(m_j+\pi h_j-m_p)}a_j(\pi)-e^{\frac{iL_p(l)}{h_p}(m_j-\pi h_j-m_p)}a_j(-\pi)\right]$$
$$+O\left(\frac{1}{L_p(l)^2}\right). \tag{5.12}$$

Hence each of the terms in the two sums on the right-hand side of (5.12) decays exponentially if $j \neq p$ and $j = p \pm 1$. Noting that $O(1/L_p(l))$ equals $O(1/l)$, if $j = p \pm 1$ then there is linear decay, and the term in G_p with $j = p$ oscillates. Clearly if $a_j(\pm\pi) = 0$ then the decay is quadratic. Similar considerations are valid for the right-hand side of (5.2a). □

Remark 7 Since $l \in \mathbf{R}^+$ it follows that the expressions on the left-hand sides of equations (5.2) taken together, define the Fourier transform of $u_p(s)$.

Remark 8 Regarding the choice of k in equation (5.3d), we note that in order to obtain the Fourier transform of $u_p(s)$, $k_p(l)$ is chosen such that

$$e_p(k_p(l), s) \equiv \exp(ils), \quad -\pi < s < \pi.$$

This yields two possible choices:

$$k_p^{\pm}(l) = -\frac{l \pm \sqrt{l^2 + 4\beta^2|h_p|^2}}{2h_p\beta}.$$

We make the choice $k_p(l) = k_p^+(l)$, which implies that $\arg k_p(l) = \pi - \arg h_j$ (the other choice $\arg k_p(l) = -\arg h_j$ implies that the right-hand sides of equations (5.2) are *unbounded* as $l \to \infty$). We also note that in order to maximize the decay of the right-hand sides of equations (5.2) as $l \to \infty$, we use the parametrization in (5.4) instead of the parametrization

$$z(s) = z_j + 2sh_j, \quad 0 < s < \pi.$$

5.1.1 The unknown values at the corners, $u_j(\pm\pi)$

Suppose that the given functions $d_j(s)$ satisfy appropriate compatibility conditions so that q_z and $q_{\bar{z}}$ are continuous at the corners, then $u_j(\pm\pi)$, $j = 1, \ldots, n$ can be determined explicitly:

$$u_j(\pi) = \frac{|h_{j+1}|\cos(\alpha_{j+1}-\alpha_j)\frac{d}{ds}d_j(\pi) - |h_j|\frac{d}{ds}d_{j+1}(-\pi)}{|h_{j+1}|\sin(\alpha_{j+1}-\alpha_j)} \tag{5.13a}$$

and

$$u_j(-\pi) = \frac{|h_j|\frac{d}{ds}d_{j-1}(\pi) - |h_{j-1}|\cos(\alpha_j - \alpha_{j-1})\frac{d}{ds}d_j(-\pi)}{|h_{j-1}|\sin(\alpha_j - \alpha_{j-1})}, \quad (5.13b)$$

where

$$\alpha_j = \arg(h_j). \quad (5.14)$$

Since the polygon is convex,

$$\alpha_{j+1} \neq \alpha_j + \pi, \quad (5.15)$$

and thus

$$\sin(\alpha_{j+1} - \alpha_j) \neq 0. \quad (5.16)$$

In order to derive (5.13) we note that (5.5) implies

$$q_z = \frac{e^{-i\alpha_j}}{2|h_j|}\left[\frac{dq_j}{ds}(s) + iu_j(s)\right], \quad s \in S_j. \quad (5.17)$$

The continuity of q_z at the corner z_j implies that the expression in (5.17) with j replaced by $j-1$ evaluated at $s = \pi$ (the right end of the side S_{j-1}) equals the expression in (5.17) evaluated at $s = -\pi$ (the left end of the side S_j), i.e.

$$\frac{e^{-i\alpha_{j-1}}}{|h_{j-1}|}\left[\frac{dq_{j-1}}{ds}(\pi) + iu_{j-1}(\pi)\right] = \frac{e^{-i\alpha_j}}{|h_j|}\left[\frac{dq_j}{ds}(-\pi) + iu_j(-\pi)\right].$$

Similarly, the continuity of $q_{\bar{z}}$ at z_j implies that

$$\frac{e^{i\alpha_{j-1}}}{|h_{j-1}|}\left[\frac{dq_{j-1}}{ds}(\pi) - iu_{j-1}(\pi)\right] = \frac{e^{i\alpha_j}}{|h_j|}\left[\frac{dq_j}{ds}(-\pi) - iu_j(-\pi)\right].$$

Solving the above two equations for $u_{j-1}(\pi)$ and $u_j(-\pi)$ and then letting $j \to j+1$ in the expression for $u_{j-1}(\pi)$ we find equations (5.13).

5.1.2 Unknown functions which vanish at the corners

Since the values of $u_p(s)$ are known at the two corners, it is possible to express the unknown function $u_p(s)$ in terms of a new unknown function, denoted by $\breve{u}_p(s)$, which *vanishes* at the corners:

$$u_p(s) = \breve{u}_p(s) + u_{\star p}(s), \quad p = 1, \ldots, n, \quad -\pi < s < \pi, \quad (5.18a)$$

with

$$u_{\star p}(s) = \frac{1}{2\pi}\left[(s+\pi)u_p(\pi) - (s-\pi)u_p(-\pi)\right], \; p = 1, \ldots, n, \; -\pi < s < \pi. \quad (5.18b)$$

The unknown functions $\{\breve{u}_j\}_1^n$ are defined by (5.18) with $u_j(\pm\pi)$ given by (5.13).

The functions $\{\breve{u}_j\}_1^n$ satisfy equations similar to (5.2) but with G_p and $\tilde{G}_p$ replaced by $G_p + U_{\star p}$ and $\tilde{G}_p + \tilde{U}_{\star p}$ respectively, where the known functions $U_{\star p}(l)$ and $\tilde{U}_{\star p}(l)$ are defined for $p = 1, \ldots, n$ and $l \in \mathbf{R}^+$ by

$$U_{\star p}(l) = \sum_{j=1}^{n} E_{jp}(k_p(l)) \int_{-\pi}^{\pi} e_j(k_p(l), s) u_{\star j}(s) ds,$$

$$\tilde{U}_{\star p}(l) = \sum_{j=1}^{n} \bar{E}_{jp}(\bar{k}_p(l)) \int_{-\pi}^{\pi} \bar{e}_j(\bar{k}_p(l), s) u_{\star p}(s) ds.$$

Hence $U_{\star p}(l)$ and $\tilde{U}_{\star p}(l)$ are defined by the following equations for $p = 1, \ldots, n$, $l \in \mathbf{R}^+$,

$$\begin{aligned} U_{\star p}(l) = \sum_{j=1}^{n} E_{jp}(k_p(l)) \Bigg\{ &[u_j(\pi) - u_j(-\pi)] \left[-\frac{i}{H_{jp}(l)} \cos(\pi H_{jp}(l)) \right. \\ &\left. + \frac{i}{\pi (H_{jp}(l))^2} \sin(\pi H_{jp}(l)) \right] \\ &+ \frac{1}{H_{jp}(l)} [u_j(\pi) + u_j(-\pi)] \sin(\pi H_{jp}(l)) \Bigg\} \end{aligned} \tag{5.19a}$$

and

$$\begin{aligned} \tilde{U}_{\star p}(l) = \sum_{j=1}^{n} \bar{E}_{jp}(\bar{k}_p(l)) \Bigg\{ &[u_j(\pi) - u_j(-\pi)] \left[\frac{i}{\bar{H}_{jp}(l)} \cos(\pi \bar{H}_{jp}(l)) \right. \\ &\left. - \frac{i}{\pi (\bar{H}_{jp}(l))^2} \sin(\pi \bar{H}_{jp}(l)) \right] \\ &+ \frac{1}{\bar{H}_{jp}(l)} [u_j(\pi) + u_j(-\pi)] \sin\left(\pi \bar{H}_{jp}(l)\right) \Bigg\}, \end{aligned} \tag{5.19b}$$

with

$$H_{jp}(l) = \frac{1}{2} \frac{h_j}{h_p} \left(l + \sqrt{l^2 + 4\beta^2 |h_p|^2} \right) - \frac{2\beta^2 \bar{h}_j h_p}{l + \sqrt{l^2 + 4\beta^2 |h_p|^2}}, \tag{5.19c}$$
$$j = 1, \ldots, n, \quad p = 1, \ldots, n, \quad l \in \mathbf{R}^+.$$

Remark 9 The function $\breve{u}_p(s)$, which is defined for $-\pi < s < \pi$, vanishes at the end points. A convenient representation for such a function is a modified sine-Fourier series expansion:

$$\breve{u}_p(s) = \sum_{m=1}^{\infty} \left[s_m^p \sin ms + c_m^p \cos\left(m - \frac{1}{2} \right) s \right], \tag{5.20a}$$

$$s_m^p = \frac{1}{\pi} \int_{-\pi}^{\pi} \breve{u}_p(s) \sin(ms) ds, \quad c_m^p = \frac{1}{\pi} \int_{-\pi}^{\pi} \breve{u}_p(s) \cos\left(m - \frac{1}{2} \right) s \, ds,$$
$$m = 1, 2, \ldots, \quad p = 1, \ldots, n. \tag{5.20b}$$

The advantage of the above expansion is that s_m^p and c_m^p are of order $1/m^3$ as $m \to \infty$, provided that $\breve{u}_p(s)$ has sufficient smoothness. The representation (5.20a) can be obtained by starting with the usual sine-Fourier series in the interval $(0, \pi)$ and using a change of variables to map this interval to $(-\pi, \pi)$. The analogue of the representation (5.20a) corresponding to the cosine-Fourier series was introduced in [IN08]. Using the techniques of [Olv08] it is possible to prove that if $\breve{u} \in C^3(-\pi, \pi)$,

$$\breve{u}_p^N(s) = \sum_{m=1}^{N} \left[s_m^p \sin ms + c_m^p \cos\left(m - \frac{1}{2} \right) s \right], \quad p = 1, \ldots, n, \tag{5.21}$$

then

$$\| \breve{u}_p - \breve{u}_p^N \|_\infty = O\left(\frac{1}{N^2} \right). \tag{5.22}$$

5.1.3 The numerical method

Proposition 6 *Let q satisfy the boundary value problem specified in Proposition 5. Assume that the values of the unknown functions $\{u_j\}_1^n$ at the corners $\{z_j\}_1^n$ are given by equations (5.13). Express $\{u_j\}_1^n$ in terms of the unknown functions $\{\breve{u}_j\}_1^n$ defined in equation (5.18) and approximate the latter functions by the functions $\{\breve{u}_j^N\}_1^n$ defined in equation (5.21). Then the constants s_m^p and c_m^p, $m = 1, \ldots, N$, $p = 1, \ldots, n$ satisfy the $2Nn$ algebraic equations*

$$\begin{aligned} 2\pi s_m^p &= i \sum_{\substack{j=1 \\ j \neq p}}^{n} \left\{ \sum_{k=1}^{N} s_k^j \left[E_{jp}(k_p(m)) S_{jp}^k(m) - \bar{E}_{jp}(\bar{k}_p(m)) \bar{S}_{jp}^k(m) \right] \right. \\ &\quad \left. + \sum_{k=1}^{N} c_k^j \left[E_{jp}(k_p(m)) C_{jp}^k(m) - \bar{E}_{jp}(\bar{k}_p(m) \bar{C}_{jp}^k(m) \right] \right\} \\ &\quad + iG_p(m) - i\tilde{G}_p(m) + iU_{\star p}(m) - i\tilde{U}_{\star p}(m) \end{aligned} \tag{5.23a}$$

and

$$2\pi c_m^p = -\sum_{\substack{j=1\\j\neq p}}^{n}\left\{\sum_{k=1}^{N} s_k^j\left[E_{jp}\left(k_p\left(m-\frac{1}{2}\right)\right)S_{jp}^k\left(m-\frac{1}{2}\right)\right.\right. \tag{5.23b}$$
$$\left. + \bar{E}_{jp}\left(\bar{k}_p\left(m-\frac{1}{2}\right)\right)\bar{S}_{jp}^k\left(m-\frac{1}{2}\right)\right]$$
$$+\sum_{k=1}^{N} c_k^j\left[E_{jp}\left(k_p\left(m-\frac{1}{2}\right)\right)C_{jp}^k\left(m-\frac{1}{2}\right)\right.$$
$$\left.\left. + \bar{E}_{jp}\left(\bar{k}_p\left(m-\frac{1}{2}\right)\right)\bar{C}_{jp}^k\left(m-\frac{1}{2}\right)\right]\right\}$$
$$- G_p\left(m-\frac{1}{2}\right) - \tilde{G}_p\left(m-\frac{1}{2}\right) - U_{\star p}\left(m-\frac{1}{2}\right) - \tilde{U}_{\star p}\left(m-\frac{1}{2}\right),$$

where the known functions G_p, $\tilde{G}_p$, $U_{\star p}$, $\tilde{U}_{\star p}$ are defined by equations (5.3) *and* (5.19), *and*

$$S_{jp}^k(m) = \frac{2ik(-1)^{k-1}\sin\left(\pi H_{jp}(m)\right)}{k^2 - \left(H_{jp}(m)\right)^2},$$
$$C_{jp}^k(m) = \frac{2\left(k-\frac{1}{2}\right)(-1)^{k-1}\cos\left(\pi H_{jp}(m)\right)}{\left(k-\frac{1}{2}\right)^2 - \left(H_{jp}(m)\right)^2},$$
$$j=1,\ldots,n, \quad p=1,\ldots,n, \quad k=1,\ldots,n, \quad m=1,\ldots,N.$$

Proof Equation (5.21) implies

$$s_m^p = \frac{1}{\pi}\int_{-\pi}^{\pi}\breve{u}_p^N(s)\sin(ms)ds.$$

Recall that the functions $\breve{u}_p$ satisfy equations similar to (5.2a) but with G_p and $\tilde{G}_p$ replaced $G_p - U_{\star p}$ and $\tilde{G}_p - \tilde{U}_{\star p}$. Replacing in the equations satisfied by $\breve{u}_p$ the function $\breve{u}_p$ with $\breve{u}_p^N$ defined in (5.21), subtracting these equations and evaluating the resulting equation at $l = m$ we find equation (5.23a). In this respect we note that the left-hand side of the resulting equation immediately yields s_m^p, whereas for the evaluation of the right-hand side of the resulting equations we use the expression,

$$\int_{-\pi}^{\pi} e^{im\frac{h_j}{h_p}s}u_j^N(s)ds = \sum_{n=1}^{N}\left\{s_n^j\int_{-\pi}^{\pi} e^{im\frac{h_j}{h_p}}\sin(ns)ds\right.$$
$$\left. + c_n^j\int_{-\pi}^{\pi} e^{im\frac{h_j}{h_p}}\cos\left(n-\frac{1}{2}\right)s\,ds\right\}$$

and then we evaluate the above integrals explicitly.

Proceeding as earlier, where we now add the equations satisfied by $\{\breve{u}_j^N\}_1^n$ and then evaluating the resulting equation at $l = m - \frac{1}{2}$, we find (5.23b). □

5.2 *Numerical results*

In order to illustrate the numerical implementation of the new collacation method to the modified Helmholtz equations, we will consider a variety of regular and irregular polygons.

We will study the modified Helmholtz equation with $\beta = 10$ and the exact solution

$$q(z, \bar{z}) = e^{11z + \frac{100}{11}\bar{z}}. \tag{5.24}$$

Analytic expressions for the known boundary functions $\{d_j(s)\}_{j=1}^n$, and the unknown boundary data $\{u_j(s)\}_{j=1}^n$ can be easily computed from (5.24).

To demonstrate the performance of the method, we use the discrete maximum relative error

$$E_\infty := \frac{||u - u^N||_\infty}{||u_n||_\infty}, \tag{5.25}$$

where

$$u_j^N(s) = \breve{u}_j^N(s) + u_{*j}(s), \quad -\pi \le s \le \pi, \; 1 \le j \le n, \tag{5.26}$$

$$||u||_\infty = \max_{1 \le j \le n} \left\{ \max_{s \in S} |u_n^{(j)}(s)| \right\}. \tag{5.27}$$

We consider 10001 evenly spaced points s_i,

$$S = \{s_i\}_{i=1}^{10001} \subset [-\pi, \pi], \tag{5.28}$$

with the points s_i, $-\pi = s_1 < s_2 < \ldots < s_{10000} < s_{10001} = \pi$, given by

$$s_i = \pi \left[-1 + \frac{2(i-1)}{10000} \right], \quad 1 \le i \le 10001. \tag{5.29}$$

We consider regular polygons with $n = 3, 4, 5, 6, 8$ sides, whose vertices lie on the circle centered at the origin with radius $\sqrt{2}$ in the complex plane (with a vertex on the positive real axis). These polygons are then rotated through an angle of $-\frac{1}{5}$ about the origin to avoid non-generic results due to alignment with the coordinate axes. Thus, we consider polygons with the vertices

$$z(j) = \sqrt{2}e^{i\left[2(j-1)\frac{\pi}{n} - \frac{1}{5}\right]}, \quad 1 \leq j \leq n. \tag{5.30}$$

We also consider irregular polygons with $n = 3, 4, 5, 6, 8$ sides, whose vertices lie on the ellipse $\left(\frac{x}{5}\right)^2 + \left(\frac{y}{2}\right)^2 = 1$ in the complex plane rotated through an angle of $\frac{1}{5}$ about the origin. The x and y coordinates of the vertices of the polygons before rotation are given (in an anticlockwise direction) in Table 1 and the polygons are shown in Table 7.1.

Triangle	$\left(-4, -\frac{6}{5}\right)$, $\left(-1, -\frac{2\sqrt{24}}{25}\right)$, $\left(3, \frac{8}{5}\right)$;
Square	$\left(1, \frac{2\sqrt{24}}{25}\right)$, $\left(-4, -\frac{6}{5}\right)$, $\left(4, -\frac{6}{5}\right)$, $\left(4, \frac{6}{5}\right)$;
Pentagon	$(0, 2)$, $(-5, 0)$, $\left(-2, -\frac{2\sqrt{21}}{25}\right)$, $\left(4, -\frac{6}{5}\right)$, $\left(3, \frac{8}{5}\right)$;
Hexagon	$\left(1, \frac{2\sqrt{24}}{25}\right)$, $\left(-\frac{9}{2}, \frac{2\sqrt{19}}{10}\right)$, $\left(-4, -\frac{6}{5}\right)$, $\left(-1, -\frac{2\sqrt{24}}{25}\right)$, $\left(2, -\frac{2\sqrt{21}}{25}\right)$, $\left(\frac{9}{2}, \frac{2\sqrt{19}}{10}\right)$;
Octagon	$\left(1, \frac{2\sqrt{24}}{25}\right)$, $\left(-2, \frac{2\sqrt{21}}{25}\right)$, $\left(-3, \frac{8}{5}\right)$, $(-5, 0)$, $\left(-4, -\frac{6}{5}\right)$, $\left(-1, -\frac{2\sqrt{24}}{25}\right)$, $\left(2, -\frac{2\sqrt{21}}{25}\right)$, $\left(3, \frac{8}{5}\right)$.

Table 1. *Vertices of irregular polygons prior to rotation*

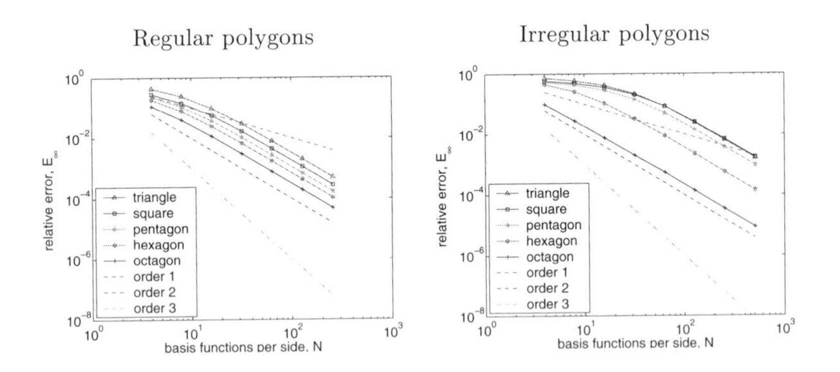

Fig. 14. E_∞ as a function of N for the modified Helmholtz equation

Fig. 14 refers to the regular and irregular polygons respectively. The red, blue and green dotted lines are the lines $\frac{1}{N}$, $\frac{1}{N^2}$ and $\frac{1}{N^3}$, indicating the slopes for first, second and third order convergence respectively. The

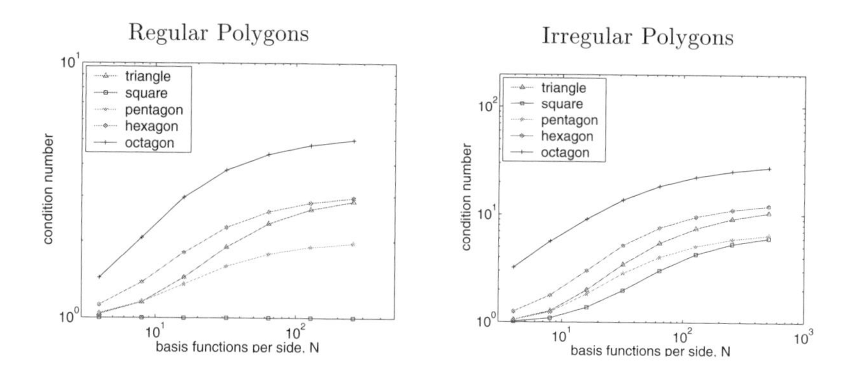

Fig. 15. The condition number of the coefficient matrix as a function of N for the modified Helmholtz equation

error lines are all asymptotically parallel to the $\frac{1}{N^2}$ line, indicating a quadratically convergent method with respect to the discrete maximum relative error. To highlight this, the order of convergence (O.o.C.) has been estimated for the triangles in Table 2. It can be seen from Fig. 15 that the condition numbers of the associated matrices are small and grow only very slowly with N.

Equilateral Triangle

N	E_∞	O.o.C.
4	4.3435e-01	–
8	2.4280e-01	0.8391
16	9.6685e-02	1.3284
32	2.9334e-02	1.7207
64	7.8274e-03	1.9060
128	1.9983e-03	1.9697
256	5.0311e-04	1.9898

Irregular Triangle

N	E_∞	O.o.C.
4	7.2627e-01	–
8	5.9449e-01	0.2889
16	4.1420e-01	0.5213
32	2.2008e-01	0.9123
64	8.3167e-02	1.4039
128	2.4321e-02	1.7738
256	6.3678e-03	1.9334
512	1.6117e-03	1.9822

Table 2. *Order of convergence (O.o.C.) for the modified Helmholtz equation in the triangles*

Remark 10 This numerical method has its origin in [FFX04] where the Laplace equation was studied. However, although the values of k were correctly chosen to be those in (5.3d), the global relations were evaluated at $l = m$ instead of $l = m$ and/or $l = m - \frac{1}{2}$. As a result, the relevant linear system possesses a large condition number and numerical computations performed in [FFX04] suggest linear convergence. The choice of $l = m$ and/or $l = m - \frac{1}{2}$was made in [SFFS07].

The implementation of the method to the modified Helmholtz equation was presented in [SSF08a]. The investigation of BVPs with boundary conditions that give rise to singularities at the corners of the polygon is work in progress.

Bibliography

[AF03] M. J. Ablowitz & A. S. Fokas (2003). *Complex Variables: Introduction and Applications*, 2nd edition, (Cambridge University Press, Cambridge).

[AF04] Y. Antipov & A. S. Fokas (2004). A transform method for the modified Helmholtz equation on the semi-strip, *Math. Proc. Camb. Phil. Soc.* **137**, 339–365.

[DF05] G. Dassios & A. S. Fokas (2005). The basic elliptic equations in an equilateral triangle, *Proc. R. Soc. Lond. A* **461**, 2721–2748.

[FF08] N. Flyer & A. S. Fokas (2008). A hybrid analytical–numerical method for solving evolution PDEs. I. The half-line, *Proc. R. Soc. Lond. A*, to appear.

[Fok97] A. S. Fokas (1997). A unified transform method for solving linear and certain nonlinear PDEs, *Proc. R. Soc. Lond. A* **453**, 1411–1443.

[Fok01] A. S. Fokas (2001). Two dimensional linear PDEs in a convex polygon, *Proc. R. Soc. Lond. A* **457**, 371–393.

[Fok02] A. S. Fokas (2002). A new transform method for evolution PDEs, *IMA J. Appl. Math.* **67**, 1–32.

[Fok08] A. S. Fokas (2008). *A Unified Approach to Boundary Value Problems* (SIAM Monographs, Philadelphia.

[FG94] A. S. Fokas & I. M. Gel'fand (1994). Integrability of linear and nonlinear evolution equations and the associated nonlinear Fourier transforms, *Lett. Math. Physics* **32**, 189–210.

[FZ02] A. S. Fokas & M. Zyskin (2002). The fundamental differential form and boundary-value problems, *Q. J. Mech. App. Maths.* **55**, 457–479.

[FFX04] S. Fulton, A. S. Fokas & C. Xenophontos (2004). An analytical method for linear elliptic PDEs and its numerical implementation, *J. of Comput. and Appl. Maths* **167**, 465–483.

[IN08] A. Iserles & S. P. Nørsett (2008). From high oscillation to rapid approximation I: Modified Fourier expansions, *IMA J. Num. Anal.* **28**, 862-887.

[GGKM67] C. S. Gardner, J. M. Greene, M. D. Kruskal & R. M. Miura (1967). Method for solving the Korteweg–de Vries equation, *Phys. Rev. Lett.* **19**, 1095.

[Lax68] P. D. Lax (1968). Integrals of nonlinear equations and solitary waves, *Commun. Pure Appl. Math.* **21**, 467–490.

[Olv08] S. Olver (2008). On the convergence rate of a modified Fourier series, DAMTP Technical Report (University of Cambridge).

[SFFS07] A. G. Sifalakis, A. S. Fokas, S. R. Fulton & Y. G. Saridakis (2007). The generalised Dirichlet–Neumann map for linear elliptic PDE's and its numerical implementation, *J. Comp. Appl. Math.* (to appear).

[SSF08a] S. A. Smitheman, E. A. Spence & A. S. Fokas (2008). A spectral collocation method for the Laplace and the modified Helmholtz equation in a convex polygon (submitted).

[SSF08b] S. A. Smitheman, E. A. Spence & A. S. Fokas (2008). (in preparation).

[SSF08c] S. A. Smitheman, E. A. Spence & A. S. Fokas (2008). A spectral collocation method for the Helmholtz equation in a convex polygon, (in preparation).

[Spe08] E. A. Spence (2008). PhD Dissertation (DAMTP, Cambridge University), to appear.